Springer Series in Game Theory

Official Series of the Game Theory Society

More information about this series at http://www.springer.com/series/8107

Carlos Alós-Ferrer • Klaus Ritzberger

The Theory of Extensive Form Games

Springer

Carlos Alós-Ferrer
Department of Economics
University of Cologne
Cologne, Germany

Klaus Ritzberger
Department of Economics and Finance
Institute for Advanced Studies
Vienna, Austria

ISSN 1868-517X ISSN 1868-5188 (electronic)
Springer Series in Game Theory
ISBN 978-3-662-49942-9 ISBN 978-3-662-49944-3 (eBook)
DOI 10.1007/978-3-662-49944-3

Library of Congress Control Number: 2016946228

Printed on acid-free paper

This Springer imprint is published by Springer Nature
The registered company is Springer-Verlag GmbH Berlin Heidelberg

In memoriam Harold W. Kuhn.

Preface

This book has grown out of a decade and a half of joint research that resulted in the articles underlying it (Alós-Ferrer and Ritzberger 2005a,b, Alós-Ferrer and Ritzberger 2008, Alós-Ferrer et al. 2011, Alós-Ferrer and Ritzberger 2013, 2015, 2016a, 2016b, 2016c). In compiling it, we have taken the chance to reorganize the material and modify and slightly expand the articles to take advantage of the book format. For example, a number of intuitive illustrations and a host of examples developed during the exploratory phase were not included in the articles due to space constraints. Those have now been reinserted. Some side results that also did not make it into the published articles have been added. A few examples from the prior literature which inspired our research are now reproduced in order to spare the reader the need to consult other sources. Many figures which we used to gain intuition and showed in research seminars, but were not included in the papers (again due to space constraints), have been inserted in the appropriate places.

The mathematical prerequisites for the current book are within the common background of advanced students and researchers in economic theory. Specifically, we assume that the reader is familiar with basic aspects of set theory and point-set topology. For those who have forgotten one or the other definition, we offer a brief mathematical appendix that summarizes the most important concepts—without any aspiration of complete coverage, of course. Other than that, the book is pretty much self-contained in terms of the mathematics.

What is assumed, however, is some basic familiarity with game theory and its applications. A host of examples are quoted throughout the book, largely from economics and related fields, without going into detail on how these examples were derived. For instance, it is assumed that the reader has seen repeated games or stochastic games before and is familiar with the basic equilibrium concepts. Hence, the target audience is advanced students who have had at least a first course on game theory, as well as researchers in the field of game theory.

We would like to thank Larry Blume, Michael Greinecker, Johannes Kern, and Bernhard von Stengel for many helpful comments. Also, we would like to thank a large number of anonymous referees, associate editors, and editors who read, commented on, and criticized different versions of the articles that gave rise to this book and who very frequently made us think harder to produce more focused and streamlined versions, more convincing examples, and clearer intuitions. Some of those referees pointed out that a consolidated reference would have helped them digest our more recent contributions more quickly—so here it is!

We also thank Bernhard von Stengel for raising the idea to produce this monograph as part of the official series of the Game Theory Society (Springer Series in Game Theory). We are indebted to Barbara Fess and Martina Bihn from Springer Verlag for their encouragement and their patience. Last but not least, we would like to thank the German Research Foundation (DFG) and the Austrian Science Fund (FWF), which provided financial support for our joint research agenda via the DACH-projects AL-1169/1-1 (C.A-F), AL-1169/1-2 (C.A-F), I338-G16 (K.R.), and I1242-G16 (K.R.).

Cologne, Germany
Vienna, Austria
Saturnalia 2015

Carlos Alós-Ferrer
Klaus Ritzberger

References

Alós-Ferrer C, Ritzberger K (2005a) Trees and decisions. Econ Theory 25(4):763–798
Alós-Ferrer C, Ritzberger K (2005b) Some remarks on pseudotrees. Order 22(1):1–9
Alós-Ferrer C, Ritzberger K (2008) Trees and extensive forms. J Econ Theory 43(1):216–250
Alós-Ferrer C, Ritzberger K (2013) Large extensive form games. Econ Theory 52(1):75–102
Alós-Ferrer C, Ritzberger K (2015) Characterizing existence of equilibrium for large extensive form games: a necessity result. Econ Theory. doi:10.1007/s00199-015-0937-0
Alós-Ferrer C, Ritzberger K (2016a) Equilibrium existence for large perfect information games. J Math Econ 62:5–18
Alós-Ferrer C, Ritzberger K (2016b) Does backwards induction imply subgame perfection? Games Econ Behav. doi:10.1016/j.geb.2016.02.005
Alós-Ferrer C, Ritzberger K (2016c) Characterizations of perfect recall. Int J Game Theory. doi:10.1007/s00182-016-0534-x
Alós-Ferrer C, Kern J, Ritzberger K (2011) Comment on 'Trees and extensive forms'. J Econ Theory 146(5):2165–2168

Contents

List of Figures

List of Symbols

$\geq$	Precedence relation (among nodes of a tree or elements of a poset)
$\subseteq$	Set inclusion
$\subset$	Strict set inclusion
$\preceq_i$	Preference relation of player i
$A_i(x)$	Set of choices available to player i at node x
C_i	Set of choices of player i
E	Set of terminal nodes of a tree
$\phi(s)$	Outcome induced by the strategy profile s
$F(N)$	Set of finite nodes of a tree
$\gamma(x,w)$	Perfect information choice at a node x which contains the outcome w
$I(N)$	Set of infinite nodes of a tree
$J(x)$	Set of players active at node x
N	Set of nodes of a tree
$P(a)$	Set of immediate predecessors of a set of plays a
$p(x)$	Immediate predecessor of a node x
$p^{-1}(x)$	Set of immediate successors of a node x
$\Pi(T)$	Single-player perfect information problem for the tree T
S_i	Set of pure strategies of player i
$S(N)$	Set of strange nodes of a tree
T	Tree
W	Set of plays of a tree
$W(x)$	Set of plays passing through the node x/set of outcomes contained in x
X	Set of moves of a tree
X_i	Set of player i's moves (decision points)
Y_t	t-th slice of a discrete tree

Chapter 1
Introduction

Consider your life. From the cradle to the grave, you have made and will make a staggering number of decisions. The alternatives you had (and the ones you will have) opened a mind-blowing number of possibilities. As you contemplate where you stand today, some of those are simply roads not taken, long gone, while others describe alternate futures which might very well keep you awake at night. Furthermore, you are not fully in control of those possibilities, for very often you make decisions without having all the information that would allow you to identify the optimal option, and even more often the decisions of others affect your path.

Now consider any economic or political problem, from repeated competition in a market to behavior in a highly regulated auction, from negotiations between labor unions and employers to international conflicts. Again, a large number of interrelated decisions by different agents, often made in the absence of full information, combine to shape the ultimate outcome. Unsurprisingly, making good decisions requires to anticipate the possible reactions of others, to understand their incentives as well as our own. In other words, the object of interest is *strategic behavior*.

The theory of games is the "theory of everything," at least when it comes to interpersonal decision problems and strategic behavior. All the situations described above are examples of so-called *extensive form games*. This book is the product of 15 years of joint research into the formalization and analysis of such structures. One objective of our research is, and has always been, to maintain a level of generality which allows game theory to unfold maximally, without artificial constraints. For, we insist, extensive form games are the model for arbitrarily complex interpersonal interactions, and unjustified constraints will only obscure this fact. At the same time, it has to be realized that, when the object of study can be as general as the one game theory covers, simplified formal models are both misleading and hopeless, and a reasonable level of mathematical sophistication is a must.

C. Alós-Ferrer, K. Ritzberger, *The Theory of Extensive Form Games*,
Springer Series in Game Theory, DOI 10.1007/978-3-662-49944-3_1

1.1 The Starting Point

1.1.1 Historical Notes

Strategic behavior and planning has attracted the attention of mathematicians, philosophers, and economists for centuries. The formal analysis of strategic interaction in economics, for instance, dates back at least to Cournot (1838) and Bertrand (1883). Historically, however, the discipline is a creature of the twentieth century. That century witnessed a first "big push" in the consolidation of the efforts to develop a general model of strategic interaction with the advent of the "Theory of Games and Economic Behavior" (von Neumann and Morgenstern 1944) in the 1940s. It was at this point that the nascent discipline received its nickname, "game theory," due to the use of parlor games as stylized versions of more serious strategic conflicts, even though it soon became clear that wars were one example of the categories of "games" that the theory should cover.

The discipline started out with the analysis of special cases, as zero-sum games. First theorems were proved, e.g., Zermelo's theorem on chess (Zermelo 1913) or the Minimax-Theorem (von Neumann 1928). Yet, the efforts remained scattered and confined to the circles of the initiated. Still, it was at this point that the first versions of the most important concepts were formulated: the normal (or strategic) form of a game, the extensive form of a game (von Neumann 1928), and the idea of a strategy.

A decade or two later, though, after World War II had come to an end, the American East coast saw an unprecedented concentration of talent and know-how gathering around the new discipline. We can do no better than to quote Aumann (1987, p. 467):

> The 1950ties were a period of excitement in game theory. The discipline had broken out of its cocoon, and was testing its wings. Giants walked the earth. At Princeton, John Nash laid the groundwork for the general non-cooperative theory, and for cooperative bargaining theory; Lloyd Shapley defined the value for coalitional games, initiated the theory of stochastic games, co-invented the core with D. B. Gillies, and, together with John Milnor, developed the first game models with continua of players; Harold Kuhn worked on behavior strategies and perfect recall; Al Tucker discovered the prisoner's dilemma; the Office of Naval Research was unstinting in its support. Three game theory conferences were held at Princeton, with the active participation of von Neumann and Morgenstern themselves. Princeton University Press published the four classical volumes of *Contributions to the Theory of Games*.

During the second half of the past century game theory experienced a period of growth and consolidation. Extensions of the theory, like games of incomplete information (Harsanyi 1967/1968), coalitional games with non-transferable utility, or the theory of the core were extensively developed and applied. The well-known asymptotic equivalence between the core of an economy and its competitive equilibrium is but one example (Debreu and Scarf 1963). The discipline soon infected places outside the ivory towers of Princeton; new centers emerged in Israel, Germany, Belgium, and the Soviet Union.

It took some time, though, until the new ideas percolated to the curriculae. In most places game theory did not properly invade the classrooms and the academic journals until the 1980s. On the other hand, this invasion had a long-lasting impact on economics, effectively starting the path to replace the then-dominant paradigm of general equilibrium theory with the language of strategic interaction. The importance of this still ongoing shift cannot be overestimated, since the former rests on the assumption that all interactions are mediated by prices, while the latter is a fully assumption-free methodology. By the end of the last century game theory had become the prevalent methodology in economics, and a prominent method in other disciplines like biology or political science. It had found profitable real-world applications beyond the planning of cold-war strategies, ranging from the design of auctions and matching mechanisms to the analysis of job markets and contracts.

Among the many remarkable aspects of this development, two are of special interest. One is that the most successful branch of game theory was its non-cooperative part—in stark contrast to the intentions of von Neumann and Morgenstern (1944). The other is that most new developments since the 1980s concerned the theory of solutions for non-cooperative games, in particular refinements of Nash equilibrium (Nash 1950), equilibrium selection (Harsanyi and Selten 1988), and an evolutionary re-interpretation of these solutions inspired by biology.

1.1.2 Extensive Form Games

In spite of the many successes of game theory, however, relatively little effort was devoted to the theory of game representations. This omission is particularly important for the theory of non-cooperative games, which happens to be the branch spanning most of the applications of game theory. For, non-cooperative games are meant to be the games with *complete* rules. A game is a complete description of a strategic, interpersonal (and maybe intertemporal) decision situation. Accordingly, one needs specific means to verify that the rules are indeed complete. The tool to do so is the representation of the game in *extensive form*, as first introduced by von Neumann (1928). This is like a "flow-chart" description of the game that specifies, for each possible contingency, who can do what and under which informational constraints. That is, a non-cooperative game has been correctly specified (and can hence presumably be analyzed) if and only if it has a representation in extensive form. Therefore, extensive form game representations are of crucial importance to the theory of non-cooperative games.

Probably many game theorists assumed that these issues had been settled in the early days, if they considered them at all. Hence the reward for reconsidering the theory of extensive form representations was implicitly considered to be small. This assumption is understandable. Indeed, since the initial contribution by von Neumann (1928) and its modification by von Neumann and Morgenstern (1944) the representation of non-cooperative games in extensive form had been generalized, in particular by Kuhn (1953), so as to cover a stunningly wide range of applications.

On the other hand, these generalizations largely concerned the informational constraints on decision making during a playing of the game and the order of moves. What by and large remained untouched since the early days were the highly restrictive cardinality assumptions. Many of the classical results in the theory of non-cooperative games concerned *finite* games, where both the number of choices at a move and the time horizon are finite.[1] This is because, at the time, games were conceptual innovations. When one explores a new idea, it is wise to begin with cases that do not pose major technical difficulties. After all, one wants to understand the problems peculiar to the novelty and not those that are raised by its interaction with the technical boundaries of the approach. And this is why finiteness was assumed in the early days.

Many applications, of course, soon transcended these finiteness restrictions. In particular economists were used to work in Euclidean spaces; many of the key models of oligopoly theory or information economics had initially been formulated with large (continuum) spaces and remained that way, even when subjected to a game-theoretic analysis. As a consequence, when game theory was applied, there was (and sometimes there still is) a gap between the pure theory and the applied models. Some contributions tried to bridge this gap with ad hoc constructions, others ignored it. In any case little effort went into generalizing the fundamental building blocks of non-cooperative games as far as finiteness assumptions were concerned.

This state of affairs is highly unsatisfactory, and very specially so for non-cooperative game theory. For, unlike other approaches in economics, game theory is more like a language than a model. Whereas in economics the model and the solution concept are often tailored to the question at hand, game theory provides an all-encompassing language to formalize and solve all sorts of problems. That this language should depend on simplifying assumptions, like finiteness, would impair its power. Of course, one can argue that in the real world "everything is finite." But such an argument ignores the usefulness of abstractions like infinity. If one does not know the precise date of the own death (or, in a less dramatic example, the last period of market competition), the appropriate model is not one with a finite horizon, but rather one with an infinite horizon where there is always a (maybe small) probability that there will be a tomorrow. If a firm can potentially decide to produce any fractional output, or undercut the competitor's price by any fraction, discretizing the action space will fail to capture the true situation. Worse, such simplifications might radically change the predictions, because one ends up analyzing the wrong game, one where key possibilities have been assumed away. There are also technical reasons to allow for infinite action spaces. An infinitesimal quantity may be hard to grasp intuitively, but it is amazingly helpful in constructing linear approximations, like partial derivatives. Moreover, there is no indication that the simplifying finiteness assumptions are indeed necessary for game theory.

[1] As always, there are exceptions. Stochastic games á la Shapley (1953), for instance, allow for an infinite time horizon; the extensive form games considered by Aumann (1964) to generalize Kuhn's theorem on perfect recall allow both for an infinite horizon and a continuum of actions.

Therefore, the purpose of this monograph is to extend the theory of extensive form games to the general case without any finiteness or other simplifying assumptions. Finite games are covered, of course, as special cases. But the inquiry into what is required in the general case will reveal a number of new insights about which structures need to be added to do what game theory traditionally does. Hence we are not trying to change the theory. Instead we aim at unifying and bringing together different approaches. The next section will illustrate the different models that have been proposed for extensive form games and explain how our work synthesizes them.

1.2 How to Model Games

Our aim is nothing less than to reconstruct the theory of non-cooperative games in extensive form with the minimal assumptions that are *necessary* for the goals of game theory.

What is meant by this admittedly bombastic phrase? With the benefit of hindsight today we can identify what the theory of non-cooperative games is capable of doing, and we can identify which building blocks are crucial for its achievements. For instance, one important step is to derive strategies from an extensive form representation of a game, in order to be able to define solution concepts like Nash equilibrium. Hence, a "minimal" assumption that is necessary for the goals of game theory is one that just makes sure that strategies can be derived from an extensive form, but not more. The way we think about such a step is to posit the derivation of strategies as a basic desideratum of the theory, and then go and try to *characterize* when this desideratum holds. A characterization is an "if and only if" statement. It, therefore, spells out conditions that are simultaneously sufficient *and* necessary for the desideratum at hand. That is, a formalism with weaker assumptions is bound to violate the desideratum.

The goal of this monograph is hence to develop a maximally general definition of an extensive form game as a framework for sequential interactive decision theories. This definition will, of course, cover finite games, but it will also encompass examples as exotic as decision problems in continuous time ("differential games"), transfinite cheap-talk games à la Aumann and Hart (2003), (infinitely) repeated games, infinite-horizon bargaining games à la Rubinstein (1982), and stochastic games à la Shapley (1953). In particular, this includes two types of generalizations with respect to other formalizations. First, no constraints will be imposed to restrict to particular classes of well-behaved games. For instance, in von Neumann games (von Neumann and Morgenstern 1944, chp. 2) the number of predecessors of every node in an information set is required to be the same. No such constraints will be imposed here. Second, there will be no extraneous finiteness or, at this point, even discreteness assumptions. In the set-up of this book immediate predecessors may not even exist (as in continuous time examples), the number of all predecessors of a node may not be finite (as in the case of terminal nodes for infinitely repeated games, but also for non-terminal nodes in more complex examples), and the number of immediate successors might be infinite (as in infinite-action games).

1.2.1 Kuhn's Graph Approach

The first question is how sequential interactive decisions should be modeled. The starting point ought to be decision theory under uncertainty, but it should be recognized that this theory is static in nature. The objects are lotteries over consequences (von Neumann and Morgenstern 1944), functions from states to consequences (Savage 1954), or functions from states to lotteries over consequences (Anscombe and Aumann 1963). Considering functions on a state space as the prime objects of analysis, as in the latter two approaches, is motivated by statistics. In probability theory and statistics random variables are functions on a state space, and an "event" is a set of states. That is, first nature rolls the dice and thereby determines a state; then this state determines the values of random variables once and for all. In such a context, decision making is about choosing a particular random variable, usually called an "act" or a "strategy." In a context where events may become known possible states are restricted to the event that occurred and probabilities are updated by Bayes' rule. Otherwise, each consecutive decision is treated like a static decision problem.

For sequential and possibly interpersonal decisions, however, a different approach is needed. The best-known domain for sequential decision making is probably the *tree* of the extensive form representation of a game, introduced in the *graph approach* due to Kuhn (1953). In finite cases a tree (see e.g. the left-hand side of Fig. 1.1) is simply a connected directed graph with no loops and a distinguished node, called the *root*, that precedes all other nodes and marks the beginning of the game. Trees serve as a transparent graphical model of how consecutive decisions are made. Each vertex (or *node*) represents a decision being made, by virtue of which one of the edges emanating from it is followed and a new vertex is reached. Further, for every vertex there exists a unique path from the root to it, which represents the chain of decisions which have to be made in order to reach that node.

Finite trees are intuitive and will most likely remain the tool for introducing extensive form games to undergraduate students. They are, however, an artificial construction, which many students struggle to relate to e.g. how potential decisions are evaluated. The latter requires a space of *outcomes* on which preferences ought to be defined. For finite examples, this can be identified with "terminal nodes" where no further decision is possible. For infinite-horizon examples they are more involved objects, specifically maximal (possibly infinite) paths along the tree, traditionally called "plays." Preferences are then defined on this derived space of plays, a constructed space with a non-intuitive structure. That is, if one follows Kuhn (1953) and starts with trees as graphs, the space of outcomes (plays) is implicit and has to be constructed.

1.2.2 Von Neumann and Morgenstern's Refined-Partitions Approach

Other authors, starting with von Neumann and Morgenstern (1944), favored a *refined-partitions approach* which more intuitively relates to how decisions refine the selection among possible outcomes, and more clearly relates to static decision theory. After all, game theory ought to be "interactive decision theory" (Aumann 1987, p. 460). The starting point is precisely the space of all possible ultimate outcomes, which is the natural space on which preferences should be defined. Initially, no outcome has been discarded, that is, the game starts at a set which contains all possible outcomes. Every decision partitions the set of outcomes, discarding a certain set thereof, and selecting a proper subset of the set of outcomes which have not yet been discarded. That is, decisions are made sequentially over (increasingly smaller) sets of outcomes. Hence, objects as e.g. preferences are defined on a natural domain. This approach generates a "set tree," where a node is a set of outcomes which might still be reached according to the chain of decisions still to be made. Further, every play (maximally long path of nodes) identifies one and only one outcome, and vice versa, hence plays and outcomes can be used interchangeably.

A well-known stylized (but illustrative) example is that of making a reservation at a restaurant. "Imagine that the only way that restaurants vary is in the menu of meals which they will serve. The individual is assumed to know the menus at all restaurants that he might select. Eventually, the individual will choose a meal, but his initial choice is of a restaurant/menu from which he will later choose his meal." (Kreps 1979, p. 565) This illustrates the modification needed when decision theory is turned sequential: Choosing a restaurant amounts to choosing a menu, which is a *set* of consequences—a set from which a further choice will be made later on. And an extension of preferences to sets opens the door to modeling further properties, e.g., the decision maker may choose a vegetarian restaurant because she foresees being tempted by steaks, or she may choose a regular restaurant because she values flexibility—which however concerns the specification of preferences. What matters at this point is how the natural extension of static decision theory to sequential decision making works. When decision theory is extended to encompass sequential decisions, consequences and states themselves become sets, and preferences and information are defined over sets of sets. This works exactly like in the theory of stochastic processes formalized as date-event trees. An *event* is a subset of a universal set of states. As information is revealed over time, the partition of the state space into events becomes progressively finer. Other objects fall in place naturally, for instance "strategies" or "acts" become functions from sets of states to sets of consequences.

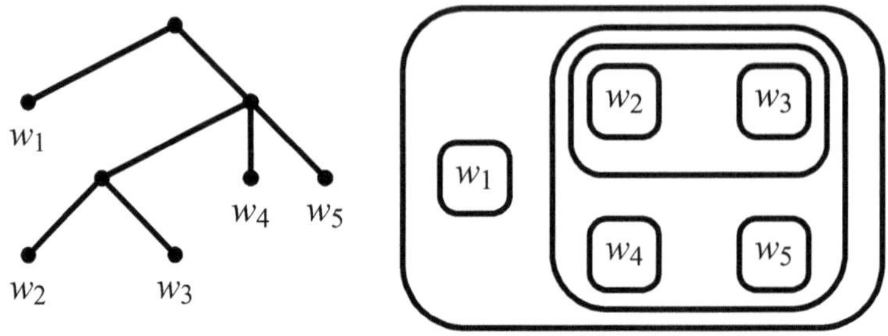

Fig. 1.1 Trees and outcomes. The *left-hand side* corresponds to the graph approach as in Kuhn (1953), the *right-hand side* to the refined-partitions approach as in von Neumann and Morgenstern (1944)

1.2.3 The Synthesis

The approach of von Neumann and Morgenstern (1944) predates the graph approach of Kuhn (1953). The set-up of von Neumann and Morgenstern (1944) was kept unnecessarily restrictive in the sense that every decision point in a given information set must have the same number of predecessors, that is, come after the exact same number of decisions. This constraint is one of the reasons which caused Kuhn (1953) to propose a game tree conceptualized as a finite directed graph (with a distinguished node, the root, that comes before any other node and marks the beginning of the game). This clearly provided a more general framework (for an explicit example see p. 199 of Kuhn 1953), but blurred the connection to decision theory. Of course, however, both approaches are related, and this relation is the starting point of our synthesis. Already von Neumann and Morgenstern (1944, pp. 60–79) discussed at length how trees emerge from successively refined information partitions, even though they confined themselves to the finite case. For small examples the relation between both approaches is indeed intuitive. Consider the left-hand side of Fig. 1.1, which represents an example of the graph approach. After all possible decisions have been made at the nodes, one of five possible outcomes is reached. Hence, the set of outcomes can be represented as $W = \{w_1, w_2, w_3, w_4, w_5\}$. This constructed set (which appears simple in this example, but will soon become complex for larger ones) is a representation of the space of outcomes, and hence preferences ought to be defined on it. The meaning of each w_k, however, has to be specified after the fact. The right-hand side of Fig. 1.1 represents the partition approach. The description of the decision problem starts with the set of outcomes. The problem is described by a collection of subsets thereof, and each decision reduces to the choice of a subset of the current set, in such a way that no intermediate subset is skipped.

In this book (or the research articles underlying it) we propose a synthesis of the decision-theoretic approach and Kuhn's graph-based alternative. In the spirit of von Neumann and Morgenstern (1944) we take the tree to be a collection of (nonempty) subsets of an underlying set of outcomes or plays, ordered by set inclusion. That is, the right-hand side of Fig. 1.1 is a tree, while the left-hand side is a handy graphical representation thereof. Yet, we do not insist on the partitional property of von Neumann and Morgenstern (1944) with respect to the players' information sets. Instead, we allow the same generality as Kuhn's graph does—and more. This results in a framework that encompasses all formalizations of extensive form games that have previously been proposed, inclusive of some exotic cases like games in continuous time. And, of course, we do not impose any cardinality assumptions at all.

More generally, the key for our approach is the equivalence (which we will explore in detail in Chap. 2) between trees as graphs and trees as set-theoretic structures. Following von Neumann and Morgenstern (1944, Section 8), we define nodes in a tree as subsets of a given set of possible ultimate outcomes, i.e., a node is the set of outcomes that may still occur conditional on the node having been reached. Hence a tree is a collection of nodes partially ordered by set inclusion, that is, a node x precedes a node y if $x \supseteq y$. A simple way to think about this approach is to start with the set of all potential outcomes that might occur (which is the space on which eventually players' preferences should be defined). The root is identified with the whole set. As the game proceeds, some outcomes are excluded, and each node is the set of outcomes which have not been discarded yet, and which might still be the ultimate outcome of play when the game proceeds through that node.

The relation to the traditional graph-approach of Kuhn (1953) is intuitive. In that approach, a node is an abstract decision point, and the tree is a partially ordered set under the precedence relation. One can simply consider the set of all "plays," that is, maximal chains of nodes from the root to the end of the game (where "end" might mean an infinite horizon). Each such play corresponds to one and only one ultimate outcome. Identify then each node with the set of plays which pass through the node, i.e., the set of plays which have not been discarded yet. The resulting object is a set representation of the original partially ordered set which corresponds to a tree viewed as a set of sets, as given above.

Our approach has many advantages. Some of them are convenient simplifications, others are of a more fundamental nature. Among the former is the fact that the set of plays (maximal chains of nodes, or, intuitively, complete histories) can be one-to-one identified with the underlying set of ultimate outcomes (as we will rigorously show in Chap. 2). A node then consists of the plays passing through it, and the underlying set can be taken to represent all plays. An element of the underlying set can thus be seen either as a possible outcome (element of some node) or as a play (maximal chain of nodes). Henceforth there is no need to distinguish between plays and outcomes. A further advantage is the fact that a clear-cut distinction is obtained between primitive concepts and derived ones. For instance, players' choices are sets of outcomes which might be chosen at certain points, and as such are primitives in the approach. In contrast, information sets are derived objects which can be deduced from choices, and strategies are simply functions from decision points to choices.

The more fundamental advantages of the approach arise from its generality. We have dropped all artificial constraints which were explicitly or implicitly present in previous approaches. As a consequence, down the road we are able to obtain rather general results and obtain insights which were obscured under other approaches. We anticipate a few examples here. In Chap. 5 we will be able to effectively draw the boundaries of game-theoretic analysis by exhibiting a characterization of which extensive form structures allow for strategies which induce unique outcomes, identifying the structural characteristics behind well-known problems in, e.g., continuous-time games. In Chap. 6 we will show that all previously known definitions of perfect recall (usually restricted to finite games) are actually equivalent even if the game has an infinite horizon, infinite action spaces, etc. Further, all of them are equivalent to a very natural but previously unknown condition which is related to the structure of players' choices. In Chap. 7, we will obtain the most general (topological) existence theorem for subgame perfect equilibria up to date, and actually we will show that our approach allows us to ask a characterization question, showing that, in a certain, natural sense, the existence theorem is as general as it can be because its hypotheses are both necessary and sufficient.

1.2.4 The Sequence Approach

The need to analyze infinite-horizon examples motivated a third approach in the literature, the *sequence approach*, which conceives of a tree as a collection of sequences of "actions." This one has been pioneered by Harris (1985) for perfect information games and extended to more general problems by Osborne and Rubinstein (1994). We are not going to provide a formal introduction to that, but the example in Fig. 1.1 can be used to illustrate how it works. Let $A = \{a_1, b_1, a_2, b_2, c_2, a_3, b_3\}$ be an abstract set of *actions*, like "move left at the beginning" (a_1) or "move right at the last decision point" (b_3). The meaning of the actions is illustrated in Fig. 1.2.

Adding a null sequence $\emptyset$, the tree would then be defined as the following set of eight sequences from A,

$$H = \{\emptyset, (a_1), (b_1), (b_1, a_2), (b_1, b_2), (b_1, c_2), (b_1, a_2, a_3), (b_1, a_2, b_3)\},$$

which readily map to the eight nodes identified on the right-hand-side of Fig. 1.1. In other words, the tree is constructed as a subset H of the set of all sequences from A, where H contains all *histories* that are consistent with the game. Though this is obvious in the finite case, the sequence approach can also cover infinite cases, provided a few conditions are met that make H a tree. Alternatively, an "action correspondence" may be specified that maps each history into available actions. In the example from Figs. 1.1 and 1.2 such a correspondence would map $\emptyset \mapsto \{a_1, b_1\}$, $(b_1) \mapsto \{a_2, b_2, c_2\}$, and $(b_1, a_2) \mapsto \{a_3, b_3\}$. More abstractly, the sequence approach amounts to embedding a subset H of possible histories into the (possibly infinite) product of the action subsets available at each decision point.

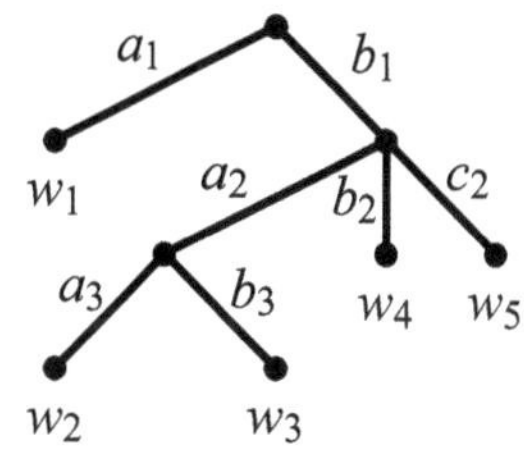

Fig. 1.2 Trees and sequences. Keeping track of the "actions" available at each node allows to identify each node with a sequence of previously taken actions

Intuitively, the correspondence between the sequence approach and the refined-partitions approach or our synthesis, as described above, is best understood by going through the graph approach. In the sequence approach, histories play the role of nodes in the graph approach. Some of such histories are maximal, that is, they correspond to (potentially infinite) plays. All others can be continued, extending them into longer histories reflecting subsequent decisions. For every such history, define the set of all plays which can be obtained by extending the given one. That set identifies the corresponding node (as a set of plays). In our approach, plays are equivalent to ultimate outcomes, and hence this identifies the corresponding tree as a set of sets of outcomes.

1.2.5 Some Illustrative Examples

The following simple example with infinite action spaces illustrates the relations between different formalizations of extensive form games.

Example 1.1 (Two Stages) Player 1 decides $a \in [0,1]$ at the root and, after observing a, player 2 decides $b \in [0,1]$. Following von Neumann and Morgenstern (1944), the root corresponds to a node encompassing the whole set of potential outcomes, $W = [0,1]^2$. At this node player 1 decides, which amounts to picking a smaller subset, a node of the form $\{a\} \times [0,1]$ where a has been selected but b remains to be decided. At each of these intermediate nodes player 2 decides, choosing a final node of the form $\{(a,b)\}$, which contains only one outcome. The game then ends. Hence, the representation of the game in our approach (or the refined-partitions approach) relies on a set N which collects all the sets (nodes) described above,

$$N = \left\{ W, (\{a\} \times [0,1])_{a \in [0,1]}, (\{w\})_{w \in W} \right\}$$

and is ordered by set inclusion. For this example, the graph approach is merely a representation of the set of sets above, with the member sets being interpreted as abstract nodes. The cardinality of the example prevents a graphical version (see Chap. 2 for details on the equivalence), but a qualitative illustration can be provided (see Fig. 1.3). The sequence approach would consider the root as a null sequence, the intermediate nodes as length-one sequences of the form (a), and the final nodes

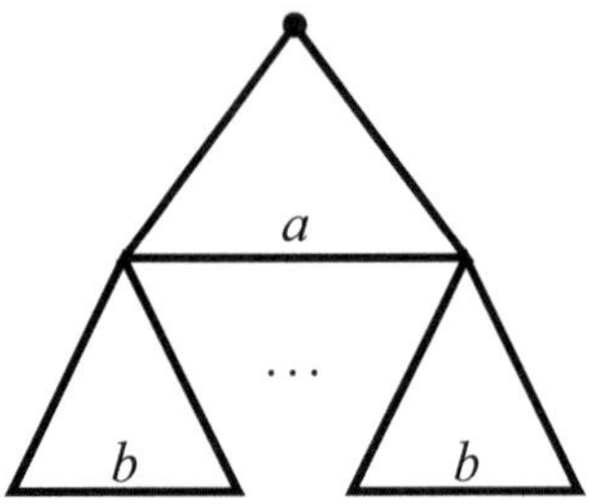

Fig. 1.3 Qualitative graphical representation of Example 1.1. Player 1 chooses a number $a \in [0,1]$. Player 2 observes a and then chooses a number $b \in [0,1]$. The game has two clearly defined "stages" and hence the equivalence between the graph approach, the refined-partitions approach, and the sequence approach is transparent

as length-two sequences (a,b). The correspondence to the other approaches is straightforward in this case.

Well-behaved examples, however, are misleading, as they create the impression that all approaches are equally well-suited for the analysis of general games. This is not the case. While the sequence approach is well-suited to apply a number of techniques from dynamic programming, it is not without problems. In particular, except for very regular structures where games have well-defined stages (as in the last example), it will typically blow up the dimension of the space to be considered by the analyst. A first intuition can be gained by a quick look at Fig. 1.2. There are clearly just five possible outcomes, $W = \{w_1, \ldots, w_5\}$, but if the possible actions are combined in a product fashion, one ends up with the set $\{a_1, b_1\} \times \{a_2, b_2, c_2\} \times \{a_3, b_3\}$, which contains 12 sequences, with the set W mapping into a strict subset thereof. The scope of the difficulties is illustrated by the following minimalistic example.

Example 1.2 (Square) Consider the following 2-player perfect information game, where player 1 picks at the root of the tree either a pair $(a,b) \in [0,1]^2$ with $a < 1$ or sets $a = 1$ and gives the move to player 2 (see Fig. 1.4 for a qualitative graphical representation). In the former case ($a < 1$) the game ends with payoffs $U_1(a,b) = ab$ and $U_2(a,b) = 1 - b$. In the latter case ($a = 1$) player 2 receives the opportunity to choose $b \in [0,1]$ and, once b has been selected, the game ends with payoffs $U_1(1,b) = b$ and $U_2(1,b) = 1 - b$. The set W of possible plays (outcomes) can be identified with the unit square $W = [0,1]^2$. Constructing the set of nodes is simple. The root and the terminal nodes are W and the collection $\{w\}$ for all $w \in W$, respectively. There is only one intermediate node (player 2's only decision point), namely the set $\{1\} \times [0,1]$ where $a = 1$ has been fixed but b remains to be decided. The set of nodes is hence

$$N = \left\{W, \{1\} \times [0,1], (\{w\})_{w \in W}\right\}$$

ordered by set inclusion.

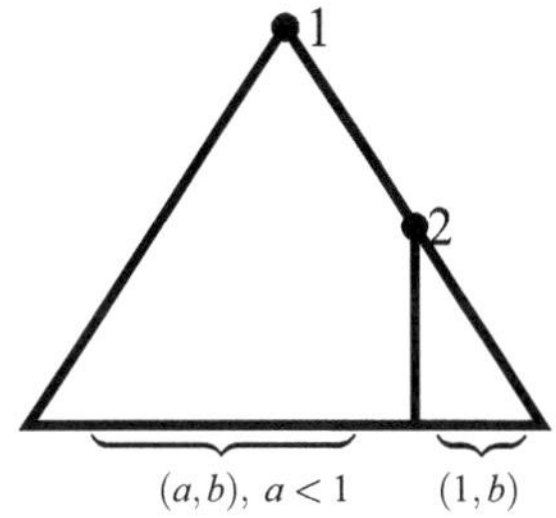

Fig. 1.4 Qualitative graphical representation of Example 1.2. Player 1 either chooses a pair (a,b) with $a<1$ or sets $a=1$ and lets player 2 decide b

The sequence approach is cumbersome in this example. Following Harris (1985), the natural choice of the action set for player 1 is $A_1=[0,1]^2$, since this player may (but need not) choose both a and b. For player 2 the action set is clearly $A_2=[0,1]$. This results in the product space $A=A_1\times A_2=[0,1]^3$. That is, the sequence approach turns a two-dimensional object into a three-dimensional one. To recover the structure of the game we must consider the set H of plays (viewed as sequences) as a subset of $[0,1]^3$. In the present case this must include all plays where player 1 has fixed both a and b, and player 2 does not get to choose. This requires the subset $H_1=\{((a,b_1),b_2)\mid b_1=b_2,\ a<1\}$ to be one part of the set of plays. The fact that player 2 does not decide on b is incorporated in H_1 by the restriction that $b_2=b_1$. The set $H\subseteq A$ must also include all plays where player 1 has chosen $a=1$ and player 2 decides on $b\in[0,1]$. This gives the set $H_2=\{((a,0),b)\mid a=1, b\in[0,1]\}$. The coordinate 0 in $(a,0)$ is an arbitrary marker indicating that player 1 does not actually decide on b. Any other marker would also do, but it would be incorrect to write (a,b), because if $a=1$ player 2's choice of b is unconstrained by player 1's decision. The set of plays finally is the subset of $[0,1]^3$ given by $H=H_1\cup H_2$.

This example (as several other examples in Chap. 7) illustrates that the sequence approach has serious drawbacks. It may be appropriate for games that have a clear-cut stage structure, like repeated games for instance, but for the general case the graph or refined-partitions approach are preferable, in particular when large games are at stake.

The graph approach and the refined-partitions approach are formally at variance. Intuitively, one should be able to use both interchangeably, and indeed for finite examples, as illustrated above, it is not difficult to move back and forth. However, once the domain of game theory is expanded to include, at the very least, infinite-horizon infinite-action games with complex information structures, it is unclear which hidden assumptions one or the other formalization might need, and whether subtle elements of the formalization may prevent an actual equivalence. The issue thus arises of whether a formalism can be given that makes clear in which sense and under which circumstances both approaches describe the same objects. To the best of our knowledge the graph approach has not been worked out in full generality, that is, without finiteness assumptions. The approach of this book is a dual one. On the one hand, we tackle the most general version of the refined-partition approach. On the other hand, we will develop equivalence results which show that our approach also captures the natural generalization of the graph approach.

1.2.6 Plan of the Book

One way to view this work is, hence, as the generalization of previous approaches to modelling extensive form games. In two respects our formalization differs from the received ones, though, because when a general enough framework is developed, it becomes obvious that certain modelling choices might not have been optimal. First, instead of emphasizing information sets, we focus on "choices," that is, on what players can do when it is their turn to decide. A *choice* is understood as a selection from the plays or outcomes that have not yet been discarded when the player chooses. The reason for this is both formal and practical. Information sets, which many textbook definitions of games take as a primitive, are just the collections of decision points at which certain menus of choices are available, thus not necessary for the formalism—they can be ignored because they can always be reconstructed, if desired. Likewise, strategies are derived objects, namely mappings assigning choices to nodes. Second, we do not insist that at each node in the tree at most one player is called upon to choose. Instead, we allow (possibly) several players to decide at the same node. To our minds this is a more clear-cut model of simultaneous decisions than the cascading information sets in the traditional set-up. Of course, nothing in our formalism excludes cascading information sets, but the flexibility to do without them allows for simpler formalizations when many players decide simultaneously.

The plan of the book is as follows. Chapter 2 examines the concept of tree and establishes the equivalence between the generalized approaches based on graphs and refined partitions. The resulting framework builds the basis for the formalization of extensive form games. Chapter 3 briefly gives the connection to the concept of pseudotree in order theory and presents original results which help understand the generality of the approach. Chapter 4 presents the general concept of an *extensive decision problem*, which captures the idea of what an extensive form game should be and is general enough to cover everything from finite games to infinite-action, infinite-horizon examples, stochastic games, and even continuous-time decision problems. Chapter 5 then relies on this framework to examine the desiderata mentioned above, namely which constraints are necessary and sufficient for every strategy profile to induce one and only one outcome. The resulting characterization identifies the proper domain of game theory and results in the concept of an *extensive forms*. Chapter 6 then adds a single further restriction, down-discreteness, and examines the resulting framework (discrete extensive forms), which essentially characterizes games where the immediate predecessor function is well defined, and which is still general enough to cover almost all economic applications. Chapter 7 turns to (subgame perfect) equilibria of perfect information games and presents the most general, topological existence theorem to date. Further, the result is formulated as a characterization, in the sense that it identifies necessary and sufficient conditions (on the topology on the outcome space of a fixed discrete extensive form) for every profile of continuous preferences to give rise to an equilibrium.

References

Anscombe F, Aumann RJ (1963) A definition of subjective probability. Ann Math Stat 34:199–205

Aumann RJ (1964) Mixed and behavior strategies in infinite extensive games. In: Advances in game theory. Annals of mathematics study, vol 52. Princeton University Press, Princeton, pp 627–650

Aumann RJ (1987) Game theory. In: Eatwell J, Milgate M, Newman P (eds) The new Palgrave, a dictionary of economics, vol 2. Macmillan, London, pp 460–482

Aumann RJ, Hart S (2003) Long cheap talk. Econometrica 71:1619–1660

Bertrand J (1883) Théorie mathématique de la richesse sociale. J des Savants 67:499–508

Cournot AA (1838) Recherches sur les principes mathématiques de la théorie des richesses. Hachette, Paris

Debreu G, Scarf H (1963) A limit theorem on the core of an economy. Int Econ Rev 4:236–246

Harris C (1985) Existence and characterization of perfect equilibrium in games of perfect information. Econometrica 53:613–628

Harsanyi JC (1967/1968) Games of incomplete information played by Bayesian players, i, ii, and iii. Manag Sci 14:159–182, 320–334, 486–502

Harsanyi JC, Selten R (1988) A general theory of equilibrium selection in games. MIT-Press, Cambridge

Kreps DM (1979) A representation theorem for "preference for flexibility." Econometrica 47:565–577

Kuhn H (1953) Extensive games and the problem of information. In: Kuhn H, Tucker A (eds) Contributions to the theory of games, vol II. Annals of mathematics study, vol 28. Princeton University Press, Princeton, pp 217–243

Nash JF (1950) Equilibrium points in n-person games. Proc Natl Acad Sci 36:48–49

Osborne MJ, Rubinstein A (1994) A course in game theory. MIT, Cambridge

Rubinstein A (1982) Perfect equilibrium in a bargaining model. Econometrica 50:97–109

Shapley L (1953) Stochastic games. Proc Natl Acad Sci 39:1095–1100

Savage L (1954) The foundations of statistics. Wiley, Hoboken

von Neumann J (1928) Zur Theorie der Gesellschaftsspiele. Mathematische Annalen 100:295–320

von Neumann J, Morgenstern O (1944) Theory of games and economic behavior. Princeton University Press, Princeton

Zermelo E (1913) Über eine Anwendung der Mengenlehre auf die Theorie des Schachspiels. In: Proceedings of the fifth international congress of mathematicians, Cambridge, vol 2, pp 501–504

Chapter 2
Game Trees

In this chapter we focus on the concept of a tree, which is the main tool to model sequential decisions.[1] In a first approximation, sequential decision theory can be seen as an extension of decision theory under uncertainty, which is static in the sense that it considers one-shot decisions. While decisions under risk can be captured by using lotteries over consequences (von Neumann and Morgenstern 1944), modeling decisions under uncertainty requires a more complex framework, where the objects of analysis are either functions from states to consequences (Savage 1954) or functions from states to lotteries over consequences (Anscombe and Aumann 1963). These approaches have a purely statistical interpretation: nature rolls the dice first and thereby determines a state; then this state determines the values of the lotteries (random variables) once and for all. An "event" is a set of states, and decision making is about choosing a particular random variable (viewed as a function on the state space), usually called an "act" or a "strategy."

In a sequential context, events may become known before a particular decision is made, giving rise to the need to model contingent plans. The events that occurred previously restrict the possible states, and the probabilities of those are updated by Bayes' rule. Otherwise, each consecutive decision can still be treated like a static decision problem. One can think of decisions as progressively refining the set of not-yet-discarded final outcomes. An illustrative example is the restaurant example already mentioned in the introduction (Kreps 1979, p. 565), where one first chooses a restaurant (viewed as a menu or set of possible meals), and only then chooses a meal. This "menu choice" approach has been generalized so as to capture features that appear relevant to sequential decisions: preference for flexibility (Kreps 1979), temptation and self-control (Gul and Pesendorfer 2001), or unforeseen contingencies (Dekel et al. 2001).[2] Such issues can be addressed by

---

[1] The material presented here is a revised and expanded version of the main part of Alós-Ferrer and Ritzberger (2005a), but some parts of that article will be dealt with in Chap. 4.

[2] For alternative approaches see Blume et al. (1991a,b).

C. Alós-Ferrer, K. Ritzberger, *The Theory of Extensive Form Games*,
Springer Series in Game Theory, DOI 10.1007/978-3-662-49944-3_2

introducing preferences over subsets, rather than elements, of the space of possible consequences.

Modeling sequential decisions over (increasingly smaller) sets of outcomes is the essence of the *refined-partitions approach* that we anticipated in the introduction. Besides this framework the traditional domain for sequential *and interactive* decision making is the *tree* of the extensive form representation of a game (Kuhn 1953). In the finite case, a tree is usually defined as a directed connected graph without loops and with a distinguished node (the "root") that precedes all other nodes. Trees serve as a graphical, transparent model of how consecutive decisions refine the selection among possible outcomes, and hence we term this alternative view the *graph approach.*

At first glance, the two approaches seem to be at odds. However, trees are closely related to collections of subsets of an underlying space of consequences or outcomes, as already observed by von Neumann and Morgenstern (1944, p. 65) for the finite case. The issue thus arises whether arbitrary trees can be recast into collections of subsets of some underlying space, thereby making them an adequate domain for sequential decision theory.

The present chapter addresses this issue. Starting from the order-theoretic concept of a tree (e.g. the traditional graphical model) but without any finiteness or discreteness assumptions it is shown that the tree can be represented as a set of sets with a particular structure. To interpret elements of these sets as consequences/outcomes requires that maximal chains of sets (called *plays*) identify elements that all sets in the chain contain. In the language of the restaurant example, a menu needs to correspond to a collection of meals, because meals are what is ultimately consumed.

Every tree has a set representation that meets this requirement. Characterizing this leads to a definition of set trees that is appropriate for sequential decision theory. For these set trees a node is like an event in probability theory, i.e. as a set of states. Furthermore, elements of these sets/nodes correspond to plays, thereby providing a framework for sequential decision theory.

The present chapter is a first step towards a general definition of an extensive form as a framework for truly sequential decision theories. This is why we start with utmost generality. For instance, in von Neumann games (von Neumann and Morgenstern 1944, chp. 2) the number of predecessors of every node in an information set is required to be the same. In our set-up immediate predecessors may not even exist, and the number of predecessors may not be finite. This allows us to consider examples as exotic as decision problems in continuous time ("differential games"), transfinite cheap-talk games à la Aumann and Hart (2003), (infinitely) repeated games, bilateral bargaining games à la Rubinstein (1982), and stochastic games à la Shapley (1953).

2.1 Preview

Every tree can be represented as a collection of (sub)sets (of some underlying set) with a particular structure (Proposition 2.1). But in a set-theoretic environment this structure can be "cleaned" without affecting the properties of the trees. We "clean" in three steps, where each step corresponds to adding structure that enables increasingly specific interpretations of the tree.

In Sect. 2.2 characterizing set representations of trees (Proposition 2.2(a)) reveals that the set-theoretic analog of the defining order-theoretic structure can be modified such that *unordered* nodes correspond to *disjoint* sets. Every tree has a set representation that satisfies this "Trivial Intersection"-property (Proposition 2.2(b)). Next, ruling out trivial structures in the tree leads to "decision trees." Characterizing set representations of these (Proposition 2.3(a)) shows that, under Trivial Intersection, the set-theoretic analog of the defining order-theoretic property of decision trees is strengthened, to "Separability" (Lemma 2.4). Every decision tree has a set representation that satisfies the two strong properties (Trivial Intersection and Separability; Proposition 2.3(b)). Moreover, every decision tree has a "canonical" set representation, where the elements of the underlying set are *plays* (Theorem 2.1) that satisfies the strong properties (by Lemma 2.5 and Corollary 2.1). This yields the first milestone: "set trees."

When can the elements of the underlying set be perceived as representing plays, as suggested by Theorem 2.1? As discussed in Sect. 2.3, this requires the underlying set to be neither too large nor too small.

In a "reduced form" (Proposition 2.4, Sect. 2.3), where redundancies in the underlying set are eliminated, the elements of the underlying set map *one-to-one* into plays (Proposition 2.5(a)). In fact, Trivial Intersection is equivalent to this property (Proposition 2.5(b)). A set tree is already in reduced form if, roughly, no element of the underlying set can be dropped: if the set tree is "irreducible." This is equivalent to the elements of the underlying set in the reduced form being the singleton sets of the originally underlying set (Proposition 2.6). Defining a "proper" order isomorphism as an order isomorphisms that preserves the "strong" properties (Lemma 2.11) yields a characterization: a collection of sets is a set tree if and only if it is properly isomorphic to its reduced form and the latter is an irreducible set tree (Theorem 2.2). If Irreducibility of the reduced form implies Irreducibility of the original set tree, the two are "doubly isomorphic." Irreducible set trees are precisely those that are doubly isomorphic to their reduced forms (Proposition 2.7). This clarifies when the underlying set is not too large.

It may still be too small. "Boundedness" of a set tree, as introduced in Sect. 2.3.4, ensures that the elements of the underlying set map *onto* plays. Achieving this involves, possibly, enlarging the underlying set. Irreducible set trees are precisely those where elements can be added so that every play is represented by a distinct element of the underlying set (Proposition 2.8). Hence, a set tree is bounded if and only if the elements of the underlying set in the reduced form represent *all* plays (Proposition 2.9). This yields the second milestone: "game trees," defined as bounded irreducible set trees.

As shown in Sect. 2.4, game trees are precisely those for which there is a bijection between elements of the underlying set and plays; equivalently, they are precisely those which are decision trees that are their own "canonical" set representation by plays (Theorem 2.3). Hence, there is no loss of generality in assuming Boundedness and Irreducibility. Yet, Boundedness does not necessarily imply that the singletons from the underlying set belong to the set of nodes. But adding the singletons as nodes to the tree does not change any essential features of the tree, provided it is a game tree (Proposition 2.11). This yields the third and last milestone: "complete game trees." A set tree is a complete game tree if and only if it is irreducible and every play has a minimum (Proposition 2.12).

2.2 Set Representations

A *preordered set* is a pair $(N, \geq)$ consisting of a nonempty set N and a reflexive and transitive binary relation $\geq$ on N.[3] A preordered set $(N, \geq)$ for which the relation $\geq$ is antisymmetric is a (partially) *ordered set* (or a *poset*). In particular, a V-*poset* is a poset $(M, \supseteq)$ where M is a collection of nonempty subsets of a given set V and $\supseteq$ is set inclusion. A nonempty subset $c \subseteq N$ of a preordered set $(N, \geq)$ is a *chain* if for all $x, y \in c$ either $x \geq y$ or $y \geq x$ (or both), i.e. if the induced preorder on c is complete.

Given a preordered set $(N, \geq)$ and an element $x \in N$ define the *up-set* (or *order filter*) $\uparrow x$ and the *down-set* (or *order ideal*) $\downarrow x$ by

$$\uparrow x = \{y \in N \,|\, y \geq x\} \text{ and } \downarrow x = \{y \in N \,|\, x \geq y\} \tag{2.1}$$

Let $\downarrow N = \{\downarrow x \,|\, x \in N\} \subseteq 2^N$ denote the set of all down-sets of $(N, \geq)$.

At this point, down-sets and up-sets are just convenient formal constructions. Intuitively, however, in the context of sequential decision problems the up-set should capture the "past" and the down-set should correspond to the "future".

An *order isomorphism* between two preordered sets $(N_1, \geq_1)$ and $(N_2, \geq_2)$ is a bijection $\varphi : N_1 \to N_2$ such that

$$x \geq_1 y \text{ if and only if } \varphi(x) \geq_2 \varphi(y) \tag{2.2}$$

for all $x, y \in N_1$.[4] This last property is referred to as "order embedding." Order isomorphism is an equivalence relation on the class of all preordered sets. Two

---

[3]See the Mathematical Appendix A.1.2 for definitions of properties of binary relations.

[4]See the Mathematical Appendix A.1.3 for definitions of properties of functions.

order-isomorphic preordered sets can be regarded as identical for all practical purposes.

Remark 2.1 If $(N_1, \geq_1)$ is a *poset*, $(N_2, \geq_2)$ a preordered set, and $\varphi : N_1 \to N_2$ an order embedding function, then φ is necessarily injective (one-to-one). For, given $x, y \in N_1$ such that $\varphi(x) = \varphi(y)$, reflexivity of $\geq_2$ implies $\varphi(x) \geq_2 \varphi(y)$ and $\varphi(y) \geq_2 \varphi(x)$ and hence $x \geq_1 y$ and $y \geq_1 x$ (by the if-part of (2.2)), together implying $x = y$ (by antisymmetry for $\geq_1$). In particular, any order-embedding surjection between two posets is an order isomorphism.

A preordered set $(N, \geq)$ *admits a set representation* if there is an order isomorphism between $(N, \geq)$ and a V-poset $(M, \supseteq)$. We say then that $(M, \supseteq)$ *represents* $(N, \geq)$. The following result shows that the class of preordered sets admitting set representations coincides with the class of posets (see also Birkhoff 1973, Theorem 21).

Proposition 2.1 *A preordered set $(N, \geq)$ admits a set representation if and only if it is a poset. In particular, any poset $(N, \geq)$ can be represented by the N-poset $(\downarrow N, \supseteq)$.*

Proof "if:" Suppose $(N, \geq)$ is a poset. Then $\varphi : N \to \downarrow N$ given by $\varphi(x) = \downarrow x$ for each $x \in N$ is onto by construction. Let $x, y \in N$ and $y \geq x$. Consider any $z \in \downarrow x$. By transitivity $y \geq x \geq z$ implies $z \in \downarrow y$, so $\varphi(y) \supseteq \varphi(x)$. Conversely, let $x, y \in N$ and $\varphi(y) \supseteq \varphi(x)$. Then $x \in \downarrow x = \varphi(x) \subseteq \varphi(y) = \downarrow y$ implies $y \geq x$. Thus, $y \geq x \Leftrightarrow \varphi(y) \supseteq \varphi(x)$ shows that φ is order embedding. By Remark 2.1 an order embedding surjection is an order isomorphism.

"only if:" Let $(N, \geq)$ be a preordered set which admits a set representation. Let $(M, \supseteq)$ be the associated poset and $\psi : N \to M$ the order isomorphism. If both $x \geq y$ and $y \geq x$ hold for some $x, y \in N$, then by (2.2) $\psi(x) = \psi(y) \in M$ implies $x = y$ because ψ is one-to-one. Hence, $\geq$ is antisymmetric. ∎

Proposition 2.1 identifies an order isomorphism, $\varphi(x) = \downarrow x$ for all $x \in N$, between N and $\downarrow N$. The resulting set representation is referred to as the *set representation by principal (order) ideals*.[5]

2.2.1 Trees and Subtrees

Definition 2.1 A **tree** is a poset $(N, \geq)$ such that $\uparrow x$ is a chain for all $x \in N$. In a tree the elements of N are called **nodes**. For nodes $x, y \in N$ say that x **precedes** (resp. follows) y if $x \geq y$ (resp. $y \geq x$) and $x \neq y$. A tree is **rooted** if there is a node $r \in N$, called the root, such that $r \geq x$ for all $x \in N$.

[5] Similar results are known, for instance, for finite arbitrary ordered sets (Davey and Priestley 1990, Theorem 8.19).

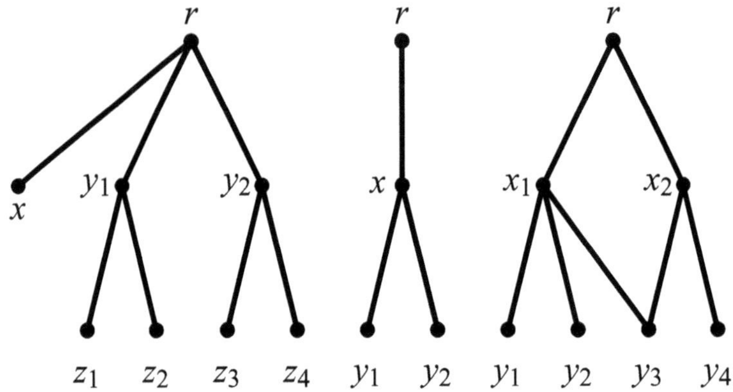

Fig. 2.1 The *left-hand-side* poset is a tree and a decision tree. The poset in the *middle* is a tree but not a decision tree. The *right-hand side* poset is not a tree

Nodes that are properly followed by other nodes are called *moves*. That is, $X = \{x \in N \mid \downarrow x \setminus \{x\} \neq \emptyset\}$ is the set of all moves. Nodes that are not properly followed by other nodes are called *terminal*, and the set of terminal nodes is denoted by $E = N \setminus X$.

The condition that $\uparrow x$ is a chain for all $x \in N$ is very intuitive. Consider the collection of all past and future events in your life as a tree. No matter how much we might regret certain past choices, the past is the past, and bygones are bygones. If we were to write the story of our lives, the collection of events and decisions before the present time would read as a linear description, with one decision neatly following another one. Alternative options that were not taken would not be part of it. The up-set $\uparrow x$ captures the past (seen from x), and hence the nodes in it should be completely ordered, making $\uparrow x$ a chain. In contrast, the down-set $\downarrow x$ captures the future (seen from x), which will in general contain many alternative paths branching out as decisions are made. Hence, each down-set will be a tree in itself: a subtree.

For finite cases, it is easy to produce graphical representations of trees or even general posets. Represent each node by a dot, start with the root at the top of a graph, and draw a line from each node down to the nodes that follow it, unless there is an intermediate node (in such a way, the notation for up- and down-sets becomes intuitive). In such a graph, $x \geq y$ if there is a sequence of lines joining x to y and x is above. Figure 2.1 presents three examples. The left-hand-side graph and the graph in the middle represent trees. As can be seen, the up-set of any node is a chain, while the down-set of any node (except the ones at the "end") is a subtree. The right-hand-side graph, however, is not a tree, because the up-set of the node y_3 is not a chain (the past is not linear).

Definition 2.1 could, of course, also be stated dually, i.e. with an element that is not followed by other nodes ("bottom" instead of "top") and $\downarrow x$, but here the opposite convention is preferred. The reason is that with this convention, we will be able to associate the order relation $\geq$ on an abstract tree with set inclusion $\supseteq$ on its set representation. With this caveat, this is the most general definition of trees in order theory (see Chap. 3 for further details).

The tree-property is preserved by order isomorphism, i.e., if a poset is order isomorphic to a tree, then it is itself a tree. By Proposition 2.1, every tree $(N, \geq)$ has a set representation by principal (order) ideals, $(\downarrow N, \supseteq)$. This is called the tree's *set representation by subtrees*, as for any $x \in N$ the ordered set $(\downarrow x, \geq)$ is itself a tree. Writing the defining property of a tree more explicitly yields:

Lemma 2.1 *A poset $(N, \geq)$ is a tree if and only if, for all $x, y, z \in N$,*

$$\text{if } y \geq x \text{ and } z \geq x \text{ then } y \geq z \text{ or } z \geq y \tag{2.3}$$

If this last property is translated into set-theoretic terms, two alternatives emerge. A V-poset $(M, \supseteq)$ satisfies *Trivial Intersection* if, for all $a, b \in M$,

$$\text{if } a \cap b \neq \emptyset \text{ then } a \subset b \text{ or } b \subseteq a \tag{2.4}$$

and it satisfies *Weak Trivial Intersection* if, for all $a, b, c \in M$,

$$\text{if } c \subseteq a \cap b \text{ then } a \subset b \text{ or } b \subseteq a \tag{2.5}$$

Of course, Trivial Intersection implies Weak Trivial Intersection. If Weak Trivial Intersection is written in terms of an abstract partial order $\geq$, property (2.3) is obtained. Trivial Intersection, on the other hand, cannot be translated back into arbitrary posets, since there is in general a difference between an intersection of two nodes being empty and not containing any other node.

Proposition 2.2 *(a) A V-poset $(M, \supseteq)$ is a tree if and only if it satisfies Weak Trivial Intersection, (2.5).*

(b) A poset $(N, \geq)$ is a tree if and only if its set representation by principal ideals $(\downarrow N, \supseteq)$ satisfies Trivial Intersection, (2.4).

Proof (a) It suffices to notice that Weak Trivial Intersection is equivalent to property (2.3) and apply Lemma 2.1.

(b) "if:" If $(\downarrow N, \supseteq)$ satisfies Trivial Intersection, then it satisfies Weak Trivial Intersection and by part (a) it is a tree. By isomorphism $(N, \geq)$ is a tree.

"only if:" Let $(N, \geq)$ be a tree and let $x, y \in N$ such that $\downarrow x \cap \downarrow y \neq \emptyset$. Next, let $z \in \downarrow x \cap \downarrow y$. It follows that $\downarrow z \subseteq \downarrow x \cap \downarrow y$. By isomorphism and part (a), $(\downarrow N, \supseteq)$ satisfies Weak Trivial Intersection, and hence either $\downarrow x \subset \downarrow y$ or $\downarrow y \subseteq \downarrow x$. ■

Any set representation of a tree is necessarily a tree, and, hence, Proposition 2.2(a) characterizes *all* set representations of trees. Proposition 2.2(b) establishes that trees can also be characterized as those posets whose set representations by principal ideals satisfy Trivial Intersection. The implication is rather natural if one observes that, for $(\downarrow N, \supseteq)$, there is no difference between an intersection of two nodes being empty and not containing any other node, i.e. Weak Trivial Intersection and Trivial Intersection are the same property for this particular set representation. Still, there may be set representations of a tree which satisfy Weak Trivial Intersection but not Trivial Intersection.

Example 2.1 Let $(M, \supseteq)$ be the $\{1,2,3\}$-poset given by

$$M = \{\{1,2,3\},\{1,2\},\{2,3\}\}$$

Trivial Intersection fails, because $\{1,2\} \cap \{2,3\} \neq \emptyset$ and neither of the nodes contains the other. But Weak Trivial Intersection holds, because its hypothesis is void (for $\{1,2\}$ and $\{2,3\}$). The set representation by subtrees is given by $M' = \{\{s_1, s_2, s_3\}, \{s_2\}, \{s_3\}\}$, where $s_1 = \{1,2,3\}$, $s_2 = \{1,2\}$, and $s_3 = \{2,3\}$. Now $\{s_2\} \cap \{s_3\} = \emptyset$, because nodes are elements of 2^M rather than M.

But this set representation is still not satisfactory. Intuitively, one would like to remove the redundant element 2 from $V = \{1,2,3\}$ and obtain a set representation in terms of a $\{1,3\}$-poset $(M'', \supseteq)$ with

$$M'' = \{\{1,3\},\{1\},\{3\}\}$$

This points to a fundamental question. In M' the primitives are the nodes, i.e., M' is formed by subsets of an underlying set (of sets). By contrast, in M'' only ultimate "outcomes" are elements of an underlying set of which the elements of M'' are subsets.

Intuitively, chains of nodes in a tree represent possible ways in which the underlying decision process may proceed. Maximal chains will then correspond to complete descriptions of all decisions taken from the beginning to the end. The classical name in game theory for such descriptions is "plays."

For a poset $(N, \geq)$ a chain $c \in 2^N$ is *maximal* if there is no $x \in N \setminus c$ such that $c \cup \{x\}$ is a chain. Every chain is contained in a maximal chain by the Hausdorff Maximality Principle (an equivalent form of the Axiom of Choice and, hence, Zorn's Lemma; see Birkhoff 1973, chp. VIII, or Hewitt and Stromberg 1965, chp. 1).

Hausdorff Maximality Principle. Let $(N, \geq)$ be a poset and $c \subseteq N$ a chain in N. Then there exists a maximal chain w in N such that $c \subseteq w$.[6]

For a tree $(N, \geq)$ a *play* w is a maximal chain in N. Denote by W the set of all plays. Given a node $x \in N$, let $W(x) = \{w \in W \,|\, x \in w\}$ be the set of all plays *passing through* x.

At this point plays are a priori complicated objects. E.g., in a V-poset, where nodes are themselves sets (subsets of V), plays are sets of subsets of V. Hence, a set of plays is a set of sets of subsets of V. Yet, there is a natural way of constructing trees in which plays, rather than being complex objects, are the primitives on which

[6] This is also known as Kuratowski's Lemma; see e.g. Hewitt and Stromberg (1965) and the Mathematical Appendix A.1.2.

decision theory can operate. To motivate this construction we begin with a few examples.

2.2.2 *Motivating Examples*

In this subsection examples are presented that illustrate the generality of the approach. Some of these will repeatedly be referred to as we proceed.

2.2.2.1 Osborne-Rubinstein Trees

A particular case of Definition 2.1 is the concept of a tree from the textbook by Osborne and Rubinstein (1994). This, in turn, is a generalization of the original concept by Kuhn (1953) of finite trees; but, by following the sequence approach, it allows for both infinitely many choices at each move and for infinitely long plays.

Let A be a set of "actions" and H a set of (finite or infinite) sequences from A such that (i) $\emptyset \in H$, (ii) if $(a_\tau)_{\tau=1}^{t} \in H$ (where t may be infinite) and $k < t$ then $(a_\tau)_{\tau=1}^{k} \in H$, and (iii) if an infinite sequence $(a_\tau)_{\tau=1}^{\infty}$ satisfies $(a_\tau)_{\tau=1}^{t} \in H$ for every positive integer t then $(a_\tau)_{\tau=1}^{\infty} \in H$. The set of plays can be identified with $W = W_\infty \cup W_F$, where

$$W_F = \left\{ h = (a_\tau)_{\tau=1}^{t} \in H \,\middle|\, t \text{ finite and } \nexists a_{t+1} \in A : (a_\tau)_{\tau=1}^{t+1} \in H \right\}$$

and $W_\infty = \{h \in H | h = (a_t)_{t=1}^{\infty}$ for some $a_t \in A, t = 1, \ldots, \infty\}$. The ordering of nodes is given by $(a_\tau)_{\tau=1}^{t} \geq (a'_\tau)_{\tau=1}^{k}$ if $t \leq k$ and $a_\tau = a'_\tau$ for all $\tau = 1, \ldots, t$, and by $\emptyset \geq h$ for all $h \in H$. Then, $(H, \geq)$ is a rooted tree. To see this, observe that the definition of the ordering on nodes implies that $\uparrow (a_\tau)_{\tau=1}^{t}$ is a chain.

2.2.2.2 Long Cheap Talk

In the Osborne-Rubinstein trees plays contain at most countably many nodes. Some games, though, involve larger trees. One example where plays are "longer" (in an ordinal sense) is the following *long cheap-talk* game by Aumann and Hart (2003). First, in the information phase, a finite two-player normal-form game is selected at random from a given set Γ. All games in Γ have the same action sets, A_1 and A_2. Second, a talk phase goes on for infinitely many periods $1, 2, \ldots$ In every period each player sends a message to the other from a given set of message M (containing at least two elements). Third, *after* these infinitely many periods of cheap-talk, the action phase takes place, where players simultaneously choose actions in the normal form game $\gamma \in \Gamma$.

The tree for this "transfinite" game cannot be defined in the Osborne-Rubinstein framework, because the "length" of the plays is $\omega+1$, where ω is the first infinite ordinal. It fits the present framework, though. Let $A=A_1\times A_2$ and $(M^2)^\infty$ be the set of all sequences of elements from $M^2=M\times M$. Denote a typical sequence $m\in(M^2)^\infty$ by $m=(m_1,\ldots,m_t,\ldots)$, where $m_t=(m_t^1,m_t^2)$. The underlying set V is given by $V=\Gamma\times\ (M^2)^\infty\times A$. The set of nodes, ordered by set inclusion, is the union $N=N_I\cup N_T\cup N_A$. Nodes in N_I model the information phase: $N_I=\{V\}\cup\{x(\gamma)\,|\,\gamma\in\Gamma\}$, where $\{V\}$ is the root and $x(\gamma)=\{(\gamma',m,a)\in V\,|\,\gamma'=\gamma\}$ are the moves reached after the normal form game $\gamma\in\Gamma$ has been selected. The first messages are sent at the nodes $x(\gamma)$. The set N_T contains all the intermediate moves, i.e. nodes of the talk phase,

$$N_T=\left\{x_t(\gamma,m)\,\middle|\,\gamma\in\Gamma,m\in(M^2)^t,t=1,2,\ldots\right\}\cup\left\{x_\infty(\gamma,m)\,\middle|\,\gamma\in\Gamma,m\in(M^2)^\infty\right\}$$

where

$$x_t(\gamma,m)=\left\{(\gamma',m',a)\in V\,|\,\gamma'=\gamma\text{ and }m'_\tau=m_\tau\text{ for all }\tau=1,\ldots,t\right\}$$

and

$$x_\infty(\gamma,m)=\left\{(\gamma',m',a)\in V\,|\,\gamma'=\gamma\text{ and }m'=m\right\}$$

The nodes $x_t(\gamma,m)$ are those where t pairs of messages have already been sent and the $(t+1)$-pair has to be chosen. At the nodes $x_\infty(\gamma,m)$ infinitely many messages have already been sent, and the actions have to be taken. Finally, the set N_A contains the "terminal nodes" selected in the action phase, $N_A=\{\{v\}\,|\,v\in V\}$. Obviously, in this well-behaved example there is a three-sided, one-to-one correspondence between the terminal nodes in N_A, the elements of the underlying set V, and the set of plays. To complete the game, payoffs can be defined either on plays, on the "terminal nodes" in N_A, or on the underlying set V.

2.2.2.3 Bilateral Bargaining

The next example corresponds to the tree of an infinite bilateral bargaining game à la Rubinstein (1982). Let $V=(\cup_{t=1}^\infty V_t)\cup V_\infty$ where, for some fixed $\delta\in(0,1)$,

$$V_t=\left\{(a_\tau)_{\tau=1}^t\,|a_\tau\in\left[0,\delta^{\tau-1}\right],\text{ for all }\tau=1,\ldots,t\right\}$$
$$V_\infty=\left\{(a_\tau)_{\tau=1}^\infty\,|a_\tau\in\left[0,\delta^{\tau-1}\right],\text{ for all }\tau\geq 1\right\}$$

and let N be given by V together with $(\{v\})_{v\in V_\infty}$ and

$$\left(\left(\{v\in\cup_{\tau=t}^\infty V_\tau\,|a_\tau=\bar a_\tau,\forall\tau=1,\ldots,t\}\right)_{(\bar a_1,\ldots,\bar a_t)\in\times_{\tau=1}^t[0,\delta^{\tau-1}]},(\{v\})_{v\in V_t}\right)_{t=1}^\infty$$

Then, $(N, \supseteq)$ is a rooted tree which has infinitely many plays of infinite length (corresponding to the elements of V_∞) and infinitely many plays of finite length (the elements of V_t). While the latter end in "terminal nodes," the former do not. Payoffs are naturally defined on the set of plays, which, again, is naturally one-to-one with the underlying set V.

2.2.2.4 Repeated Games

In a repeated game a constituent one-shot game is repeated infinitely often. Representing this in extensive form gives rise to a large tree. Let I be a player set and A_i an action set for player $i \in I$. Players choose actions in discrete time, $t = 0, 1, 2, \ldots$ A possible set representation for the tree is as follows. Denote $A = \times_{i \in I} A_i$ and let $V = A^\infty = \{a : \mathbb{N} \to A\}$ be the set of all possible sequences of decisions (plays) from "time" $t = 0$ to infinity. Define, for all $t \geq 1$,

$$N_t = \{x \subseteq V \,|\, a(\tau) = b(\tau) \text{ for all } \tau = 0, 1, \ldots, t \text{ for all } a, b \in x\}$$

With $N = \{V\} \cup (\cup_{t=1}^{\infty} N_t)$ the A^∞-poset $(N, \supseteq)$ is the rooted tree of the repeated game.[7] To see that it is a tree, it is enough to verify Trivial Intersection, (2.4), and apply Proposition 2.2. Let $x, y \in N$ be such that $x \cap y \neq \emptyset$. If either x or y are the root V, the conclusion follows. If not, choose $a \in x \cap y$ and let t, τ be such that $x \in N_t$ and $y \in N_\tau$. If $t = \tau$ then $b(s) = a(s)$ for all $s = 0, 1, \ldots, t$ for all $b \in x \cup y$, and $x = y$ follows from the definition. If $t \neq \tau$, assume without loss of generality that $t < \tau$. Then, $b(s) = a(s)$ for all $s = 0, 1, \ldots, \tau$ for all $b \in y$, and $a \in x$ implies $y \subseteq x$.

The interpretation of this tree is straightforward. A node $x \in N_t \subseteq N$ is a specification of all the actions taken up to "time" t. No future actions have yet been discarded. At this node, all players simultaneously choose actions and this results in a new node $y \in N_{t+1}$ being reached. In contrast to the Aumann-Hart example above, in repeated games there are no "terminal nodes" on which payoff can be defined. There is still a natural one-to-one correspondence between the elements of the underlying set $V = A^\infty$ and the set of plays, so that payoffs can be defined on these.

2.2.2.5 Stochastic Games

The trees for stochastic games (see Shapley 1953) are also covered by the present framework. Following Friedman (1990), consider a stochastic game to be a situation in which an n-player normal form game γ from a given set Γ is selected in every period $t = 0, 1, 2, \ldots$ according to a transition function $q(\gamma' \,|\, \gamma, s)$, where $q(\gamma' \,|\, \gamma, s)$

[7] In this specification players who choose simultaneously are active at the same node; see Chap. 4 for details.

denotes the probability that next period's game is γ', given that in the current period game γ is played and players have chosen the strategy combination $s \in S(\gamma)$ in γ. (The set $S(\gamma)$ is the set of strategy combinations in the game γ.)

Of course, the transition function does not affect the underlying tree. The latter describes *possibilities*. In an extensive form representation of the stochastic game the transition function becomes merely a (fixed) strategy for a distinguished player, called "nature." The tree can be constructed in analogy to repeated games with the underlying set

$$V = \left\{ a : \mathbb{N} \to \Gamma \times \cup_{\gamma \in \Gamma} S(\gamma) \,|\, a_t = (\gamma_t, s_t) \text{ with } s_t \in S(\gamma_t) \right\}.$$

Nature will decide at nodes where all past decisions (including the selected games) up to time t are given. Personal players decide at nodes following those, where the next game γ_{t+1} has been decided, but the action profile is still open.

2.2.2.6 Differential Games

Up to this point trees had a discrete structure. The purpose of the following is to illustrate that the present concepts go further. They include the trees of so-called "differential games" (decision problems in continuous time). The reader is referred to Dockner et al. (2000) for an introduction to the topic.

Let V be the set of functions $f : \mathbb{R}_+ \to A$, where A is some given set of "actions," containing at least two elements, and let

$$N = \{x_t(g) \,|\, g \in V,\ t \in \mathbb{R}_+\} \text{ where}$$
$$x_t(g) = \{f \in V \,|\, f(\tau) = g(\tau),\ \forall \tau \in [0,t)\}$$

for any $g \in V$ and $t \in \mathbb{R}_+$. Intuitively, at each point in time $t \in \mathbb{R}_+$ a decision $a_t \in A$ is taken. The "history" of all decisions in the past (up to, but exclusive of, time t) is a function $f : [0,t) \to A$, i.e. $f(\tau) = a_\tau$ for all $\tau \in [0,t)$. A node at "time" t is the set of all functions that coincide with f on $[0,t)$, all possibilities still open for their values thereafter.

We claim that $(N, \supseteq)$ is a V-poset satisfying Trivial Intersection, thus a tree by Proposition 2.2(a). To verify this, let $x_t(g)$ and $x_\tau(h)$ be two nodes, with $g, h \in V$ and $t, \tau \in \mathbb{R}_+$. If $x_t(g) \cap x_\tau(h) \neq \emptyset$, then there is some $f \in V$ such that $f(s) = g(s)$ for all $s \in [0,t)$ and $f(s) = h(s)$ for all $s \in [0,\tau)$. If, say, $\tau \leq t$, then $g(s) = f(s) = h(s)$ for all $s \in [0,\tau)$, implying that $x_t(g) \subseteq x_\tau(h)$ as required.

There is no "point in time" where the decision between two distinct nodes $x_t(g)$ and $x_t(h)$ for which $g(\tau) = h(\tau)$ for all $\tau \in [0,t)$, but $g(t) \neq h(t)$, is actually "taken."[8]

[8] Suppose the convention in the specification of nodes would be changed such that for two functions to belong to the same node they would have to agree on the closed interval $[0,t]$. Then there would be no "point in time" when the decision actually "becomes effective."

Still, the definition is operational: in each node $x_t(g)$ the decision that an agent has to take is clear and the history up to that point is clearly specified. Ultimately, a function $f \in V$ becomes a complete description of all decisions taken from the beginning to the end, i.e. it represents exactly one play. Though this equivalence is transparent in the example, it is not clear in general.

This example has the additional value of putting differential games into a clear game-theoretic context. Whether differential games, and continuous-time decision processes in general, can actually be seen as "proper games" has permeated this branch of the literature. Recall, for instance, the "chattering problem" arising when one attempts to define strategies in continuous time as limits of discrete-time strategies as the time steps shrink to zero; see e.g. Davidson and Harris (1981) and Fudenberg and Levine (1986) for more details.

The point of all examples above is that the framework is sufficiently general to capture even very large decision problems. In all cases the problem can be represented such that there is a one-to-one correspondence between plays and elements of the underlying set. This is not a coincidence, as the subsequent analysis shows.

2.2.3 Decision Trees

In finite cases, it is easy to see that each terminal node (node without successors) is associated to exactly one play (maximal chain) in the set of nodes. Each such maximal chain correspond to a complete possible history of play in the tree, from the beginning to the end, and hence the name "play." A cursory look at Fig. 2.1 shows the intuition for the two trees, and it also illustrates that this simple property fails if the poset fails to be a tree (right-hand-side example). For infinite cases, there might not be terminal nodes, but the set of plays will still be fundamental for the analysis. Given a tree one can associate to every node x the set of plays (maximal chains) that pass through this node, given by $W(x)$ as defined above. The next result identifies key properties of this mapping $W : N \twoheadrightarrow W$ that associates to each node $x \in N$ the plays passing through x.

Lemma 2.2 *For any tree* $(N, \geq)$ *and all nodes* $x, y \in N$*:*

(a) The set $W(x)$ *of plays passing through* x *is nonempty,*
(b) if $x \geq y$ *then* $W(x) \supseteq W(y)$.

Proof (a) Take the chain $\{x\} \subseteq N$ and apply the Hausdorff Maximality Principle.

(b) First it is proven that if $w \in W(x)$ then there exists a chain $c \in 2^N$ such that $w = c \cup \uparrow x$ and $c \subseteq \downarrow x \setminus \{x\}$. Note that $\uparrow x \subseteq w$ by maximality of the play w. For, given $y \in \uparrow x$ and $z \in w$, either $y \geq x \geq z$, implying $y \geq z$ by transitivity, or $y, z \uparrow x$, implying that either $y \geq z$ or $z \geq y$ because $\uparrow x$ is a chain. Hence, $w \cup \{y\}$ is a chain, and it follows that $y \in w$ because w is a play. Finally, let $c = w \setminus \uparrow x$. Since w is a chain, it follows that $c \subseteq \downarrow x \setminus \{x\}$.

Now, let $x, y \in N$ and $x \geq y$. Consider any $w \in W(y)$. Since $x \in \uparrow y$, by the previous statement it follows that $x \in w$, i.e. $w \in W(x)$. Since $w \in W(y)$ is arbitrary, $W(y) \subseteq W(x)$. ■

Any chain is a tree. But in a chain nodes that follow a given one do not represent alternatives. To model decisions, a given node should be followed by several others that are not related by $\geq$. The idea is that $\geq$ expresses "history," while nodes not related by $\geq$ model decisions among *alternative* "histories." Consider the tree depicted in the middle of Fig. 2.1. Since the root r is followed only by the node x or successors of x, there is actually no decision to make at r. The decision maker who is implicitly active at r only has one alternative. There is hence no reason to include the node x in the model. Whatever the decision problem captured by this example, it can be more simply captured by a tree where r is followed directly by y_1 and y_2, without reference to a node x. If every node represents a decision, the following is obtained.

Definition 2.2 A **decision tree** is a tree $(N, \geq)$ such that for all $x, y \in N$

$$\text{if } W(x) = W(y) \text{ then } x = y \tag{2.6}$$

A decision tree is a tree without irrelevant nodes, where a node is irrelevant if it is followed only by one other node. The presence of irrelevant nodes would make it hard to recover nodes as sets of plays, since the plays passing through two different nodes may be identical.[9] In the example in the middle of Fig. 2.1, r and x are different nodes, bur $W(r) = W(x)$. Hence this example is not a decision tree, while the example on the left-hand-side of the figure is. Obviously, irrelevant nodes serve no purpose for decision theory, and they cause added difficulties since one is not able to associate a node x to the set of plays $W(x)$. Hence, Definition 2.2 rules out irrelevant nodes. A decision tree is a tree where every node is followed by at least two different alternatives, as made explicit in the following characterization.

Lemma 2.3 *A tree $(N, \geq)$ is a decision tree if and only if for all $x, y \in N$*

$$\begin{array}{c} \text{if } x \geq y \text{ and } y \not\geq x \text{ then there is } z \in N \\ \text{such that } x \geq z,\ y \not\geq z, \text{ and } z \not\geq y \end{array} \tag{2.7}$$

Proof "if:" Let $(N, \geq)$ be a tree and $x, y \in N$ such that $W(x) = W(y)$. Then, for any $w \in W(x) = W(y)$, that $x, y \in w$ implies either $x \geq y$ or $y \geq x$ (or both), because $w \in W$ is a chain. Assume, without loss of generality, that $x \geq y$. Suppose $y \not\geq x$. Then, by (2.7) there exists $z \in \downarrow x$ such that $z \not\geq y$ and $y \not\geq z$. Since $W(z) \subseteq W(x)$ by Lemma 2.2(b), $y \notin w$ for all $w \in W(z)$. Hence, $W(y) \subseteq W(x) \setminus W(z)$ contradicts

[9]One may use objects other than plays to obtain a set representation, though, as will be shown in Chap. 3.

$W(x) = W(y)$. Thus, also $y \geq x$ must hold, so that $x = y$ (by antisymmetry) verifies (2.6).

"only if:" Let $(N, \geq)$ be a decision tree, and let $x, y \in N$ such that $x \geq y$ and $y \not\geq x$. By Lemma 2.2(b), $W(x) \supseteq W(y)$. By (2.6), $W(x) \supset W(y)$, i.e. there exists $w \in W(x) \setminus W(y)$. For any $z \in w$, either $z \geq x$ or $x \geq z$. If $z \geq x$, transitivity implies $z \geq y$. Hence, there must be some $z \in w$ such that $x \geq z$ and both $z \not\geq y$ and $y \not\geq z$ hold. For, otherwise for all $z \in w$ either $z \geq y$ or $y \geq z$, which implies that $w \cup \{y\}$ is a chain. Maximality of $w \in W$ implies that $y \in w$ and $w \in W(y)$, a contradiction. ■

Since property (2.7) is given in terms of the relation $\geq$, the property of being a decision tree is preserved by order isomorphism. That is, if a poset is order isomorphic to a decision tree, it is itself a decision tree. The translation of (2.7) into set-theoretic terms gives rise to the following definitions.

A V-poset $(M, \supseteq)$ satisfies *Separability* if, for all $a, b \in M$,

$$\text{if } b \subset a, \text{ then there is } c \in M \text{ such that } c \subseteq a \text{ and } b \cap c = \emptyset \tag{2.8}$$

and it satisfies *Weak Separability* if, for all $a, b \in M$,

$$\text{if } b \subset a, \text{ there is } c \in M \text{ such that } c \subseteq a \text{ but } c \setminus b \neq \emptyset \text{ and } b \setminus c \neq \emptyset \tag{2.9}$$

Clearly, Separability implies Weak Separability. If the latter is written in terms of an abstract partial order $\geq$, property (2.7) is obtained. Separability, on the other hand, cannot be translated back into arbitrary posets. The difference exists, however, only in the absence of Trivial Intersection.

Lemma 2.4 *Let $(M, \supseteq)$ be a V-poset satisfying Trivial Intersection, and let $b, c \in M$. Then,*

$$b \setminus c \neq \emptyset \textit{ and } c \setminus b \neq \emptyset \textit{ if and only if } b \cap c = \emptyset$$

In particular, under Trivial Intersection, Weak Separability holds if and only if Separability holds.

Proof The "if"-part is trivial. To see the "only if"-part, suppose $b \cap c \neq \emptyset$ for $b, c \in M$ as in (2.9). Then by Trivial Intersection (2.4) either $c \subset b$, in contradiction to $c \setminus b \neq \emptyset$, or $b \subseteq c$, in contradiction to $b \setminus c \neq \emptyset$. ■

It follows that Separability and Weak Separability are equivalent for the set representation by subtrees, but not necessarily for arbitrary set representations of trees.

Proposition 2.3 *(a) A V-poset $(M, \supseteq)$ is a decision tree if and only if it satisfies Weak Trivial Intersection and Weak Separability.*

(b) A poset $(N, \geq)$ is a decision tree if and only if its set representation by subtrees $(\downarrow N, \supseteq)$ satisfies Trivial Intersection and Separability.

Proof (a) It suffices to notice that Weak Separability is equivalent to property (2.7) and apply Lemma 2.3 and Proposition 2.2(a).

(b) "if:" If $(\downarrow N, \supseteq)$ satisfies Trivial Intersection and Separability, then it satisfies Weak Trivial Intersection and Weak Separability and by part (a) it is a decision tree. By isomorphism $(N, \geq)$ is a decision tree.

"only if:" Let $(N, \geq)$ be a decision tree. By Proposition 2.2(a), $(\downarrow N, \supseteq)$ satisfies Trivial Intersection. But, by isomorphism, $(\downarrow N, \supseteq)$ is a decision tree and by part (a) satisfies Weak Separability. By Lemma 2.4, we have that $(\downarrow N, \supseteq)$ satisfies also Separability. ■

This result is the analogue to Proposition 2.2 for decision trees. Any set representation of a decision tree is a decision tree and, hence, Proposition 2.3(a) characterizes *all* set representations of decision trees.

Since the set (of sets) M in the set representation of a decision tree can be arbitrary, there may be set representations of a decision tree for which Separability fails, but then, by Lemma 2.4, Trivial Intersection must also fail.

Example 2.2 Consider again the $\{1,2,3\}$-poset $(M, \supseteq)$ from Example 2.1. Separability (2.8) does not hold, because $\{1,2\} \subset \{1,2,3\}$ and yet the only other node contained in $\{1,2,3\}$, that is, $\{2,3\}$, has a nonempty intersection with $\{1,2\}$. However, Weak Separability (2.9) holds. Neither of the nodes $\{1,2\}$ and $\{2,3\}$ contains the other.

Example 2.3 (Differential game) The differential game tree (Sect. 2.2.2.6) is a decision tree. For, let $x_t(g)$ and $x_\tau(h)$ be two nodes, with $g, h \in V$ and $t, \tau \in \mathbb{R}_+$, such that $x_t(g) \subset x_\tau(h)$. Then, $\tau < t$. Choose any $f \in V$ such that $f(s) = h(s)$ for all $s \in [0, \tau)$ and $f(\tau) \neq g(\tau)$. Then, for any s with $\tau < s < t$, we have that $x_s(f) \subseteq x_\tau(h)$ but $x_s(f) \cap x_t(g) = \emptyset$, verifying Separability. Since this example also satisfies Trivial Intersection, it follows from Proposition 2.3(a) that this tree is a decision tree.

2.2.4 Representation by Plays

The arbitrariness of the V-poset representing a tree makes it hard to interpret the elements of V. Yet, every decision tree $(N, \geq)$ admits a set representation $(M, \supseteq)$ where $M \subseteq 2^W$ is a collection of nonempty sets of plays, i.e., every decision tree can be represented by a W-poset.

Intuitively, one should be able to take plays and nodes alternatively as the primitives. If nodes are the primitives, plays are derived as maximal chains. If plays are the primitives, nodes are recovered as sets of plays sharing a common history. Formally, for a tree $(N, \geq)$ its *image in plays* is the tree $(W(N), \supseteq)$, where

$$W(N) = \{W(x)\}_{x \in N} = \left\{a \in 2^W \,|\, \exists x \in N : a = W(x)\right\}.$$

This is illustrated in Fig. 2.2, which reconsiders the example in Fig. 1.1.

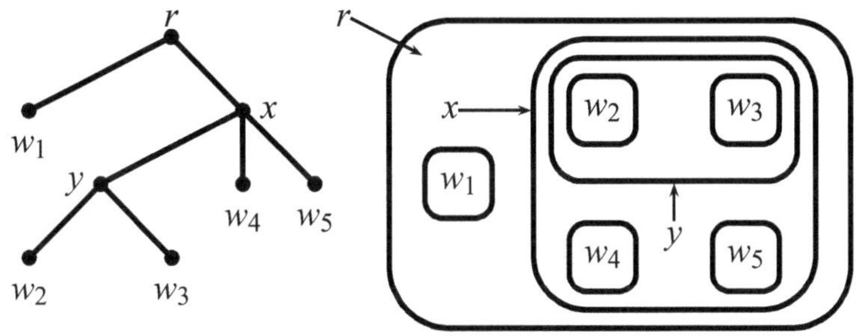

Fig. 2.2 Image in plays of a tree. Each node in the *left-hand side* tree is mapped to the set of plays passing through the node

It is easy to see that a tree's image in plays satisfies Trivial Intersection, and, hence, is itself a tree (by Proposition 2.2(a)).

Lemma 2.5 *If* $(N, \geq)$ *is a tree, its image in plays satisfies Trivial Intersection.*

Proof We have to prove that, for all $x, y \in N$,

$$\text{if } W(x) \cap W(y) \neq \emptyset \text{ then either } W(x) \subset W(y) \text{ or } W(y) \subseteq W(x).$$

Suppose there is $w \in W(x) \cap W(y)$. Since w is a chain with $x \in w$ and $y \in w$, either $x \geq y$ or $y \geq x$ or both. In the first case $W(y) \subseteq W(x)$ and in the second case $W(x) \subseteq W(y)$ by Lemma 2.2(b). ■

Say that a tree $(N, \geq)$ can be *represented by plays* if the mapping[10] $W : N \twoheadrightarrow W(N)$ is an order isomorphism between $(N, \geq)$ and its image in plays $(W(N), \supseteq)$. The latter is then called the tree's (set) *representation by plays.* The image in plays is the natural candidate for a "canonical" set representation. An arbitrary tree, though, need not be order isomorphic to its image in plays.

Theorem 2.1 *A tree* $(N, \geq)$ *can be represented by plays if and only if it is a decision tree.*

Proof "if:" Let W be the set of plays. The set $W(N)$ and its elements are nonempty by Lemma 2.2(a). The mapping $W : N \to W(N)$ is one-to-one by (2.6) and onto by construction. Next, it is verified that the bijection W is order embedding.

Let $x, y \in N$. If $y \geq x$, then by Lemma 2.2(b) $W(x) \subseteq W(y)$. Conversely, suppose $W(x) \subseteq W(y)$. Choose $w \in W(x) \subseteq W(y)$. Since $x, y \in w$, either $x \geq y$ or $y \geq x$. In the first case the previous argument would imply $W(x) = W(y)$ and, therefore, $x = y$, because W is one-to-one. Since $\geq$ is reflexive, in both cases $y \geq x$. Hence, $y \geq x \Leftrightarrow W(x) \subseteq W(y)$ for all $x, y \in N$, i.e. $W(\cdot)$ is an order isomorphism.

[10] No confusion should arise between the mapping $W(\cdot)$ assigning to each node x the set of plays passing through x and the set W of all plays.

"only if:" If $W(x) = W(y)$ for $x, y \in N$, then $x = y$ because the mapping W is one-to-one. ■

The set representation by plays of a decision tree is itself a decision tree (by isomorphism) which satisfies Trivial Intersection by Lemma 2.5 and Weak Separability by Proposition 2.3(a). Hence, it also satisfies Separability (by Lemma 2.4).

Corollary 2.1 *If $(N, \geq)$ is a decision tree then its image in plays, $(W(N), \supseteq)$, satisfies Trivial Intersection, (2.4), and Separability, (2.8).*

Hence, the set representation by plays of a decision tree satisfies Trivial Intersection and Separability, and is order-isomorphic to the decision tree.

These results can also be understood as follows. The properties that characterize set representations of decision trees, Weak Trivial Intersection and Weak Separability, have order-theoretic analogues, that are preserved by order isomorphisms. Trivial Intersection and Separability, on the other hand, make sense only for V-posets and, hence, are not preserved by order isomorphisms. However, both the set representation by subtrees and the set representation by plays of a decision tree satisfy Trivial Intersection and Separability and can be taken as "canonical." The former gives a particularly "bulky" representation, while the latter conforms to our intuition, according to which we should be able to take either nodes or plays as primitives.

Example 2.4 Let $(M, \supseteq)$ be a decision tree with $M = \{\{1,2,3\}, \{1,2\}, \{3\}\}$. This tree satisfies Trivial Intersection and Separability, but a decision between 1 and 2 is never taken, i.e., there is a redundant element in the underlying set. Its set representation by subtrees is given by

$$M' = \{\{s_1, s_2, s_3\}, \{s_2\}, \{s_3\}\}$$

where $s_1 = \{1,2,3\}$, $s_2 = \{1,2\}$, and $s_3 = \{3\}$. In this representation there is also an irrelevant element in the underlying set, because no decision is ever taken to select s_1. The representation by plays of $(M, \supseteq)$ is given by $M' = \{\{w_1, w_2\}, \{w_1\}, \{w_2\}\}$, where $w_1 = \{\{1,2,3\}, \{1,2\}\}$ and $w_2 = \{\{1,2,3\}, \{3\}\}$. In a sense, the redundant element 2 has disappeared.

This example shows that the representation by plays "reduces" the underlying set, eliminating irrelevant elements. What is still missing is a further separation property which guarantees that, given two elements of the underlying set, there is always a decision to distinguish between them.

Example 2.5 (Differential game, continued) Since the differential game tree (Sect. 2.2.2.6) is a decision tree, it can be represented by plays. Given a node $x_t(g)$, the set of plays passing through it is given by

$$W(x_t(g)) = \left\{ \{x_\tau(f)\}_{\tau \in [0,\infty)} \,|\, f \in V \text{ with } f(\tau) = g(\tau) \, \forall \tau \in [0,t) \right\}$$

Example 2.6 (Osborne-Rubinstein trees) As originally defined, Osborne-Rubinstein trees (Sect. 2.2.2.1) are *not* decision trees. The problem is that the formalism allows for a node to fail to represent a decision, in the sense that it might be followed only by one other node and the successors of the latter (recall the middle graph in Fig. 2.1). This is, however, a rather minor point. Osborne-Rubinstein trees are decision trees if the following additional condition holds: If $(a_\tau)_{\tau=1}^t \in H$ then there is $a'_t \neq a_t$ such that $\left((a_\tau)_{\tau=1}^{t-1}, a'_t\right)$ belongs to H. Under this condition $(H, \geq)$ is a decision tree and can be represented by plays as follows. Each sequence in H is identified with the subset of W given by $W\left((a_\tau)_{\tau=1}^t\right) = \{(b_\tau)_{\tau=1}^r \in W \mid r \geq t \text{ and } b_\tau = a_\tau \, \forall\, \tau \leq t\}$ if t is finite, and $W(h) = \{h\}$ if $h \in W_\infty$.

2.3 Set Trees

Definition 2.3 A V-poset $(M, \supseteq)$ is a V**-set tree** if it satisfies Trivial Intersection, (2.4), and Separability, (2.8). A V-set tree is **rooted** if $V \in M$.

All V-set trees are decision trees, but not all V-posets that are decision trees are also V-set trees (see Proposition 2.3(a)). However, given a decision tree $(N, \geq)$, there are two set representations that are V-set trees: the set representation by subtrees, by Proposition 2.3(b), and the image in plays, by Theorem 2.1, which is a W-set tree by Lemma 2.5 and Corollary 2.1.

Intuitively, Separability for a V-set tree $(M, \supseteq)$ ensures that there are *no redundant nodes* in M. Yet, there may still be *redundant elements* in V. Roughly, an element $v \in V$ is redundant, if it can be deleted without affecting the structure of the tree. But there are two meanings for when an element of V is redundant.

Example 2.7 Let $V = \{1,2,3,4,5\}$ and $M = \{\{1,2,3,4\},\{1,2\},\{3\}\}$. Then $(M, \supseteq)$ satisfies Trivial Intersection (2.4) and Separability (2.8). For, if $b \subset a$ then $a = \{1,2,3,4\}$, so that there always is $c \in M \setminus \{a,b\}$ such that $c \subset a$ and $b \cap c = \emptyset$. On the other hand, V contains redundant elements for two reasons.

First, $4 \notin \{1,2\} \cup \{3\}$, but $4 \in a \in M$ implies $a = \{1,2,3,4\}$ so that $\{1,2\} \cup \{3\} \subset a$; hence, there is no $b \in M$ with $v \in b \setminus a$ for $v = 1,2,3$. Intuitively, element $4 \in V$ is *not separable*. Similarly, since there is no $a \in M$ with $5 \in a$, there are no $a, b \in M$ such that $5 \in a \setminus b$ and $v \in b \setminus a$ for $v = 1,2,3,4$. Second, $1 \neq 2$, but $1 \in c \in M$ if and only if $2 \in c \in M$. Intuitively, elements $1,2 \in V$ are *duplicates*.

In this example we attribute the first redundancy to the two elements $4,5 \in V$ not being *separable*. The structure of the tree $(M, \supseteq)$ would not be affected by eliminating elements 4 and 5 from V. The second redundancy we attribute to elements $1,2 \in V$ being *duplicates*. If one of them were eliminated (or they would be identified), the structure of the tree $(M, \supseteq)$ would not be affected.

Let $(M, \supseteq)$ be a V-poset, $v \in V$, and define $\uparrow\{v\} = \{a \in M \mid v \in a\}$. If $\{v\} \in M$, this coincides with the previously defined up-set. With this convention, the aforementioned redundancies can be tackled. Starting with duplicates, define the equivalence relation $\sim$ on V by

$$v \sim v' \text{ if } \uparrow\{v\} = \uparrow\{v'\} \tag{2.10}$$

that is, if, for all $a \in M$, $v \in a \Leftrightarrow v' \in a$. Note that $v \in a \Leftrightarrow [v] \subseteq a$ for all $a \in M$ and all $v \in V$, where $[v]$ denotes the equivalence class (with respect to $\sim$) to which v belongs. In Example 2.7 we have $1 \sim 2$, so $[1] = [2] = \{1,2\}$, but $[v] = \{v\}$ for $v = 3,4,5$.

By definition, it is now justified to write $\uparrow[v] = \uparrow\{v\}$. Obviously, any V-poset $(M, \supseteq)$ can be identified with a $(V/\sim)$-poset, where $V/\sim$ is the quotient set, and this representation will contain no duplicate elements.

Turning to separable elements, consider the subset $S(V)$ of the quotient space $V/\sim$ defined by

$$S(V) = \left\{[v] \in V/\sim \;\middle|\; \cap_{a \in \uparrow[v]} a = [v]\right\} \tag{2.11}$$

which will be referred to as the set of *separable* equivalence classes. In Example 2.7 we have $\uparrow[4] = \{\{1,2,3,4\}\}$ and $\uparrow[5] = \emptyset$, so $[4], [5] \notin S(V)$, while $\cap_{a \in \uparrow[v]} a = [v]$ for $v = 1,2,3$. The following justifies the use of the word "separable" for these classes, by characterizing them as those equivalence classes that can be "separated" from other classes by nodes.

Lemma 2.6 *Let $(M, \supseteq)$ be a V-poset. The equivalence class $[v] \in V/\sim$ is separable, i.e. $[v] \in S(V)$, if and only if for all $v' \in V \setminus [v]$ there is $a \in M$ such that $[v] \subseteq a$ and $v' \notin a$, i.e. $V \setminus [v] = V \setminus \cap_{a \in \uparrow[v]} a = \cup_{a \in \uparrow[v]} (V \setminus a)$.*

Proof First, let $[v] \in V/\sim$ be such that $\uparrow[v] = \emptyset$. Then, $\cap_{a \in \uparrow[v]} a = \emptyset$, i.e. $[v]$ is not separable. Since there is no $a \in M$ such that $[v] \subseteq a$, the property is false, verifying the equivalence in this case.

Let now $[v] \in V/\sim$ be such that $\uparrow[v] \neq \emptyset$. Note that $[v] \subseteq \cap_{a \in \uparrow[v]} a$ whenever $\uparrow[v] \neq \emptyset$. Then, $[v]$ is separable if and only if $[v] = \cap_{a \in \uparrow[v]} a$, or, equivalently, $V \setminus [v] = V \setminus \cap_{a \in \uparrow[v]} a = \cup_{a \in \uparrow[v]} (V \setminus a)$, which proves the claim. ■

Lemma 2.7 *Let $(M, \supseteq)$ be a V-poset. If $a, b \in M$ are such that $a \cap b \neq \emptyset$ (not necessarily $a \neq b$) then there is $[v] \in S(V)$ such that $[v] \subseteq a \cap b$.*

Proof For all $v \in a \cap b$ we have $[v] \subseteq \cap_{c \in \uparrow[v]} c \neq \emptyset$. If there is some $v \in a \cap b$ such that $\cap_{c \in \uparrow[v]} c \subseteq [v]$ the statement is verified. Hence, suppose that $[v] \subset \cap_{c \in \uparrow[v]} c$ for all $v \in a \cap b$. But then that $a \cap b = \cup_{v \in a \cap b} [v] \subset \cup_{v \in a \cap b} \left(\cap_{c \in \uparrow[v]} c\right) \subseteq a \cap b$ yields a contradiction. ■

The *reduced form* of a V-poset $(M, \supseteq)$ is the $S(V)$-poset $(M^*, \supseteq)$, where

$$M^* = \{a^* \subseteq S(V) \,|\, \exists a \in M : [v] \in a^* \Leftrightarrow [v] \subseteq a\}$$

For instance, $S(V) = \{[1], [3]\}$ and $M^* = \{\{[1], [3]\}, \{[1]\}, \{[3]\}\}$ in Example 2.7.

Proposition 2.4 *If the V-poset* $(M, \supseteq)$ *is a V-set tree, then it is order isomorphic to its reduced form with order isomorphism* $\varphi : M \to M^*$ *given by* $\varphi(a) = \{[v] \in S(V) \,|\, [v] \subseteq a\}$.

Proof We first show that the mapping φ, as defined in the statement, is onto. Let $a^* \in M^*$. Then there is $a \in M$ such that $[v] \in a^*$ if and only if $[v] \subseteq a$, i.e. $a^* = \varphi(a)$ and φ is onto.

Let $a, b \in M$ be such that $a \subseteq b$. Then $a \supseteq [v] \in S(V)$ implies $b \supseteq [v] \in S(V)$, so $\varphi(a) \subseteq \varphi(b)$. Conversely, if $a, b \in M$ are such that $\varphi(a) \subseteq \varphi(b)$, then $[v] \in \varphi(a)$ implies $[v] \in \varphi(b)$, so $a \supseteq [v] \in S(V)$ implies $b \supseteq [v]$, hence, $a \cap b \neq \emptyset$. By Trivial Intersection, either $a \subseteq b$ or $b \subset a$. If $b \subset a$, then by Separability there is $c \in M$ such that $c \subseteq a$ and $b \cap c = \emptyset$. By Lemma 2.7, we can choose $[v'] \in S(V)$ such that $[v'] \subseteq c$. Then $\varphi(c) \subseteq \varphi(a) \subseteq \varphi(b)$ implies $[v'] \subseteq b$, in contradiction to $b \cap c = \emptyset$. Hence, $a \subseteq b$ must hold and φ is order embedding. By Remark 2.1, the statement is verified. ∎

That the hypothesis of a V-set tree (rather than a V-poset) is *necessary* for Proposition 2.4 is illustrated by the following example.

Example 2.8 Reconsider Example 2.1. There, $\uparrow\{1\} = \{\{1,2,3\},\{1,2\}\}$, $\uparrow\{2\} = \{\{1,2,3\},\{1,2\},\{2,3\}\} = M$, and $\uparrow\{3\} = \{\{1,2,3\},\{2,3\}\}$, so all equivalence classes with respect to $\sim$ are singletons, but only $2 \in V$ is separable, i.e. $S(V) = \{[2]\}$, and $[2] \subseteq a$ for all $a \in M$. Therefore, $M^* = \{\{2\}\}$ cannot be order isomorphic to $(M, \supseteq)$. This is due to a violation of Trivial Intersection.

Yet, this example does not mean that Proposition 2.4 can be strengthened to a characterization. The next example shows that there are V-posets (in fact, trees) that are order isomorphic to their reduced form, but are not V-set trees. The crucial point is the step from Weak Trivial Intersection to Trivial Intersection.

Example 2.9 Let $V = \{1,2,3\}$ and $M = \{\{1,2,3\},\{1,2\},\{2,3\},\{1,3\}\}$. Then,

$$\uparrow\{1\} = \{\{1,2,3\},\{1,2\},\{1,3\}\} \text{ and } \cap_{a\in\uparrow\{1\}} a = \{1\} = [1],$$
$$\uparrow\{2\} = \{\{1,2,3\},\{1,2\},\{2,3\}\} \text{ and } \cap_{a\in\uparrow\{2\}} a = \{2\} = [2],$$
$$\uparrow\{3\} = \{\{1,2,3\},\{2,3\},\{1,3\}\} \text{ and } \cap_{a\in\uparrow\{3\}} a = \{3\} = [3],$$

so all equivalence classes are singletons and all elements of V are separable, i.e., $S(V) = \{[1],[2],[3]\}$. Therefore,

$$M^* = \{\{[1],[2],[3]\},\{[1],[2]\},\{[2],[3]\},\{[1],[3]\}\}$$

so $(M, \supseteq)$ and $(M^*, \supseteq)$ are order isomorphic by $\varphi(a) = \{[v] \in S(V) \,|\, [v] \subseteq a\}$.

This example also shows that without Trivial Intersection (but still with Separability) it may not be possible to find a subset $V' \subseteq V$ such that Trivial Intersection holds for $(M', \supseteq)$ with $M' = \{a' \subseteq V' \mid \exists a \in M : a' = a \cap V'\}$ and $(M', \supseteq)$ is order isomorphic to $(M,\supseteq)$. For, if $1 \in V'$ then $1 \in \{1,2\} \cap V'$ and $1 \in \{1,3\} \cap V'$. So, if $(M',\supseteq)$ satisfies Trivial Intersection, then $1 \notin V'$. Therefore, if $\{1,2\} \cap V' \neq \emptyset$, it follows that $2 \in V'$ and $2 \in \{1,2\} \cap V'$. But then $2 \in \{2,3\} \cap V'$ contradicts Trivial Intersection on $(M',\supseteq)$. Hence, $(M',\supseteq)$ cannot satisfy Trivial Intersection and be order isomorphic to $(M,\supseteq)$ at the same time.

2.3.1 *Reduced Form and Plays*

By Lemma 2.7, for a V-poset $(M,\supseteq)$ every element $a \in M$ contains at least one separable equivalence class. Conversely, if for a V-poset $(M,\supseteq)$ and $v \in V$ there exists some $a \in M$ such that $a \subseteq [v]$, then $[v] \in S(V)$. For, $a \subseteq [v]$ implies $a = [v]$ (because $v' \in [v]$ implies $v' \sim v$ and, therefore, $v' \in a$ and $[v] \subseteq a$) so that $a \in \uparrow [v]$ and $a = [v] \subseteq b$ for all $b \in \uparrow [v]$ imply that $a = \cap_{b \in \uparrow [v]} b = [v]$, as required. That there is $a \in M$ such that $a \subseteq [v]$ is, therefore, sufficient for $[v] \in S(V)$. But it is not necessary, as the next example shows.

Example 2.10 Let $V = [0,1]$ and $M = \left\{ (\{v\})_{v \in (0,1]}, (x_t)_{t=1}^{\infty} \right\}$, where $x_t = [0, 1/t]$ for all $t = 1,2,\ldots$ Then $\cap_{a \in \uparrow [0]} a = \{0\} = [0]$, but there is no $a \in M$ such that $a = [0]$.

Hence, there are more separable equivalence classes than those which coincide with a node without a successor. (A Cantor-set construction would allow for an example with infinitely many such classes.)

Proposition 2.5 *Let $(M,\supseteq)$ be a V-poset.*

(a) If $\uparrow [v] = \uparrow [v']$, then $[v] = [v']$, for all $[v], [v'] \in S(V)$.
(b) $(M,\supseteq)$ satisfies Trivial Intersection if and only if $\uparrow [v] \in W$ for all $[v] \in S(V)$, where W denotes the set of plays (maximal chains) for $(M,\supseteq)$.

Proof (a) If $\uparrow [v] = \uparrow [v']$, then $[v] = \cap_{a \in \uparrow [v]} a = \cap_{a \in \uparrow [v']} a = [v']$.

(b) "if:" Let W be the set of maximal chains for $(M,\supseteq)$ and assume that $\uparrow [v] \in W$ for all $[v] \in S(V)$. If $a, b \in M$ are such that $a \cap b \neq \emptyset$ then by Lemma 2.7 there is $[v] \in S(V)$ such that $[v] \subseteq a \cap b$, i.e., $a, b \in \uparrow [v]$. But then $\uparrow [v] \in W$ implies either $a \subset b$ or $b \subseteq a$, verifying Trivial Intersection.

"only if:" By Trivial Intersection $\uparrow [v]$ is a chain for all $[v] \in S(V)$. Suppose there is $a \in M \setminus \uparrow [v]$ such that $\uparrow [v] \cup \{a\}$ is a chain. If there would be some $b \in \uparrow [v]$ such that $b \subseteq a$, then $[v] \subseteq a$ in contradiction to $a \notin \uparrow [v]$. Thus, if $\uparrow [v] \cup \{a\}$ is a chain, then $a \subset b$ for all $b \in \uparrow [v]$. Then $a \subseteq \cap_{b \in \uparrow [v]} b = [v]$, i.e. $a = [v]$, again in contradiction to $a \notin \uparrow [v]$. ■

Part (a) of Proposition 2.5 states that the mapping $[v] \mapsto \uparrow [v]$ is one-to-one (injective). Obviously, this holds even for arbitrary V-posets on the separable equivalence classes $S(V)$. Moreover, if $\uparrow [v] \in W$ for all $[v] \in S(V)$, then $(M,\supseteq)$ satisfies Trivial Intersection, by Proposition 2.5 (b).The latter says that Trivial

Intersection is equivalent to $\{\uparrow[v] \mid [v] \in S(V)\} \subseteq W$. The reverse inclusion is not true, though, as a slight modification of the last example shows.

Example 2.11 Let $V = (0,1]$ and $M = \left\{(\{v\})_{v\in(0,1]}, (x_t)_{t=1}^{\infty}\right\}$, where $x_t = (0,1/t]$ for all $t = 1,2,\ldots$ Then $w = \{x_t\}_{t=1}^{\infty}$ is a play that corresponds to no separable class.

The problem in this example is that V itself is not large enough, since intuitively a play fails to lead to an ultimate outcome (even in the limit).

2.3.2 Irreducible Set Trees

Proposition 2.5 suggests that separable equivalence classes in $S(V)$ can be used to represent plays for a V-set tree. If all separable equivalence classes were singletons in V, this would yield an interpretation of the elements of V as representatives of plays.

A V-poset $(M,\supseteq)$ is *irreducible* (or satisfies Irreducibility) if, for all $v, v' \in V$,

$$\text{if } v \neq v' \text{ then } \exists a,b \in M \text{ such that } v \in a \setminus b \text{ and } v' \in b \setminus a \tag{2.12}$$

By Lemma 2.4 it follows that, if a V-poset $(M,\supseteq)$ satisfies Trivial Intersection, then Irreducibility holds if and only if *Strong Irreducibility* holds[11]: for all $v, v' \in V$

$$\text{if } v \neq v' \text{ then } \exists a,b \in M : v \in a,\ v' \in b, \text{ and } a \cap b = \emptyset \tag{2.13}$$

Lemma 2.8 *If a V-poset $(M,\supseteq)$ satisfies Trivial Intersection and Irreducibility, then it satisfies Separability.*

Proof Let $a,b \in M$. If $b \subset a$ then there are $v' \in b$ and $v \in a \setminus b$ such that $v \neq v'$. If Irreducibility holds, there are $a', b' \in M$ such that $v \in a' \setminus b'$ and $v' \in b' \setminus a'$. By Trivial Intersection $v \in a \cap a'$ implies either $a \subseteq a'$ or $a' \subset a$. Yet, $a \subseteq a'$ is impossible, because $v' \in a \setminus a'$, so $a' \subset a$. If $a' \cap b \neq \emptyset$ then, again by Trivial Intersection, either $b \subset a'$ in contradiction to $v' \in b \setminus a'$ or $a' \subseteq b$ in contradiction to $v \in a' \setminus b$. Therefore, $a' \cap b = \emptyset$, as required by Separability. ■

The converse of Lemma 2.8 is not true. The trivial V-set tree $(\{V\},\supseteq)$ satisfies Separability, because the hypothesis is void, but it fails Irreducibility, whenever V is not a singleton set.

The set representation by plays of a decision tree satisfies Trivial Intersection by Lemma 2.5. It is easy to show that it also satisfies Irreducibility and, therefore, is an

[11] Observe the formal analogy of Strong Irreducibility with the definition of a Hausdorff space in topology.

irreducible set tree.[12] Hence, every decision tree $(N, \geq)$ is order isomorphic to an irreducible W-set tree. But the hypothesis of a *decision* tree is only required to make the image in plays a set representation.

Lemma 2.9 *If $(N, \geq)$ is a tree, its image in plays $(W(N), \supseteq)$ is an irreducible tree.*

Proof By Lemma 2.5, $(W(N), \supseteq)$ satisfies Trivial Intersection. We choose to prove Strong Irreducibility (2.13), i.e., for all $v, w \in W$

$$\text{if } v \neq w \text{ then there are } x_v, x_w \in N \text{ such that}$$
$$v \in W(x_v),\ w \in W(x_w),\ \text{and } W(x_v) \cap W(x_w) = \emptyset.$$

If $v \neq w$, since both v and w are maximal chains, neither $v \subseteq w$ nor $w \subseteq v$. Choose $x_v \in v \setminus w$ and $x_w \in w \setminus v$. Then $v \in W(x_v)$ and $w \in W(x_w)$ and it remains to show that $W(x_v) \cap W(x_w) = \emptyset$. Suppose otherwise. Then by Trivial Intersection either $W(x_v) \subseteq W(x_w)$ or $W(x_w) \subseteq W(x_v)$. In the first case $v \in W(x_w)$ implies the contradiction $x_w \in v$. In the second case $w \in W(x_v)$ implies the contradiction $x_v \in w$. ■

Recall that, by Lemma 2.6, separable equivalence classes can be "separated" from other classes by choosing appropriate elements of M. This implies:

Lemma 2.10 *Let $(M, \supseteq)$ be a V-poset. Its reduced form $(M^*, \supseteq)$ is irreducible.*

Proof We have to show that, if $[v], [v'] \in S(V)$ such that $[v] \neq [v']$, then there are $a', b' \in M^*$ such that $v \in a' \setminus b'$ and $v' \subseteq b' \setminus a'$, or, equivalently,

$$\text{if } [v], [v'] \in S(V),\ [v] \neq [v'], \text{ then there are } a, b \in M \text{ such that}$$
$$[v] \subseteq a \setminus b \text{ and } [v'] \subseteq b \setminus a.$$

By Lemma 2.6, there exist $a, b \in M$ such that $[v] \subseteq a$ and $[v'] \subseteq b$, but $v \notin b$ and $v' \notin a$. It follows that $[v] \subseteq a \setminus b$ and $[v'] \subseteq b \setminus a$, as required. ■

Proposition 2.6 *A V-poset is irreducible if and only if $S(V) = \{\{v\}\}_{v \in V}$.*

Proof "if:" If $S(V) = \{\{v\}\}_{v \in V}$ then by Lemma 2.10 Irreducibility holds for all $v, v' \in V$. (The set tree and its reduced form must then be identical.)

"only if:" To see the converse, let $(M, \supseteq)$ be an irreducible V-poset and consider any $v \in V$. By Irreducibility for any $v' \in V \setminus \{v\}$ there are $a, b \in M$ such that $v \in a \setminus b$ and $v' \in b \setminus a$, implying that $[v] \neq [v']$ and, therefore, $[v] = \{v\}$ for all $v \in V$. That is, all equivalence classes are singletons. We still have to show that they are separable.

Consider any class $[v] = \{v\}$. If $[v] \subset \bigcap_{a \in \uparrow[v]} a$ then there exists $v' \neq v$ such that $v' \in a$ for all $a \in \uparrow \{v\}$. This contradicts Irreducibility, because the latter requires the existence of some $a \in \uparrow \{v\}$ and $b \in M$ such that $v' \in b \setminus a$. Hence, $[v] = \bigcap_{a \in \uparrow[v]} a$, i.e.

[12] The set representation by subtrees of a decision tree *cannot* satisfy Irreducibility. For, if $x \neq y$ and $x \geq y$ then $\downarrow y \subseteq \downarrow x$, so that $x \in \downarrow z$ and $y \in \downarrow z'$ for $z, z' \in N$ implies $\downarrow x \subseteq \downarrow z$ and $\downarrow y \subseteq \downarrow z'$ and, therefore, $y \notin \downarrow z' \setminus \downarrow z$.

$[v]=\{v\} \in S(V)$. This implies $\{\{v\}\}_{v\in V} \subseteq S(V)$. Since all equivalence classes are singletons, the reverse inclusion follows. ■

For irreducible V-set trees the set $S(V)$ simply consists of the collection of all singleton subsets of V. Thus, for an irreducible V-set tree the mapping $v \mapsto \uparrow\{v\}$ on V is a one-to-one function into the set W of plays (by Proposition 2.5).

Example 2.12 (Differential game and Irreducibility) The differential game tree (Sect. 2.2.2.6) satisfies (Strong) Irreducibility. For, if $g,f \in V$ and $g \neq f$, then there is some $t \in \mathbb{R}_+$ such that $f(t) \neq g(t)$. For any τ with $t < \tau$ it then follows that $x_\tau(f) \cap x_\tau(g) = \emptyset$ (because elements of V are functions), verifying (2.13). Moreover, every $g \in V$ uniquely induces the play $\uparrow\{g\} = \{x_t(g)\}_{t\in[0,\infty)}$.

2.3.3 Proper Order Isomorphism

When are the "strong" properties, Trivial Intersection and Separability, preserved by order isomorphisms? Let $(M,\supseteq)$ be a V-poset, $(M',\supseteq)$ a V'-poset, and $\varphi : M \to M'$ an order isomorphism between the two. The order isomorphism φ is *proper* if

$$\varphi(a) \cap \varphi(b) = \emptyset \text{ implies } a \cap b = \emptyset \text{ for all } a,b \in M \tag{2.14}$$

Note that properness need not be symmetric. That $(M,\supseteq)$ is properly isomorphic to $(M',\supseteq)$ does not necessarily imply that $(M',\supseteq)$ is properly isomorphic to $(M,\supseteq)$. Yet, a proper order isomorphism is necessary and sufficient to preserve the strong properties.

Lemma 2.11 *Let $(M,\supseteq)$ be a V-poset, $(M',\supseteq)$ a V'-poset, and $\varphi : M \to M'$ an order isomorphism between the two.*

(a) *If $(M,\supseteq)$ satisfies Trivial Intersection, then φ is proper.*
(b) *If $(M',\supseteq)$ satisfies Trivial Intersection, then $(M,\supseteq)$ satisfies Trivial Intersection if and only if φ is proper.*
(c) *If $(M',\supseteq)$ is a V'-set tree, then $(M,\supseteq)$ is a V-set tree if and only if φ is proper.*

Proof (a) Assume that $(M,\supseteq)$ satisfies Trivial Intersection and let $a,b \in M$ be such that $a \cap b \neq \emptyset$. Then either $a \subseteq b$ or $b \subset a$; hence, by order isomorphism, (2.2), either $\varphi(a) \subseteq \varphi(b)$ or $\varphi(b) \subset \varphi(a)$, i.e. $\varphi(a) \cap \varphi(b) \neq \emptyset$.

(b) Suppose $(M',\supseteq)$ satisfies Trivial Intersection. If is φ proper and $a \cap b \neq \emptyset$, for $a,b \in M$, then by (2.14) $\varphi(a) \cap \varphi(b) \neq \emptyset$. Therefore, either $\varphi(a) \subseteq \varphi(b)$ or $\varphi(b) \subset \varphi(a)$; by order isomorphism, (2.2), either $a \subseteq b$ or $b \subset a$. The "only if"-part follows from (a).

(c) By (b) and Lemma 2.4 it is enough to establish Weak Separability, under the hypothesis that $(M',\supseteq)$ is a V'-set tree. But Weak Separability is preserved by order isomorphism. Hence, this is immediate. ■

Consider a decision tree $(M,\supseteq)$. Its image in plays, $(W(M),\supseteq)$, is order isomorphic to $(M,\supseteq)$ by Theorem 2.1. Still, the order isomorphism is not necessarily proper. To see why, recall that decision trees are characterized only by Weak Trivial Intersection and Weak Separability (Proposition 2.3(a)), while $(W(M),\supseteq)$ is a W-set tree by Lemma 2.5 and Corollary 2.1. Hence, the concept of proper order isomorphism adds further structure.

On the other hand, if both $(M,\supseteq)$ and $(M',\supseteq)$ satisfy Trivial Intersection, then all order isomorphism between them *and their inverses* are trivially proper. Thus, the following characterization says that set trees are precisely those posets (of sets) that are properly order isomorphic to their reduced forms, provided the latter are irreducible set trees.

Theorem 2.2 *A V-poset $(M,\supseteq)$ is a V-set tree if and only if its reduced form $(M^*,\supseteq)$ is an irreducible $S(V)$-set tree and $(M,\supseteq)$ is properly (order) isomorphic to $(M^*,\supseteq)$.*

Proof The "if"-part is immediate from Lemma 2.11. For the "only if"-part, note that Irreducibility of $(M^*,\supseteq)$ follows from Lemma 2.10. That $(M,\supseteq)$ is order isomorphic to its reduced form follows from Proposition 2.4. That the order isomorphism $\varphi(a) = \{[v] \in S(V) \mid [v] \subseteq a\}$ from Proposition 2.4 is proper follows from Lemma 2.7. For, if $a \cap b \neq \emptyset$, for some $a,b \in M$, then by Lemma 2.7 there is an equivalence class $[v] \in S(V)$ such that $[v] \subseteq a \cap b$, hence $[v] \in \varphi(a) \cap \varphi(b)$ by the construction of φ. ■

Intuitively, Theorem 2.2 shows that any V-set tree is properly order-isomorphic to an irreducible set tree, obtained by appropriately shrinking the underlying set V. Moreover, by Proposition 2.5 (b) for any V-set tree the mapping $[v] \mapsto \uparrow[v]$ on $S(V)$ is an injection into the set W of plays (by Proposition 2.5 (a)). If this mapping were onto, W and $S(V)$ could be identified. (Recall, though, Example 2.11.) In particular, if this mapping were onto for an *irreducible* V-set tree, then there would be no distinction between W and V (rather than $S(V)$), due to Proposition 2.6.

This motivates the following definitions: Let $(M,\supseteq)$ be a V-poset and $(M',\supseteq)$ a V'-poset. An order isomorphism $\varphi : M \to M'$ is an *isomorphic embedding* if there is an injection $f : V \to V'$ such that

$$f(a) = \{v' \in V' \mid \exists v \in a : v' = f(v)\} \subseteq \varphi(a) \text{ for all } a \in M \tag{2.15}$$

If $(M,\supseteq)$ is isomorphically embedded in $(M',\supseteq)$ and, moreover, f is also onto (surjective) and satisfies $f(a) = \varphi(a)$ for all $a \in M$, then $(M,\supseteq)$ and $(M',\supseteq)$ are *doubly* (order) *isomorphic*.

Every isomorphic embedding is a proper order isomorphism. For, if $(M,\supseteq)$ is isomorphically embedded in $(M',\supseteq)$ and $v \in a \cap b$ with $a,b \in M$, then $f(v) \in \varphi(a) \cap \varphi(b)$ implies $\varphi(a) \cap \varphi(b) \neq \emptyset$. The converse is false: By Theorem 2.2 a V-set tree is properly order isomorphic to its reduced form, the latter is irreducible, but the former may not be. Still, only irreducible V-posets can be isomorphically embedded in irreducible V'-posets.

Lemma 2.12 *Let $(M, \supseteq)$ be a V-poset which is isomorphically embedded in a V'-poset $(M', \supseteq)$. If $(M', \supseteq)$ satisfies Irreducibility, (2.12), then so does $(M, \supseteq)$.*

Proof Assume that $(M, \supseteq)$ is isomorphically embedded in $(M', \supseteq)$. Let $v, w \in V$ such that $v \neq w$. Since the mapping f is one-to-one, $f(v) \neq f(w)$. By Irreducibility for $(M', \supseteq)$, there are $a', b' \in M'$ such that $f(v) \in a' \backslash b'$ and $f(w) \in b' \backslash a'$. Let $a, b \in M$ such that $a' = \varphi(a)$ and $b' = \varphi(b)$, where φ is the order isomorphism. Since $f(v) \in a'$ and f is one-to-one, it follows that $v \in a$. Since $f(v) \notin b'$, it follows that $v \notin b$. Hence, $v \in a \backslash b$ and, analogously, $w \in b \backslash a$. ■

In other words, like other "strong" properties, Irreducibility is inherited by V-posets isomorphically embedded in irreducible V'-posets.

Proposition 2.7 *A V-set-tree is irreducible if and only if it is doubly isomorphic to its reduced form.*

Proof The "if"-part follows from Lemmata 2.10 and 2.12. For the "only if"-part, let $(M, \supseteq)$ be an irreducible V-set tree. By Theorem 2.2 it is properly order isomorphic to its reduced form. By Proposition 2.6 $S(V) = \{\{v\}\}_{v \in V}$ and the mapping f, given by $f(v) = [v]$, is a bijection such that $\varphi(a) = \{[v] \in S(V) \,|\, [v] \subseteq a\} = f(a)$ for all $a \in M$. ■

This clarifies the status of Irreducibility. It is equivalent to the property that a set tree is not only properly order isomorphic to its reduced form, but the underlying sets, V for the set tree and $S(V)$ for its reduced form, also "look alike."

2.3.4 Bounded Set Trees

Proposition 2.5 says that certain plays for a V-set tree $(M, \supseteq)$ can be represented by elements of $S(V)$. According to Proposition 2.6 equivalence classes in $S(V)$ have to be used to represent plays, because V may be "too large." If the V-set tree were irreducible, V could be used directly. But even Irreducibility does not ensure that *all* plays for $(M, \supseteq)$ can be represented by elements of V. The problem is that for some play $w \in W$ the set $\{v \in V \,|\, v \in a,\ \forall a \in w\}$ may be empty, so that not every play is represented by some $v \in V$, i.e., the given set V may be "too small." This was the case in Example 2.11 and is so in the following.

Example 2.13 (Infinite Centipede) Let $V = \{1, 2, \ldots\}$ be the set of natural numbers, define $a_t = \{\tau \in V \,|\, t \leq \tau\}$ for all $t = 1, 2, \ldots$, and let $M = \{(\{t\})_{t=1}^{\infty}, (a_t)_{t=1}^{\infty}\}$. See Fig. 2.3 for a graphical representation. Since $a_\tau \subseteq a_t \Leftrightarrow \tau \geq t$ and $\{\tau\} \subseteq a_t \Leftrightarrow t \leq \tau$, for all $\tau, t = 1, 2, \ldots$, Trivial Intersection holds. Moreover, $\tau \neq t$, say, $t < \tau$, implies that $t \in \{t\}$, $\tau \in a_\tau$, and $\{t\} \cap a_\tau = \emptyset$, so Irreducibility also holds. Therefore, $(M, \supseteq)$ is an irreducible V-set tree.

The set W of plays for $(M, \supseteq)$ consists of sets of the form $\{\{a_\tau\}_{\tau=1}^{t}, \{t\}\}$ for all $t = 1, 2, \ldots$ plus the play $\{a_t\}_{t=1}^{\infty}$. Every play of the form $\{\{a_\tau\}_{\tau=1}^{t}, \{t\}\}$ can be represented by the natural number t, for all $t = 1, 2, \ldots$, but the play $\{a_t\}_{t=1}^{\infty}$ cannot

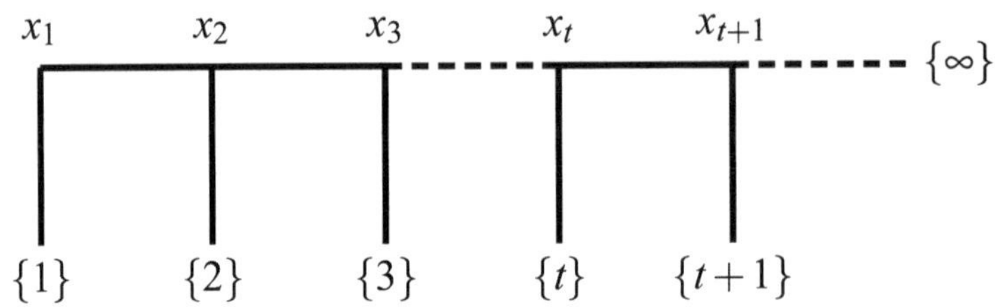

Fig. 2.3 The infinite centipede. $W = \{1, 2, \ldots, \infty\}$, $x_t = \{t, t+1, \ldots, \infty\}$

be represented by a natural number. Yet, if the element "∞" were added to V, the latter play could be represented by this added element.

This suggests that for an irreducible V-set tree the underlying set V could be used to represent all plays, provided V is "large enough." The following makes precise what "large enough" V means. A V-poset $(M, \supseteq)$ is *bounded* (from below) (or satisfies Boundedness) if every chain in M has a lower bound in V, i.e., if

$$\text{for all chains } c \in 2^M \text{ there is } v \in V \text{ such that } v \in a \text{ for all } a \in c \tag{2.16}$$

The image in plays of a tree $(N, \geq)$ is bounded. For, if $c \subseteq W(N)$ is a chain, then there is a chain $c' \subseteq N$ such that $x \in c'$ implies $W(x) \in c$. By the Hausdorff Maximality Principle there is a play $w \in W$ for $(N, \geq)$ such that $c' \subseteq w$. Therefore, $w \in W(x)$ for all $W(x) \in c$. It follows from Theorem 2.1 that the set representation by plays of any decision tree is bounded (and irreducible by Lemma 2.9).

Example 2.14 (Differential game and Boundedness) The differential game tree (Sect. 2.2.2.6) is bounded. For, consider any chain $c \in 2^N$ and let $x_t(g), x_\tau(h) \in c$. If $\tau \leq t$, then (since c is a chain) $x_t(g) \subseteq x_\tau(h)$, and it follows that $g(s) = h(s)$ for all $s \in [0, \tau)$. Hence, the mapping $f_c : \mathbb{R}_+ \to A$ defined in (2.17) below is a well defined function.

$$f_c(t) = \begin{cases} f(t) & \text{if there exists } x_\tau(f) \in c \text{ with } \tau > t \\ a_o \in A & \text{otherwise} \end{cases} \tag{2.17}$$

By construction, $f_c \in x_\tau(f)$ for all $x_\tau(f) \in c$, which proves the claim.

Lemma 2.13 *A V-poset $(M, \supseteq)$ satisfies Trivial Intersection and Boundedness if and only if*

$$c \in 2^M \textit{ is a chain if and only if } \exists v \in V : v \in a,\ \forall a \in c \tag{2.18}$$

Proof "if:" Suppose that $(M, \supseteq)$ satisfies that $c \in 2^M$ is a chain if and only if there is $v \in V$ such that $v \in a$ for all $a \in c$. Then, for any chain $c \in 2^M$ the "only if" part implies that there is $v \in V$ which forms a lower bound on c. Furthermore, if $a, a' \in M$

are such that $a \cap a' \neq \emptyset$, then that there is $v \in a \cap a'$ implies from the "if" part that $\{a,a'\} \in 2^M$ is a chain, i.e., either $a' \subset a$ or $a \subseteq a'$, verifying Trivial Intersection.

"only if:" Suppose $(M, \supseteq)$ satisfies Trivial Intersection and every chain in M has a lower bound in V. Then if $c \in 2^M$ is a chain, there is $v \in V$ such that $v \in a$ for all $a \in c$. On the other hand, if $v \in a$ for all $a \in c$ for some $c \in 2^M$, then if $a, a' \in c$ that $v \in a \cap a'$ implies from Trivial Intersection either $a' \subset a$ or $a \subseteq a'$, i.e., $c \in 2^M$ is a chain. ■

Condition (2.18) implies that $\bigcap_{a \in c} a \neq \emptyset$ for any chain c, thus preventing the situation in Examples 2.11 and 2.13. It will now be shown that there is no loss of generality in assuming that an irreducible V-set tree is bounded. To do this, the underlying set V gets enlarged to a superset V_B giving rise to a new V_B-set tree, that is bounded and properly order isomorphic to the original tree. The idea of enlarging is captured by an isomorphic embedding, where $f : V \to V_B$ is simply the identity.

Proposition 2.8 *A V-poset $(M, \supseteq)$ is an irreducible V-set tree if and only if it is isomorphically embedded in some bounded irreducible V_B-set tree $(M_B, \supseteq)$.*

Proof The "if"-part follows from Lemmata 2.11 and 2.12. To show the "only if"-part, let $(M, \supseteq)$ be an irreducible V-set tree. By Proposition 2.6 $S(V) = \{\{v\}\}_{v \in V}$ and by Proposition 2.5 the mapping $v \mapsto \uparrow \{v\}$ on V is an injection into the set W of plays. Let $W^* = \{\uparrow \{v\} \,|\, v \in V\}$ and define the superset V_B of V by $V_B = V \cup (W \setminus W^*)$. By construction, this union is disjoint.

For any $a \in M$ let $\Phi(a) = a \cup (W(a) \setminus W^*) \subseteq V_B$. Hence $\Phi(a) \cap V = a$. Let

$$M_B = \Phi(M) = \left\{ a' \subseteq V_B \,\middle|\, \exists a \in M : a' = \Phi(a) \right\}$$

It follows that $\Phi : M \to M_B$ is onto and also one-to-one, because if $a, b \in M$ are such that $\Phi(a) = \Phi(b)$, then $a = \Phi(a) \cap V = \Phi(b) \cap V = b$. Moreover, $\Phi(a) \subseteq \Phi(b)$ implies $a = \Phi(a) \cap V \subseteq \Phi(b) \cap V = b$ for all $\Phi(a), \Phi(b) \in M_B$. Conversely, if $a, b \in M$ are such that $a \subseteq b$ then by Lemma 2.2(b) $W(a) \subseteq W(b)$ which implies (by the construction of Φ) that $\Phi(a) \subseteq \Phi(b)$. Hence, Φ is an order isomorphism. Condition (2.15) holds with $f : V \to V_B$ given by $f(v) = v$ for all $v \in V$, so that Φ is an isomorphic embedding.

To establish that $(M_B, \supseteq)$ is a bounded irreducible V_B-set tree, the following is needed.

Claim. If $a \cap b = \emptyset$ then $\Phi(a) \cap \Phi(b) = \emptyset$ for all $a, b \in M$.

To see this, recall that $V_B = V \cup (W \setminus W^*)$ and the union is disjoint. If $\Phi(a) \cap \Phi(b) \neq \emptyset$ and $a \cap b = \emptyset$, then there exists $w \in W(a) \cap W(b)$, in contradiction to $a \cap b = \emptyset$.

Clearly, Φ has inverse $\Phi^{-1} : M_B \to M$ given by $\Phi^{-1}(a') = a' \cap V$ for all $a' \in M_B$. The Claim implies that Φ^{-1} is a proper order isomorphism and hence Lemma 2.11(b) implies that $(M_B, \supseteq)$ satisfies Trivial Intersection. Next, we verify Irreducibility on V_B. If $v', w' \in V_B$ are such that $v' \neq w'$, there are three possibilities.

If $v', w' \in V$, Irreducibility for $(M, \supseteq)$ implies that there are $a, b \in M$ such that $v' \in a \setminus b$ and $w' \in b \setminus a$. It follows that $v' \in \Phi(a) \setminus \Phi(b)$ and $w' \in \Phi(b) \setminus \Phi(a)$.

If $v' \in V$ and $w' \in V_B \backslash V$ (and analogously for the reciprocal case), then w' is a play for $(M, \supseteq)$. If $v' \in a$ for all $a \in w'$, then $w' = \uparrow \{v'\} \in \cup_{v \in V} \uparrow \{v\}$, a contradiction. Hence, there exists $a \in w'$ such that $v' \notin a$. Therefore, $w' \neq \uparrow \{v'\}$, i.e., they are two different plays in $(M, \supseteq)$. Let $a, b \in M$ such that $a \in \uparrow \{v'\} \backslash w'$ and $b \in w' \backslash \uparrow \{v'\}$. Then $v' \in a$, because $a \in \uparrow \{v'\}$. And $v' \notin b$, because $b \in w' \backslash \uparrow \{v'\}$, i.e., $v' \in a \backslash b = a \backslash \Phi(b) \subset \Phi(a) \setminus \Phi(b)$. On the other hand, because $b \in w' \in V_B \backslash V$, it follows that $w' \in W(b) \setminus (\cup_{v \in V} \uparrow \{v\}) \subset \Phi(b)$. Since $a \notin w'$, we have that $w' \notin \Phi(a)$ and hence $w' \in \Phi(b) \setminus \Phi(a)$.

Finally, if $v', w' \in V_B \backslash V = W \setminus (\cup_{v \in V} \uparrow \{v\})$, then v' and w' are two different plays in $(M, \supseteq)$. Let $a, b \in M$ such that $a \in v' \backslash w'$ and $b \in w' \backslash v'$. It follows that $v' \in \Phi(a) \setminus \Phi(b)$ and $w' \in \Phi(b) \setminus \Phi(a)$.

This shows that Irreducibility holds for all $v', w' \in V_B$. By Lemma 2.8, it follows that the V_B-poset $(M_B, \supseteq)$ is an irreducible V_B-set tree. To establish that $(M_B, \supseteq)$ is also bounded from below, consider any chain c' in M_B. By Irreducibility and the Claim above,

$$c = \Phi^{-1}(c') \equiv \{a \in M \,|\, \exists a' \in c' : a' = \Phi(a)\}$$

is a chain in M. By the Hausdorff Maximality Principle there is a play w in M such that $c \subseteq w$. If $w \in \cup_{v \in V} \uparrow \{v\}$, let $v \in V \subset V_B$ be such that $w = \uparrow \{v\}$. Then v is a lower bound for c and hence for c'. If $w \in W \setminus \cup_{v \in V} \uparrow \{v\} \subset V_B$, then $w \in W(a) \setminus (\cup_{v \in V} \uparrow \{v\}) = \Phi(a)$ for all $a \in c$ and hence is a lower bound for c'. ∎

Combining this with Theorem 2.2 it follows that any V-set tree is properly order isomorphic to an irreducible set tree which itself is isomorphically embedded in a bounded irreducible set tree. In this sense, there is no loss of generality in assuming Irreducibility and Boundedness when working with set trees.

Proposition 2.9 *A V-set tree $(M, \supseteq)$ is bounded if and only if the map $[v] \mapsto \uparrow [v]$ on $S(V)$ is onto W, where W is the set of plays for $(M, \supseteq)$.*

Proof "if:" Suppose the mapping $[v] \mapsto \uparrow [v]$ on $S(V)$ is onto W. Let $c \in 2^M$ be a chain in M. By the Hausdorff Maximality Principle there is a play $w \in W$ such that $c \subseteq w$. By hypothesis there is $[v] \in S(V)$ such that $w = \uparrow [v]$. Therefore, $v \in a$ for all $a \in c \subseteq w$.

"only if:" Let $(M, \supseteq)$ be a bounded V-set tree and $w \in W$ a play. By hypothesis there is $v \in V$ such that $v \in a$ for all $a \in w$. By Trivial Intersection and the fact that w is a maximal chain, $\uparrow [v] = w$ and $[v] = \cap \{a \,|\, a \in w\}$, and, hence, $[v] \in S(V)$. ∎

Combining the last result with Proposition 2.6 it follows that for a bounded irreducible V-set tree the mapping $v \mapsto \uparrow \{v\}$ is a bijection from V onto the set W of plays. This result is particularly transparent in the differential game (Sect. 2.2.2.6).

2.4 Game Trees

Theorem 2.2 and Proposition 2.8 show that every set tree can be made into a bounded irreducible set tree by appropriately shrinking and enlarging the underlying set. Then, the underlying set V and the set W of plays can be identified (by Propositions 2.5 and 2.9) and the mapping W from nodes to (sets of) plays passing through them becomes the identity.

Definition 2.4 A **game tree** is a W-poset $(N, \supseteq)$ that satisfies

(GT1) $c \in 2^N$ is a chain if and only if $\exists w \in W$ such that $w \in x \, \forall x \in c$.
(GT2) If $w, w' \in W$ with $w \neq w'$, then $\exists x, y \in N$ such that $w \in x \setminus y$ and $w' \in y \setminus x$.

By Lemma 2.13 condition (GT1) holds if and only if Trivial Intersection holds and every chain in N has a lower bound in W. Since condition (GT2) is simply Irreducibility, a W-poset $(N, \supseteq)$ is a game tree if and only if it is a bounded irreducible W-set tree.

It will now be shown that a game tree is its own set representation by plays. Yet, when precisely is a set tree its own set representation by plays? The idea is that, not only the set of plays W and the underlying set V are bijective, but, additionally, this bijection can be used to reconstruct the plays passing through a node from the elements which the node contains. This is formalized as a double isomorphism.

Definition 2.5 A V-poset $(M, \supseteq)$ is **its own set representation by plays** if there exists a bijection $\psi : V \to W$ such that $\psi(a) \equiv \{\psi(v)\}_{v \in a} = W(a)$ for all $a \in M$, where $W : M \to W(M)$ is the mapping assigning plays passing through a node, given by $W(a) = \{w \in W \,|\, a \in w\}$.

Remark 2.2 Let $2^\psi : 2^V \to 2^W$ be the trivial extension of ψ to the power set given by $2^\psi(a) = \psi(a)$ for all $a \subseteq V$. Let $i_M : M \to 2^V$ be the immersion of M into the power set of V (given by $i_M(x) = x$ for all $x \in M$) and $i_W : W(M) \to 2^W$ the analogous immersion of $W(M)$ into the set of sets of plays. Then, the previous definition amounts to the diagram in Fig. 2.4 being commutative.

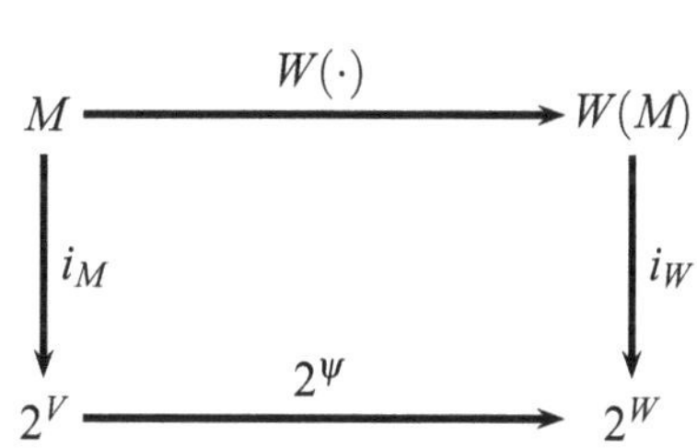

Fig. 2.4 Own representation by plays. Interpretation of Definition 2.5 in terms of a commutative diagram

Lemma 2.14 *A tree $(M, \supseteq)$ that is its own representation by plays is a decision tree.*

Proof Suppose $a, b \in M$ with $a \neq b$ but $W(a) = W(b)$. Without loss of generality, we can assume $a \backslash b \neq \emptyset$. Let $v \in a \backslash b$. Let $\psi : V \to W$ be the bijection given in Definition 2.5. Then, $\psi(v) \in W(a) = W(b) = \{\psi(v')\}_{v' \in b}$ and, hence, there is $v' \in b$ (in particular, $v' \neq v$) such that $\psi(v) = \psi(v')$, a contradiction. ■

Combining Theorem 2.1 with Lemma 2.14 shows that a tree is its own representation by plays if and only if it is doubly isomorphic to its image in plays via the natural order isomorphism.

The Infinite Centipede from Example 2.13 shows that Trivial Intersection and Irreducibility are *not* sufficient for $(M, \supseteq)$ to be its own set representation by plays. This is purely due to the fact that the underlying set V is given, and is unrelated to properties of the decision tree. That the underlying set V is large enough to contain a lower bound for every chain in M is expressed by adding the converse to the implication in Trivial Intersection, as in condition (GT1). The "only if"-part of condition (GT1) is purely a convention on how the set representation is chosen. That is, by contrast to its "if"-part (viz. Trivial Intersection) and Irreducibility (GT2) it has no impact on the corresponding decision tree $(N, \geq)$, by Proposition 2.8.

With these insights at hand we are now ready to prove the main theorem of this chapter.

Theorem 2.3 *For any V-poset $(M, \supseteq)$ the following statements are equivalent.*

(a) $(M, \supseteq)$ is a game tree.
(b) $(M, \supseteq)$ is a tree and its own representation by plays.
(c) $\psi(v) = \uparrow\{v\}$ defines a bijection from V to the set W of maximal chains in M.

Proof "(a) implies (b):" By Proposition 2.6 $S(V) = \{\{v\}\}_{v \in V}$, and by Proposition 2.5, for every $v \in V$, the set $\uparrow\{v\} \equiv \{a \in M | v \in a\} \subseteq M$ is a play (maximal chain).

Define $\psi : V \to W$ by $\psi(v) = \uparrow\{v\}$ for all $v \in V$. By Proposition 2.5(a) this function is one-to-one and by Proposition 2.9 it is onto. Hence, it is a bijection. Moreover, it follows that $\psi^{-1}(w)$ is given by the only element in $\cap_{a \in w} a = \{v \in V | v \in a, \forall a \in w\}$.

Next, it is verified that $\psi(a) \equiv \{\psi(v)\}_{v \in a} = W(a)$ for all $a \in M$. For $a \in M$ the plays passing through are $W(a) = \{w \in W | a \in w\}$, as usual. Let $w' \in W(a)$ and $v' = \psi^{-1}(w')$. Since $a \in w' = \psi(v') = \{a' \in M | v' \in a'\}$, it follows that $v' \in a$. Hence $w' \in \psi(a)$. Conversely, let $w' \in \psi(a)$. Then there is $v' \in a$ such that $\psi(v') = w'$. Since $w' = \psi(v') = \{a' \in M | v' \in a'\}$, it follows that $a \in w'$. Hence, $w' \in W(a)$. In summary, $\psi(a) = W(a)$ for all $a \in M$, as required by Definition 2.5.

"(b) implies (c):" Suppose that $(M, \supseteq)$ is a decision tree and that there exists some bijection $\tilde{\psi} : V \to W$ such that $\tilde{\psi}(a) \equiv \{\tilde{\psi}(v)\}_{v \in a} = W(a)$ for all $a \in M$. Since $(W(M), \supseteq)$ is a decision tree by Lemma 2.9, it follows from Lemmata 2.11 and 2.12 that $(M, \supseteq)$ is an irreducible V-set tree.

Consequently, by Propositions 2.5 and 2.6, $\psi(v) = \uparrow \{v\}$ is a play for all $v \in V$. It suffices to show that $\psi(v) = \tilde{\psi}(v)$ for all $v \in V$. Let $v \in V$ and consider any $a \in \psi(v) = \uparrow \{v\}$. Then $v \in a$ implies $\tilde{\psi}(v) \in \tilde{\psi}(a) = W(a)$ and, hence, $a \in \tilde{\psi}(v)$. It follows that $\psi(v) \subseteq \tilde{\psi}(v)$ and, since both $\psi(v)$ and $\tilde{\psi}(v)$ are plays, $\psi(v) = \tilde{\psi}(v)$ by maximality.

"(c) implies (a):" Suppose there is $v \in V$ such that $v \in a$ for all $a \in c$ for some $c \in 2^M$. Then $c \subseteq \psi(v) = \uparrow \{v\} \in W$ implies that c is a chain. This verifies the "if"-part of (GT1).

If $c \in 2^M$ is chain, by the Hausdorff Maximality Principle there is a maximal chain $c' \in W$ such that $c \subseteq c'$. Since ψ is onto, there is $v \in V$ such that $\psi(v) = c' \in W$. Then $v \in b$ for all $b \in c'$ implies $v \in a$ for all $a \in c$. This verifies the "only if"-part of (GT1).

Let $v, v' \in V$ be such that $v \neq v'$. Then $\psi(v)$ and $\psi(v')$ are distinct plays, because ψ is one-to-one. If for every $a \in \psi(v)$ also $v' \in a$ would hold, then $\psi(v) \subseteq \psi(v')$ would imply the contradiction $\psi(v) = \psi(v')$ by maximality. Hence, there is $a \in \psi(v)$ such that $v' \notin a$. A symmetric argument shows that there is $b \in \psi(v')$ such that $v \notin b$. Thus $v \in a \setminus b$ and $w \in b \setminus a$ verifies (GT2). ■

Along the way it has been shown that it is justified to call the mapping $\psi : V \to W$ defined by $\psi(v) = \uparrow \{v\}$ the *canonical mapping*. After all, the proof of "(b) implies (c)" shows that the bijection from the underlying set to the set of plays is unique.

Corollary 2.2 *For any game tree $(N, \supseteq)$ the bijection in Definition 2.5 from V to the set W of plays is unique and given by $\psi(v) = \uparrow \{v\}$ for all $v \in V$.*

In a game tree, by Theorem 2.3, the sets V and W can be identified. The bijection $\psi : V \to W$ becomes the identity on all of $V = W$, and then $\psi(a) = W(a)$ for all $a \in M$, i.e., a node can be identified with the set of plays that pass through it. Game trees are those trees for which it is inconsequential whether nodes or plays are taken to be the primitives. Hence, one can use the concepts of "plays" and "ultimate outcomes" interchangeably.

2.4.1 *Complete Game Trees*

As the Infinite Centipede from Example 2.13 shows, even for a game tree some plays may not end at nodes. Yet, it will be shown that terminal nodes (singletons) can be added without altering the structure of the tree. Hence, having all plays ending at nodes or not becomes a modelling decision.

Definition 2.6 A game tree $(N, \supseteq)$ is **complete** if $\{w\} \in N$ for all $w \in W$ (where W is the underlying set).

If a game tree contains all singleton sets, i.e. if it is complete, then Irreducibility, (GT2), holds trivially for all $w, w' \in W$. This simple observation leads to the following characterization of a complete game tree that is important for further reference.

Proposition 2.10 *A W-poset $(N, \supseteq)$ is a complete game tree if and only if (GT1) holds and $\{w\} \in N$ for all $w \in W$.*

Theorem 2.3 implies that every game tree can be *completed* by adding all singletons $\{w\}$ for $w \in W$ to N without affecting the set of plays:

Proposition 2.11 *If a V-poset $(N, \supseteq)$ is a game tree and $Z = \{\{v\} \mid v \in V\}$ is the collection of singleton sets, then $(N \cup Z, \supseteq)$ is a complete game tree. Moreover, $\Phi(w) = w \cup \{\{\psi^{-1}(w)\} \cap Z\}$ defines a bijection between the set of plays for $(N, \supseteq)$ and the set of plays for $(N \cup Z, \supseteq)$.*

Proof Let W be the set of plays for the game tree $(N, \supseteq)$ and W' the set of plays for $(N', \supseteq)$ where $N' = N \cup Z$. Let $\psi : V \to W$ be the canonical mapping, defined by $\psi(v) = \{x \in N | v \in x\} = \uparrow \{v\}$. Define $\Phi : W \to W'$ by $\Phi(w) = w \cup \{\{\psi^{-1}(w)\} \cap Z\}$. We show that Φ is a bijection.

Let $w' \in W'$ and $w = w' \cap N$. Then $w \in W$, $\psi^{-1}(w) \in V$, and $\Phi(w) = w' \in W'$. Hence, Φ is onto (surjective). If $\Phi(w) = \Phi(\hat{w})$ for $w, \hat{w} \in W$ then

$$w = \Phi(w) \cap N = \Phi(\hat{w}) \cap N = \hat{w}$$

Thus, Φ is one-to-one (injective). It follows that W and W' are set-isomorphic (bijective). By Theorem 2.3 (applied to $(N, \supseteq)$), ψ is bijective and hence $\psi' \equiv \Phi \circ \psi : V \to W'$ is also bijective. But

$$\begin{aligned} \psi'(v) &= \{x \in N | v \in x\} \cup \{\{v' \in V | v' \in x, \forall x \in \psi(v)\} \cap Z\} \\ &= \{x' \in N' | v \in x'\} \end{aligned}$$

because $\{v' \in V | v' \in x, \forall x \in \psi(v)\} = \{\psi^{-1}(\psi(v))\} = \{v\}$ by Theorem 2.3. Therefore, Theorem 2.3 now implies that $(N', \supseteq)$ is a complete game tree. ∎

Proposition 2.11 is about adding singletons. In Chap. 4 below it will be shown that a particular class of terminal nodes may also be removed from a game tree without affecting its structure. For the moment observe that in Proposition 2.11, *all* singletons are added to N. But some of the singletons may already have to belong to N, by Irreducibility (GT2). The next example illustrates this.

Example 2.15 (Twins example) Let $V = [0,1]$ and $N = \{(\{v\})_{v \in V}, (x_t)_{t=1}^{\infty}\}$, where

$$x_t = \left[0, \frac{1}{t+1}\right] \cup \left[\frac{t}{t+1}, 1\right] \text{ for all } t = 1, 2, \ldots$$

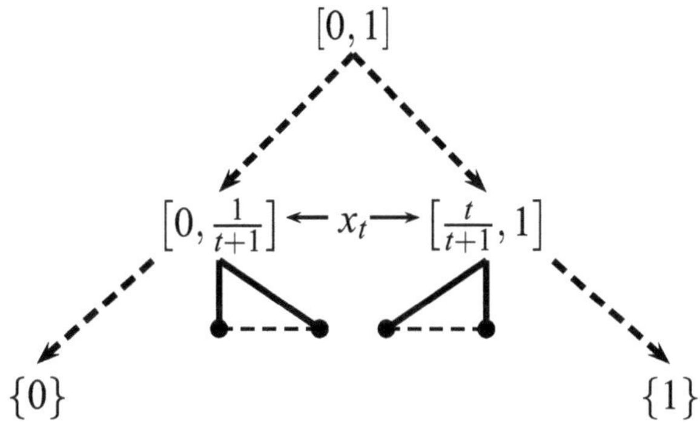

Fig. 2.5 The Twins Example, qualitative graphical representation. The node x_t, for $t = 1, 2, \ldots$, is the union of the two intervals $\left[0, \frac{1}{t+1}\right]$ and $\left[\frac{t}{t+1}, 1\right]$, but in the limit it becomes two separate singleton nodes, $\{0\}$ and $\{1\}$

A qualitative representation can be seen in Fig. 2.5. Because $\{v\} \in N$ for all $v \in V$, the poset $(N, \supseteq)$ is a complete game tree. The set $c_\infty = \{x_t \in N \,|\, t = 1, 2, \ldots\}$ is not a play (because $c_\infty \cup \{0\}$ is a chain), nor is any set $c_t = \{x_\tau \in N \,|\, \tau = 1, \ldots, t\}$ (for the same reason). The set of plays is given by

$$W = \left\{ \left((c_t \cup \{v\})_{v \in \left(\frac{1}{t+2}, \frac{1}{t+1}\right]}, (c_t \cup \{v\})_{v \in \left[\frac{t}{t+1}, \frac{t+1}{t+2}\right)} \right)_{t=1}^{\infty}, c_\infty \cup \{0\}, c_\infty \cup \{1\} \right\}$$

Hence, V and W are naturally isomorphic by the bijection $\psi : V \to W$, where $\psi(v) = c_{t(v)} \cup \{v\}$ and $t(v)$ is the largest integer such that $t(v) \leq (1 - v)/v$ for all $v \in (0, 1/2]$, $\psi(v) = c_{t(v)} \cup \{v\}$ and $t(v)$ is the largest integer such that $t(v) \leq v/(1 - v)$ for all $v \in (1/2, 1)$, $\psi(0) = c_\infty \cup \{0\}$, and $\psi(1) = c_\infty \cup \{1\}$.

The singletons $\{0\}$ and $\{1\}$ could *not* have been added, as in Proposition 2.11, to a game tree. For, if originally, say $\{0\}$ would not be a node, then the original tree fails Irreducibility (GT2). This is so, because without $\{0\}$ there would not be any node that separates $0 \in V$ from $1 \in V$ (i.e. that $0 \in V$ belongs to a node would imply that $1 \in V$ belongs to this node). Only if the underlying set is modified to become $V' = (0, 1]$, the resulting ordered set $(N \setminus \{\{0\}\}, \supseteq)$ would be a game tree; but then $\{0\}$ could not be added. Hence, the class of singletons that can truly be added (i.e. without already being there), as in Proposition 2.11, forms a particular subset of the set of all singletons. To unveil what special subset that is, however, takes extra concepts that are deferred to Chap. 4 (based on Alós-Ferrer and Ritzberger 2008).

By Lemma 2.5 the representation by plays of a decision tree is an irreducible set tree. By Proposition 2.8 every irreducible set tree can be modified—by adding elements to the underlying set V—to become a bounded irreducible tree, i.e. a game tree. (Hence, every decision tree is order isomorphic to a game tree.) By Proposition 2.11 every game tree can be modified—by adding nodes—to become a complete game tree. Neither of these modifications changes any essential features of the tree.

Proposition 2.12 *Let* $(N, \supseteq)$ *be a* V*-set tree. Then,* $(N, \supseteq)$ *is a complete game tree if and only if it is irreducible and every play has a minimum.*

Proof "if:" Let $(N, \supseteq)$ be an irreducible V-set tree for which every play has a minimum. Then $(N, \supseteq)$ is a game tree by Theorem 2.3. For any $v \in V$ let $\psi(v) = \uparrow\{v\} = w \in W$ be the play associated to v by the canonical mapping. By hypothesis w has a lower bound $z \in w$ and, by the definition of w, we have that $v \in z$.

If there is $v' \in z$ with $v' \neq v$, then by Irreducibility there are $x, x' \in N$ such that $v \in x \setminus x'$ and $v' \in x' \setminus x$. Since $v \in x$, we have $x \in w$. Since $v' \in z$ and $z \subseteq x$ (because $z \subseteq y$ for all $y \in w$), it follows that $v' \in x$, a contradiction. Therefore, $z = \{v\} \in N$.

"only if:" If $(N, \supseteq)$ is a complete game tree, then it is irreducible, because $\{v\} \in N$ for all $v \in V$. If $w \in W$ then by (GT1) there is $v \in V$ such that $v \in x$ for all $x \in w$. By definition $x \supseteq \{v\} \in N$ for all $x \in w$, so $w \subseteq \uparrow\{v\}$. Since $\uparrow\{v\} \in W$ by Theorem 2.3(c), maximality implies $w = \uparrow\{v\}$. Therefore, if $z \in N$ is such that $z \subseteq x$ for all $x \in w$, then, in particular, $z \subseteq \{v\}$ implies $z = \{v\}$. Since, moreover, $\{v\} \in w$ the play $w = \uparrow\{v\}$ has the minimum $\{v\} \in N$. ■

Say that a game tree (or a V-poset) $(N, \supseteq)$ has *finite horizon* if all chains in N are finite. In this important special case, a complete game tree is characterized by a much simpler condition than the combination of (GT1) and (GT2) from Definition 2.4.

Proposition 2.13 *If* $(N, \supseteq)$ *is a* V*-poset with finite horizon, then* $(N, \supseteq)$ *is a complete game tree if and only if it satisfies Trivial Intersection and* $\{v\} \in N$ *for all* $v \in V$*.*

Proof "if:" If $v, v' \in V$ are such that $v \neq v'$ then, because $\{\hat{v}\} \in N$ for all $\hat{v} \in V$, $v \in \{v\} \setminus \{v'\}$ and $v' \in \{v'\} \setminus \{v\}$ verifies (GT2). If $c \in 2^N$ is a chain, then by hypothesis it is finite, so that $\bigcap_{x \in c} x \neq \emptyset$ implies that there is $v \in \bigcap_{x \in c} x \subseteq V$. Since the chain c is arbitrary, the tree is bounded. By Lemma 2.13, (GT1) holds, and $(N, \supseteq)$ is a game tree.

"only if:" If $(N, \supseteq)$ is a complete game tree, then Trivial Intersection follows from (GT1) and Lemma 2.13. For any $v \in V$ the chain $\uparrow\{v\}$ is finite by hypothesis and, therefore, contains a smallest node $x^* \in N$ such that $v \in x^*$ (i.e. $v \in y \in N \Rightarrow x^* \subseteq y$). If there is $v' \in V \setminus \{v\}$ such that $v' \in x^*$, then by (GT2) there are $x, y \in N$ such that $v' \in x \setminus y$ and $v \in y \setminus x$. But $v \in y$ implies $v \in x^* \subseteq y$ in contradiction to $v' \notin y$. Hence, there is no $v' \in V \setminus \{v\}$ such that $v' \in x^*$ and, therefore, $\{v\} = x^* \in N$. ■

Note that all *finite game trees*, i.e. game trees with underlying *finite* set V, are necessarily complete game trees. Without finiteness, however, a game tree as in Definition 2.6 is quite general. It includes the trees for all examples from Sect. 2.2.2.6.

Example 2.16 (Differential game and Completeness) As shown before, the differential game tree is irreducible and bounded, and hence a game tree, where the plays

$w \in W$ can be identified with the underlying elements (functions) $f \in V$. To turn this game tree into a complete game tree, it is enough to add the singletons as nodes (applying Proposition 2.11), e.g. simply allowing t to be infinite in the definition of $x_t(g)$ and N.

No play for these trees has a lower bound in N (but only a lower bound in V), unless the singletons have been added. Infima of chains, on the other hand, are easy to identify. Let c be a chain. If, for all t, there exists $x_\tau(f) \in c$ with $\tau > t$ (which is true for instance for plays), then $\{f_c\}$, where f_c is defined in (2.17), is an infimum for c. If there exists t such that the chain contains no node $x_\tau(f)$ with $\tau > t$, then let t^* be the supremum over all such t. The node $x_{t^*}(f_c)$ is an infimum for the chain. To see this, take any node $x_\tau(f) \in c$; necessarily, $\tau \leq t^*$ and $x_{t^*}(f_c) \subseteq x_\tau(f)$.

However, $x_{t^*}(f_c) \notin c$ in general, i.e., the chain may not have a minimum. Consider a fixed $f \in V$ and take the chain $\{x_t(f) \,|\, t < t^*\}$ for a given t^*. Obviously, $x_{t^*}(f_c)$ is the infimum of the chain, but is not an element of the chain. Intuitively, this says that no node other than the root itself can be reached in a finite number of "steps" from the root.

The last observation raises the issue whether or not a game tree has enough structure to serve as the "objective" description of what may happen in the course of an extensive form game. Chapter 5 will tackle this issue by imposing a number of desiderata.

2.5 Summary

This chapter studies how arbitrary trees can be represented by a collection of sets. The purpose of such a representation is to provide a domain for sequential decision theory. To do this requires two things: First, a node should be an event in the sense of probability theory, i.e. a set of states. Second, the elements of the nodes/sets should have meaning as representatives of ultimate outcomes. We show that both desiderata can be met without loss of generality by the current definition of a game tree: a collection of subsets of an underlying set (of plays) such that (i) a family of those subsets is a chain if and only if all its elements (nodes) contain a common element (play), and (ii) for any two distinct elements (plays) there are two sets (nodes) such that the first set (node) contains the first element (play), but not the second, and the second set (node) contains the second element (play), but not the first.

This definition is essentially equivalent to the notion of a decision tree, i.e., all requirements on top of being a set representation of an (order-theoretic) decision tree are purely modelling conventions. At the same time it captures the intuition that nodes and plays could both serve as the primitives for the model. Intuitively, Theorem 2.3 says that game trees are characterized by

$$\text{“}x \in w \Leftrightarrow w \in x\text{”}$$

for nodes $x \in N$ and plays $w \in W$ (abusing notation, of course).

Even though game trees are the appropriate domain for extensive form games, the definition of the latter remains to be supplied in later chapters. On the way a number of problems will need to be clarified. For instance, with the present generality nodes need not have immediate predecessors (even if choices do), as in the differential game example (Sect. 2.2.2.6). This poses a problem with alternating moves, as they appear in games of perfect information. In the differential game example with two players, one could let A_2 be the set of functions from A_1 to an ultimate action space for player 2, modelling that player 2's decision conditions on what player 1 has chosen. But, in the tree, the two players' decision would formally be taken simultaneously. Ideally, one would like to let player 1 move at the immediate predecessor of player 2's decision points. In continuous time, however, immediate predecessors do not exist. On the other hand, alternating moves could be viewed as a property of the situation that is to be modeled by the tree, so that the tree inherits a discrete structure from what it models.

Passing from an appropriate definition of extensive form games to "solutions" in terms of strategies will raise even further issues. At this point of the development it is an open issue under which conditions strategy combinations induce outcomes. (This shall be addressed in Chap. 5.) Even if pure strategies do, there also remains, of course, a measurability issue whether behavior strategy combinations induce well-defined probability distributions on plays.[13]

[13]For instance, consider the well-known problem with the law of large numbers arising if a player tosses a coin repeatedly in continuous time (e.g. Alós-Ferrer 2002; Judd 1985).

References

Alós-Ferrer C (2002) Individual randomness with a continuum of agents. Mimeo, Vienna

Alós-Ferrer C, Ritzberger K (2005a) Trees and decisions. Econ Theory 25(4):763–798

Alós-Ferrer C, Ritzberger K (2008) Trees and extensive forms. J Econ Theory 43(1):216–250

Anscombe F, Aumann RJ (1963) A definition of subjective probability. Ann Math Stat 34:199–205

Aumann RJ, Hart S (2003) Long cheap talk. Econometrica 71:1619–1660

Birkhoff G (1973) Lattice theory, vol XXV, 3rd edn. American Mathematical Society Colloquium Publications, Providence

Blume LE, Brandenburger A, Dekel E (1991a) Lexicographic probabilities and choice under uncertainty. Econometrica 59:61–79

Blume LE, Brandenburger A, Dekel E (1991b) Lexicographic probabilities and equilibrium refinements. Econometrica 59:81–98

Davey B, Priestley H (1990) Introduction to lattices and order. Cambridge University Press, Cambridge

Davidson R, Harris R (1981) Nonconvexities in continuous-time investment. Rev Econ Stud 48:235–253

Dekel E, Lipman B, Rustichini A (2001) Representing preferences with a unique subjective state space. Econometrica 69:891–934

Dockner E, Jörgensen S, Long NV, Sorger G (2000) Differential games in economics and management science. Cambridge University Press, Cambridge

Friedman JW (1990) Game theory with applications to economics, 2nd edn. Oxford University Press, Oxford

Fudenberg D, Levine D (1986) Limit games and limit equilibria. J Econ Theory 38:261–279

Gul F, Pesendorfer W (2001) Temptation and self-control. Econometrica 69:1403–1435

Hewitt E, Stromberg K (1965) Real and abstract analysis, graduate texts in mathematics, vol 25. Springer, Berlin/Heidelberg/New York

Judd K (1985) The law of large numbers with a continuum of i.i.d random variables. J Econ Theory 35:19–25

Kreps DM (1979) A representation theorem for "preference for flexibility." Econometrica 47:565–577

Kuhn H (1953) Extensive games and the problem of information. In: Kuhn H, Tucker A (eds) Contributions to the theory of games, vol II. Annals of mathematics study, vol 28. Princeton University Press, Princeton, pp 217–243

Osborne MJ, Rubinstein A (1994) A course in game theory. MIT, Cambridge

Rubinstein A (1982) Perfect equilibrium in a bargaining model. Econometrica 50:97–109

Savage L (1954) The foundations of statistics. Wiley, Hoboken

Shapley L (1953) Stochastic games. Proc Natl Acad Sci 39:1095–1100

von Neumann J, Morgenstern O (1944) Theory of games and economic behavior. Princeton University Press, Princeton

Chapter 3
Pseudotrees and Order Theory

In this chapter we will make a quick excursion into order theory with the purpose of tying up a few loose ends and illustrating the connection to the mathematical literature on trees.[1] Even though these considerations yield interesting insights, the reader interested exclusively in game theory may safely skip the chapter in a first reading.

The first section looks at how the concept of tree has been treated and generalized in order theory, and clarifies the correspondence to the concepts used there and related order-theoretic structures. In doing so, new results which were first derived in Alós-Ferrer and Ritzberger (2005b) are presented. One of the purposes of this excursion is quite simply to alert the reader to the existence of the connections, in the hope that future researchers will not reproduce work being done in other disciplines.

The second part looks back at one of the main results from the last chapter, Theorem 2.1. That result stated that trees can be represented by plays if and only if they are decision trees. Decision trees are trees where irrelevant nodes (those with a single immediate successor) have been excluded. While this is a convenient (and, we believe, fully justified) modelling decision, it is perfectly legitimate to ask whether appropriate representations could also be obtained for trees where irrelevant nodes are allowed. The answer is positive.

To make explicit the relation of the concepts discussed in this book and order theory, the notation in this chapter slightly deviates from the one in other chapters. This should also help emphasize that the objects under scrutiny in this chapter are not quite the same as in others, even though the differences are sometimes subtle. In particular, we alert the reader to the difference between a poset $(P,\geq)$ and its dual $(P,\leq)$. While we have been relying on the notation $(P,\geq)$ for trees, the order-theoretic literature relies on the dual notation $(P,\leq)$, which we will follow in this

[1] This chapter is a revised and slightly expanded version of Alós-Ferrer and Ritzberger (2005b).

C. Alós-Ferrer, K. Ritzberger, *The Theory of Extensive Form Games*,
Springer Series in Game Theory, DOI 10.1007/978-3-662-49944-3_3

chapter. The difference is of course superficial, but the reader should be warned to avoid confusion.

Recall that a subset A of a partially ordered set (poset) $(P, \leq)$ is a *chain* if it is totally ordered under the restricted order. Equivalently, A is a chain if every finite, nonempty subset F of A has an upper bound in F (hence a maximum).[2]

3.1 Pseudotrees

A poset is totally (or linearly) ordered if and only if it is a chain. In the theory of Boolean Algebras, the concept of tree has been treated as a generalization of the concept of linearly ordered set. The following definition is taken from Koppelberg and Monk (1992)

Definition 3.1 A **pseudotree** is a poset $(T, \leq)$ such that, for each $t \in T$, the down-set (or principal order ideal, see Davey and Priestley 1990, p. 185)

$$\downarrow t = \{x \in T | x \leq t\}$$

is a chain.

A *dual pseudotree* is a poset $(T, \geq)$ such that, for each $t \in T$, the up-set (or principal order filter)

$$\uparrow t = \{x \in T | t \leq x\}$$

is a chain, i.e. if $(T, \leq)$ is a pseudotree. That is, in the rest of the book we actually work with dual pseudotrees. Of course, working with pseudotrees or dual pseudotrees is purely a matter of convention, as one can be converted to the other by reversing the partial order. While in the rest of the book we follow the dual notation in order to have an intuitive connection between the relation $\geq$ and set inclusion, in this chapter we follow the notation of Koppelberg and Monk (1992).

Except for the use of the dual notation and the fact that no root is assumed, our notion of tree (Definition 2.1) corresponds to the concept of pseudotree in order theory.[3] However, in order theory the word "tree" is usually reserved for posets such that, additionally, the down-sets are well-ordered, i.e. for all $t \in T$ every subset of $\downarrow t$ has a first element (a minimum) (see, e.g., Koppelberg 1989, chp. 6). This condition (which we will refer to as "up-discreteness" in Chap. 5) implies that immediate successors of non-terminal nodes are well-defined, but is unrelated to the existence of immediate predecessors (as will be clarified in Chap. 6 on discrete

[2] See the Mathematical Appendix A.1.2 for definitions of properties of binary relations.

[3] Other authors also call such a poset simply a tree, e.g. Davey and Priestley (1990, Exercise 1.7).

trees). Koppelberg and Monk (1992) dropped the well-ordered requirement and called the resulting concept a *pseudotree*, which corresponds to the definition above.

A poset $(P, \leq)$ has a *bottom element* if there exists $x_0 \in P$ such that $x_0 \leq x \, \forall x \in P$. A *top element* is defined dually. A pseudotree is *rooted* if it has a bottom element; notice, though, that general pseudotrees need not be rooted.

Order-theoretic analysis of pseudotrees has concentrated on the various (set) Boolean algebras that they give rise to. Koppelberg and Monk (1992) study the *pseudotree algebra* of a pseudotree $(T, \leq)$, i.e. the set algebra generated by $\{\uparrow x \,|\, x \in T\}$. Baur and Heindorf (1997) and Baur (2000) study the *initial chain algebra*, i.e. the algebra generated by $\{\downarrow x \,|\, x \in T\}$. The previous chapter studies the set representation of dual pseudotrees which "branch at every node" (decision trees). In the following section, we tackle the order-theoretic characterization of the concept of pseudotree.

3.2 Directed Sets and Pseudotrees

Apart from the concept of pseudotree, in order theory there are two different, notable generalizations of the concept of (nonempty) chains. Again, a set A is a chain if every finite, nonempty subset F of A has an upper bound in F (hence a maximum). The two generalizations differ only with respect to which set contains the upper bound from the previous sentence.

Definition 3.2 Let $(P, \leq)$ be a poset. A nonempty subset A of P is

(a) **directed** if every finite subset of A has an upper bound *in* A.
(b) **consistent** if every finite subset of A has an upper bound *in* P.

Note that all chains are directed sets, and that all directed sets and hence all chains are consistent.

Often a result holds for all directed sets in a poset if and only if it holds for all chains. An outstanding example is the following. A poset $(P, \leq)$ is a *CPO* (short for Complete Partially Ordered set) if it has a bottom element and every directed subset $D \subseteq P$ has a supremum (join) $\bigvee D$ (see e.g. Davey and Priestley 1990). A poset is *chain-complete* if every chain, including the empty chain, has a supremum. It is a well-known but nontrivial result (whose proof uses the machinery of ordinals, see Markowsky 1976) that a poset is a CPO if and only if it is chain-complete.

Results like this raise the question of how much is added by the generalization of the concept of chain to that of directed sets for the understanding of general posets. A complementary question could be phrased as follows: For which posets are all directed sets chains? In such posets, results like the one quoted above would collapse to trivial statements.

Why is this relevant? The reason is that, as the next (straightforward) proposition shows, any poset such that all its directed sets are chains is necessarily a pseudotree. The converse is even stronger, since in a pseudotree, all consistent sets are chains.

Proposition 3.1 *Let $(P,\leq)$ be a poset. The following statements are equivalent:*

(a) $(P,\leq)$ *is a pseudotree.*
(b) *All consistent subsets of P are chains.*
(c) *All directed subsets of P are chains.*

Proof "(a) implies (b):" Let $(P,\leq)$ be a pseudotree. Let C be a consistent subset $(P,\leq)$ and $d,d' \in C$. Since C is consistent, there exists an upper bound in $(P,\leq)$ for $\{d,d'\}$, i.e. $x \in P$ such that $d \leq x$ and $d' \leq x$. It follows that $d,d' \in \downarrow x$. Since $\downarrow x$ is a chain, either $d \leq d'$ or $d' \leq d$, proving that C is a chain.

"(b) implies (c):" Immediate.

"(c) implies (a):" Let $(P,\leq)$ be a poset such that all its directed subsets are chains and consider $x \in P$. Since $x \in \downarrow x$ is an upper bound for all subsets of $\downarrow x$, it follows that $\downarrow x$ is directed and hence a chain. It follows that $(P,\leq)$ is a pseudotree. ■

This result offers a characterization of pseudotrees as the class of posets where the generalization of the concept of chain to directed (or consistent) sets is vacuous. Pseudotrees provide an example of posets where all consistent sets are actually directed, i.e. the generalization of the concept of directed sets to consistent sets is vacuous. One may ask whether there are other examples of posets satisfying this property, i.e. posets where all consistent sets are directed but some directed sets are not chains. The next proposition shows that the answer is negative.

Proposition 3.2 *A poset is a pseudotree if and only if all its consistent sets are directed.*

Proof The "only if"-part follows from Proposition 3.1. To see the converse, let $(P,\leq)$ be a poset such that all its consistent subsets are directed.

Choose $x \in P$. We need to show that $\downarrow x$ is a chain. Let $y,z \in \downarrow x$. Since $y \leq x$ and $z \leq x$, the set $\{y,z\}$ is consistent. By hypothesis it must then be directed. Since $\{y,z\}$ is a finite subset of itself, it follows that either $y \leq z$ or $z \leq y$, showing that $\downarrow x$ is a chain. ■

Given a poset $(P,\leq)$, consider the three classes of subsets given by nonempty chains, directed sets, and consistent sets. Propositions 3.1 and 3.2 show that it is enough that any two of these three classes coincide to conclude that $(P,\leq)$ is a pseudotree. Reciprocally, it follows that, for any poset which is not a pseudotree, there must exist at least a consistent subset which is not directed, and a directed subset which is not a chain.

While it is hardly surprising that the three "nested" concepts above collapse in the case of pseudotrees, Proposition 3.2 and the simplicity of its proof were somewhat unexpected. The results above illustrate that, from the point of view of order theory, (pseudo)trees are very powerful structures.

3.3 Pseudotrees as (Semi)Lattices

In view of Proposition 3.1, we obtain the following, immediate Corollary.

Corollary 3.1 *A pseudotree is a CPO if and only if it is chain-complete.*

Of course, this is just a restatement of the celebrated result by Markowsky (1976), which holds for arbitrary posets. We merely remark here that the result is trivial for pseudotrees. This observation, though, points out that chain-complete pseudotrees have a strong algebraic structure. Indeed, we will argue that such pseudotrees are a subclass of *semilattices*, and they turn into lattices under minimal additional requirements.

A poset $(P,\leq)$ is a *lattice* if any two elements $x,y\in P$ have a least upper bound (supremum), denoted $x\vee y$ (the *join* of x and y) and a greatest lower bound (infimum), denoted $x\wedge y$ (the *meet* of x and y). A lattice is *complete* if every subset $S\subseteq P$ has a supremum $\bigvee S$ and an infimum $\bigwedge S$. Since $S=\emptyset$ is allowed, any complete lattice has both top and bottom elements.

A *meet-semilattice* is a poset such that any two elements have an infimum. A complete meet-semilattice is a CPO satisfying any of the equivalent conditions of the following, well-known Lemma (see Davey and Priestley 1990, Lemma 3.20).

Lemma 3.1 *Let $(P,\leq)$ be a CPO. The following conditions are equivalent:*

- *(SL1)* $(P,\leq)$ *is* consistently complete, *i.e. the join $\bigvee S$ exists for every consistent subset S of P.*
- *(SL2)* *The join $\bigvee S$ exists whenever $S\subseteq P$ has an upper bound.*
- *(SL3)* *The meet $\bigwedge S$ exists whenever $S\subseteq P$ is nonempty.*
- *(SL4)* $\downarrow x$ *is a complete lattice for all $x\in P$.*
- *(SL5)* $P\bigoplus 1$ *is a complete lattice, where $P\bigoplus 1$ denotes the poset formed by adding a new element (the 1) to P and extending the order on P in such a way that 1 is a top element, i.e. $x\leq 1$ for all $x\in P$.*

In view of this it is immediate to observe that chain-complete pseudotrees (i.e. pseudotrees which are CPOs) coincide with complete meet-semilattices.

Proposition 3.3 *A pseudotree $(T,\leq)$ is chain-complete if and only if it is a complete meet-semilattice.*

Proof Suppose $(T,\leq)$ is chain-complete. By Proposition 3.1(b) that every chain has a supremum implies condition (SL1) in Lemma 3.1.

Suppose now $(T,\leq)$ is a complete meet-semilattice. Then it is a CPO by definition, hence chain-complete by Corollary 3.1. ■

Hence, conditions (SL1–5) hold for all chain-complete pseudotrees. In particular, adjoining a top to a chain-complete pseudotree yields a complete lattice. We now show that a weaker property characterizes the class of rooted pseudotrees which turn into (not necessarily complete) lattices when a top element is adjoined.

Baur and Heindorf (1997) introduce the concept of *well-met pseudotrees*: A pseudotree is *well-met* if any two compatible elements[4] have a greatest lower bound (infimum). They show (Baur and Heindorf 1997, Corollary 1.3) that any pseudotree can be monomorphically embedded in a well-met pseudotree.

In a rooted pseudotree all elements are compatible. Hence, in a well-met rooted pseudotree, any two elements have an infimum. Thus, a rooted pseudotree is well-met if and only if it is a meet-semilattice. It is well-known, though, that adjoining a top element to a general meet-semilattice does *not* necessarily yield a lattice.[5]

Proposition 3.4 *Let $(T,\leq)$ be a rooted pseudotree. Then T is well-met (i.e. a meet-semilattice) if and only if $T\bigoplus 1$ is a lattice.*

Proof "if:" Suppose that $T\bigoplus 1$ is a lattice and let $x,y\in T$ (which are compatible, because T is rooted). Since $T\bigoplus 1$ is a lattice, there exists an infimum z for x and y in $T\bigoplus 1$. If $z=1$, it follows that $x=z\notin T$, a contradiction, and hence $z\in T$, proving the claim.

"only if:" Consider any two elements $x,y\in T$. Either they are related by $\leq$ or not. If $x\leq y$, y is the supremum of $\{x,y\}$, and reciprocally if $y\leq x$. If x,y are unrelated by $\leq$, then there cannot exist an upper bound. For, if $x\leq z$ and $y\leq z$, then $x,y\in\downarrow z$, which is a chain, a contradiction. Thus, once the top element is added, it becomes the least upper bound of any pair of unrelated elements. ■

Finally, we observe that the class of well-met, rooted pseudotrees is larger than the class of chain-complete pseudotrees. This result also follows indirectly from Proposition 3.4 and (SL5).

Proposition 3.5 *Every chain-complete pseudotree is well-met.*

Proof If $x,y\in T$ are compatible, the set $S=\downarrow x\cap\downarrow y$ is nonempty. Since both $\downarrow x$ and $\downarrow y$ are chains, S has a supremum, z. Since both x and y are upper bounds for S, it follows that $z\in S$ and hence it is a greatest lower bound for x and y. ■

The converse is not true, as the pseudotree $(\mathbb{Z},\leq)$ shows. Notice that, in the proof of the last Proposition, we only make use of the property that every *nonempty* chain in T has a supremum in T. The difference with chain-completeness is small but significant, since a supremum for the empty chain is necessarily a bottom element, i.e. a root for the pseudotree.

[4] Two elements x,y of a poset $(P,\leq)$ are compatible if there exists some $z\in P$ such that $z\leq x$ and $z\leq y$.

[5] This is known if all chains have finite length (see Birkhoff 1973, p.23), but, as the next Proposition shows, this condition is not necessary.

3.4 Representation of Pseudotrees

Trees and pseudotrees are of particular interest to economics and game theory, where they serve as a transparent model of sequential decision making. The emphasis in game theory is on *plays* and outcomes, as those provide a domain for decision-makers' preferences. Intuitively, an outcome is the final result of the chain of decisions taken in the course of a game, while a play (or maximal history) is a full record of all decisions taken. Following von Neumann and Morgenstern (1944), we may think of decisions taken along the trees as sampling subsets of a universal set of outcomes. We should then be able to identify each node (element of the tree) with the set of outcomes (or plays) which have still not been discarded at the decision point represented by the node.

This intuition was first formalized in Alós-Ferrer and Ritzberger (2005a) and has been described in the previous chapter. From the point of view of order theory, that chapter contains representation theorems for certain classes of pseudotrees, very specially Theorem 2.1 on decision trees. Focusing on decision trees entails the position that nodes (immediately) followed by exactly one other node represent no decision and, hence, do not belong to the description of a decision problem. This is not a necessary position, though. If trivial structures are allowed for, a result analogous to Theorem 2.1 obtains: General trees can be represented by *extensible* and maximal chains just like in Theorem 2.1. We now recall the necessary concepts and present this generalization which, in our view, sheds light on the concept of a pseudotree.

Let $(P,\leq)$ be a nonempty poset. A chain $w\subseteq P$ is maximal if there exists no $x\in P\setminus w$ such that $w\cup\{x\}$ is a chain. That is, a maximal chain is a *play* in the game-theoretic sense. Let W denote the set of maximal chains of $(P,\leq)$ and, for each $x\in P$, denote by $W(x)=\{w\in W|x\in w\}$ the set of maximal chains "passing through" x. By the Hausdorff Maximality Principle, $W(x)$ is nonempty for all $x\in P$ (see e.g. Birkhoff 1973, Chapter VIII). We can define the poset $(W(P),\supseteq)$, where $W(P)=\{W(x)|x\in P\}$ and $\supseteq$ is set inclusion, and consider $W(\cdot)$ as a mapping from P to $W(P)$. Then, the following result holds:

Lemma 3.2 *Let $(T,\leq)$ be a pseudotree. If $x,y\in T$ are such that $x\leq y$, then $W(x)\supseteq W(y)$.*

Proof Let $w\in W(y)$. Notice that $\downarrow y\subseteq w$ by maximality of w. To see this, fix $z\in\downarrow y$ and consider any $t\in w$. Since w is a chain and $y\in w$, either $y\leq t$ or $t\leq y$. In the first case $z\leq y\leq t$ implies $z\leq t$. In the second case $z,t\in\downarrow y$ implies (since $\downarrow y$ is a chain) that either $z\leq t$ or $t\leq z$. This proves that $w\cup\{z\}$ is a chain, which by maximality of w implies that $z\in w$.

Now, suppose $x\leq y$ and consider any $w\in W(y)$. By the previous statement $x\in w$, i.e. $w\in W(x)$, proving the claim. ■

A *decision pseudotree* is a pseudotree $(T,\leq)$ such that if $W(x)=W(y)$ for two elements $x,y\in T$, then necessarily $x=y$ (recall Definition 2.2). In a decision pseudotree, every element has to be followed (in the sense of $\leq$) by at least two other elements.

It is straightforward to extend Theorem 2.1 to show that a pseudotree $(T, \leq)$ is a decision pseudotree if and only if the mapping $W(\cdot)$ is an order isomorphism (recall that pesudotrees are not necessarily rooted, hence the result is slightly more general than that Theorem). This fact will actually become a corollary of the results proved below. The pseudotree $(W(T), \supseteq)$ is then a set representation of $(T, \leq)$.

While, as we have argued before, decision (pseudo)trees should be the objects of interest for game theory, from the theoretical point of view the question remains whether a similar characterization can be given for general pseudotrees. This question is also of interest for game theory, because certain definitions of extensive form games (notably the one in Osborne and Rubinstein 1994, recall Sect. 2.2.2.1 and Example 2.6) make use of trees which are in principle not decision trees.

In this section such a characterization is provided. Of course, in a tree which is not a decision tree, there might exist distinct elements x, y such that $W(x) = W(y)$. Intuitively, there is no decision to be taken at x, and y immediately follows x (or the other way around). Thus, a representation theorem cannot be based on maximal chains alone. Instead, it must rely on a weaker concept.

Given a poset $(P, \leq)$, say that a chain $w \subseteq P$ is *extensible* if there exists $x \in P$ such that $w = \downarrow x$ and

$$\bigcap \{w' \in W \mid w \subseteq w'\} \setminus w \neq \emptyset.$$

In words, a chain of the form $w = \downarrow x$ is extensible if all maximal chains that contain it have a common part outside w. In decision-theoretic terms, this common part comes forcefully "after" w. Hence, a decision at x cannot "discard" anything, but becomes a trivial "decision." For the class of pseudotrees the following characterization of extensible chains obtains.

Lemma 3.3 *Given a pseudotree $(T, \leq)$, a chain $w \subseteq T$ is an extensible chain if and only if there exists $x \in T$ and $y \in \uparrow x \setminus \{x\}$ such that $w = \downarrow x$ and $W(x) = W(y)$.*

Proof Suppose $w = \downarrow x$ is extensible and let $y \in \cap\{w' \in W | w \subseteq w'\}$ such that $y \notin w$. It follows that $y \in \uparrow x \setminus \{x\}$. By Lemma 3.2 $W(y) \subseteq W(x)$ and, thus, it suffices to show that $W(x) \subseteq W(y)$.

Let $w' \in W(x)$, i.e. $x \in w' \in W$. Since w' is a maximal chain, $w = \downarrow x \subseteq w'$. By the choice of y we have that $y \in w'$, thus, $w' \in W(y)$. This completes the proof of the "only if"-part.

Suppose now $w = \downarrow x$ and $W(x) = W(y)$ with $y \in \uparrow x \setminus \{x\}$, $x \in T$. Let $w' \in W$ such that $w \subseteq w'$. Since $w = \downarrow x$, we have that $w' \in W(x) = W(y)$, thus, $y \in w'$. Since $y \notin w$, this proves that w is extensible. ■

Call a chain of a poset $(P, \leq)$ *elementary* if it is either extensible or a maximal chain. Denote by Ω the set of chains of $(P, \leq)$ which are either maximal or extensible (i.e. the elementary chains) and, given $x \in P$, let $\Omega(x) = \{w \in \Omega | x \in w\}$. Obviously, $W(x) \subseteq \Omega(x)$. We can consider $\Omega(\cdot)$ as a mapping from $(P, \leq)$ to the poset $(\Omega(P),\supseteq)$, where $\Omega(P)=\{\Omega(x)|x\in P\}$ and $\supseteq$ is set inclusion. Then, the next Theorem shows that this latter poset yields a set representation of $(P,\leq)$.

Example 3.1 Let $P=\{x,y,z_1,z_2\}$ and $\leq$ be given by $x\leq y,z_1,z_2$ and $y\leq z_1,z_2$. $(P,\leq)$ is a pseudotree. There are only two maximal chains, $w_1=\{x,y,z_1\}$ and $w_2=\{x,y,z_2\}$. Further, $W(x)=\{w_1,w_2\}=W(y)$ and, thus, $W(P)$ does not yield a set representation of $(P,\leq)$. However, there is an extensible chain $w_0=\{x\}$, thus, $\Omega(x)=\{w_0,w_1,w_2\}\supset\{w_1,w_2\}=\Omega(y)$, and $\Omega(P)$ yields the desired set representation.

Theorem 3.1 *A poset* $(P,\leq)$ *is a pseudotree if and only if the mapping* Ω *is an order isomorphism.*

Proof "if:" Suppose that $\Omega(\cdot)$ is an order isomorphism. Let $z\in P$. We have to show that $\downarrow z$ is a chain. Let $x,y\in\downarrow z$. By order isomorphism it follows that $\Omega(z)\subseteq\Omega(x)\cap\Omega(y)$. Let $w\in\Omega(z)$. Then $x,y\in w$ and the conclusion follows from the fact that w is a chain.

"only if:" Suppose $(P,\leq)$ is a pseudotree. The mapping $\Omega(\cdot)$ is surjective by construction. Since any order-embedding surjection is an order isomorphism, it is enough to show that $\Omega(\cdot)$ is order-embedding, i.e. $x\leq y$ if and only if $\Omega(x)\supseteq\Omega(y)$.

Let $x,y\in P$ be such that $x\leq y$ and $w\in\Omega(y)$. If $w\in W(y)$, then by Lemma 3.2 $w\in W(x)\subseteq\Omega(x)$. If $w\notin W(y)$, then by Lemma 3.3 there exists $z\in P$ such that $w=\downarrow z$. Since $x\leq y\in w$, it follows that $x\leq z$ and, thus, $x\in\downarrow z=w$, implying that $w\in\Omega(x)$. Hence, $\Omega(x)\supseteq\Omega(y)$.

Suppose now $\Omega(x)\supseteq\Omega(y)$ for some $x,y\in P$. In particular, this implies that $W(x)\supseteq W(y)$. Let $w\in\Omega(y)$; then $x,y\in w$ and, since w is a chain, either $x\leq y$ or $y\leq x$. Suppose $y\leq x$ and $x\neq y$. That $y\leq x$ implies $W(x)=W(y)$ by Lemma 3.2 and the hypothesis. Then, that $x\in\uparrow y\setminus\{y\}$ implies that $w^*=\downarrow y$ is an extensible chain. But $w^*\in\Omega(y)$ implies $w^*\in\Omega(x)$ by the hypothesis, in contradiction to $x\notin w^*=\downarrow y$. ■

As a consequence of Theorem 3.1, a characterization of decision pseudotrees is obtained.

Corollary 3.2 *A pseudotree is a decision pseudotree if and only if there are no extensible chains.*

Proof "if:" Let $(T,\leq)$ be a pseudotree. Let $x,y\in T$ be such that $W(x)=W(y)$. If there are no extensible chains, $\Omega(x)=\Omega(y)$ and, by Theorem 3.1, $x=y$.

"only if:" Let $(T,\leq)$ be a decision pseudotree, and let w be an extensible chain. Then by Lemma 3.3 there exist $x,y\in T$ such that $w=\downarrow x$, $y\in\uparrow x\setminus\{x\}$, and $W(x)=W(y)$. Since $(T,\leq)$ is a decision pseudotree, we must have $x=y$, a contradiction. ■

That $(T,\leq)$ is a decision pseudotree if and only if the mapping $W(\cdot)$ is an order isomorphism is an obvious corollary of Theorem 3.1 and Corollary 3.2. This is Theorem 2.1 from the previous chapter.

3.5 Summary

This chapter considers the concept of "pseudotrees" studied in order theory, which essentially correspond to the (dual of the) concept of tree in the previous chapter. In particular, in (dual) pseudotrees the set of predecessors is a chain, but a node may be followed by only one other node. Hence, they are not decision trees. We show that a poset is a pseudotree if and only if all its consistent subsets are chains, or, equivalently, all its directed subsets are chains (Proposition 3.1). In fact, a poset is a pseudotree if and only if all its consistent subsets are directed (Proposition 3.2).

Since pseudotrees are not decision trees, the representation by plays does not work. But if one adds to the plays all extensible chains, an analogous result to Theorem 2.1 holds: A poset is a pseudotree if and only if it can be represented by plays and extensible chains (Theorem 3.1). Of course, a (rooted) pseudotree is a decision tree if and only if there are no extensible chains.

References

Alós-Ferrer C, Ritzberger K (2005a) Trees and decisions. Econ Theory 25(4):763–798

Alós-Ferrer C, Ritzberger K (2005b) Some remarks on pseudotrees. Order 22(1):1–9

Baur L (2000) Cardinal functions on initial chain algebras on pseudotrees. Order 17:1–21

Baur L, Heindorf L (1997) Initial chain algebras on pseudotrees. Order 14:21–38

Birkhoff G (1973) Lattice theory. American mathematical society, vol XXV, 3rd edn. Colloquium Publications, Providence

Davey B, Priestley H (1990) Introduction to lattices and order. Cambridge University Press, Cambridge

Koppelberg S (1989) General theory of Boolean algebras. In: Monk J, Bonnet R (eds) Handbook of boolean algebras, vol 1. Elsevier, Amsterdam/New York, pp 741–773

Koppelberg S, Monk J (1992) Pseudo-trees and Boolean algebras. Order 8:359–577

Markowsky G (1976) Chain-complete posets and directed sets with applications. Algebra Universalis 6:53–68

Osborne MJ, Rubinstein A (1994) A course in game theory. MIT, Cambridge

von Neumann J, Morgenstern O (1944) Theory of games and economic behavior. Princeton University Press, Princeton

Chapter 4
Extensive Decision Problems

This chapter takes advantage of the findings in Chap. 2, in particular of the general concept of a game tree (Definition 2.4), and presents its main application.[1] For the formalization of non-cooperative games the game tree captures the "objective" description of what may happen in the course of the game—an outside observer's viewpoint, as it were. To make a tree into a game takes a second, "subjective," layer that brings in the decision makers (players). The first step on this second level is a specification of who can when do what, and under which informational constraints. An "Extensive Decision Problem" (EDP), as defined below, formalizes these specifications in full generality.

An EDP is a preliminary concept, though. Due to the generality of its underlying game trees, it captures all game specifications that have been proposed in the literature. In particular, it also captures exotic cases like "differential games" (see Sect. 2.2.2.6). Yet, as of now there is no guarantee that such exotic games can also be played, in the sense that strategies yield outcomes that players can evaluate. In fact, it will be shown in Chap. 5 that without further restriction such exotic games may not be playable. These considerations provide the reason why we do not refer to an EDP as an "extensive form." The latter term remains reserved for game specifications that can indeed be played.

[1]This chapter is based on an expanded version of Alós-Ferrer and Ritzberger (2008), with the addition of some parts of Alós-Ferrer and Ritzberger (2005a).

C. Alós-Ferrer, K. Ritzberger, *The Theory of Extensive Form Games*,
Springer Series in Game Theory, DOI 10.1007/978-3-662-49944-3_4

4.1 Preview

To identify which further restrictions guarantee that a game is in fact playable takes a framework in which this problem can be posed. And it is this framework that an EDP, as in Definition 4.1, provides. It is shown below that the definition of an EDP consists of independent properties, and yields a specification such that every play from the game tree can be selected by appropriate choices on the part of the players (Theorem 4.1). On the other hand, an EDP, like the original definition by Kuhn (1953) of an extensive form, rules out cases of absent-mindedness (Proposition 4.1), where a play passes through an information set more than once.

With the definition of an EDP at hand it remains to be seen, though, whether or not an EDP can be defined on an arbitrary game tree. It is shown below that this is a matter of the tree (Proposition 4.3): An EDP can be defined on a game tree if and only if the tree has "available (perfect information) choices." (Theorem 4.2).

This insight derives from a deeper inquiry into properties of game trees. In particular, this chapter introduces three such properties: Weak up-discreteness, coherence, and regularity (see Definition 4.2 below). And it provides a full classification of nodes in a game tree. The classification of nodes enables a result complementary to Proposition 2.11 by identifying the nodes that may or may not be present in a game tree without any effect on its structure (Proposition 4.2 below). Of the three properties of game trees weak up-discreteness turns out to be sufficient for a game tree to allow "available (perfect information) choices" (Definition 4.4 below). Hence, weak up-discreteness guarantees that an EDP can be defined on the game tree.

Since ultimately what makes an EDP playable turns out to be properties of its underlying game tree, the results in the present chapter will be used later on, when playable EDPs are at stake. This chapter thus prepares for the analysis in Chap. 5.

4.2 Definition of Extensive Decision Problems

In this Section it is shown that game trees lend themselves naturally to the formalization of sequential multi-person decision problems. Since the aim is only to provide an extensive form *representation*—rather than a vehicle for solving games—the proposed definition concerns only the rules of the game, not the players' preferences.[2]

An extensive form requires a specification of the players' choices on top of the specification of the tree. This is also true in the traditional definition of an extensive form: choices of players (and, thereby, the information structures of players) have to be specified separately. The concept of an "information set" (the set of nodes where a certain menu of choices is available), however, need not make sense in the general

[2]In the older literature this is often referred to as a "game form."

setting: information sets may not exist. Therefore, the objects added to turn a tree into a game form are not information structures, but "choices"—alternatives for the selection among plays.

Let $T = (N, \supseteq)$ be a game tree with set of plays W and

$$X = \{x \in N \,|\, \exists z \in N : z \subset x\}$$

the set of *moves*. At each move, one or more players will choose certain sets of plays. In order to model such decisions, one needs a notion of which sets of plays are available at which moves. For a set $a \subseteq W$ of plays let $\downarrow a = \{x \in N \,|\, x \subseteq a\}$ be its *down-set* and define the set of *immediate predecessors* of a as

$$P(a) = \{x \in N \,|\, \exists y \in \downarrow a : \uparrow x = \uparrow y \setminus \downarrow a\} \tag{4.1}$$

Say that a set a of plays is *available at* the move $x \in X$ if $x \in P(a)$.

The reason for this apparently complex definition is that, in the general setting considered here, immediate predecessors of nodes might not exist (e.g., in the case of decisions in continuous time), and hence one cannot appeal to a shortcut through immediate predecessors of nodes. Indeed, since every node $x \in N$ is a set of plays, nodes too may, but need not, have immediate predecessors according to the definition above.

The idea behind the definition is as follows. Consider Fig. 4.1, which depicts a set of plays a which is a subset of x (for simplicity; in general a need not be contained in x). The discontinuous lines depict chains of nodes. The node x is a predecessor of a because there exists some other node y, contained in a, such that x is the minimum among those predecessors of y that are not contained in a. That is, there is a chain of nodes containing x whose members are eventually contained in the set of plays a, but the node x is the "last" one along the chain which is *not* contained in a. Intuitively, the set of plays a "leads towards" a certain chain of successors of x. Clearly, if $x \in P(a)$, then there is a play $w \in x \cap a$ that passes through x and a and another $w' \in x \setminus a$ that passes through x but not through a, hence a represents some "choice".

Effectively, $P(a)$ is the collection of nodes that are minimal with respect to a nontrivial intersection with $a \subseteq W$; intuitively, "the last event after which a becomes known." For, if $x \cap a \neq \emptyset \neq x \setminus a$ and $y \subseteq a$ for all $y \in \downarrow x \setminus \{x\}$, then for any $y \in \downarrow x \setminus \{x\}$ it holds that $y \in \downarrow a$ and $\uparrow x = \uparrow y \setminus \downarrow a$, hence $x \in P(a)$. Conversely, if $x \in P(a)$, then for any $y \in \downarrow x \setminus \{x\}$ it holds that $y \subseteq a$, hence $x \cap a \neq \emptyset$, and $x \notin \downarrow a$ implies $x \setminus a \neq \emptyset$.

Fig. 4.1 Immediate predecessors of a set of plays. The set of plays a is available at the node x because x is the minimum among those predecessors of some node $y \subseteq a$ that are *not* contained in the set a

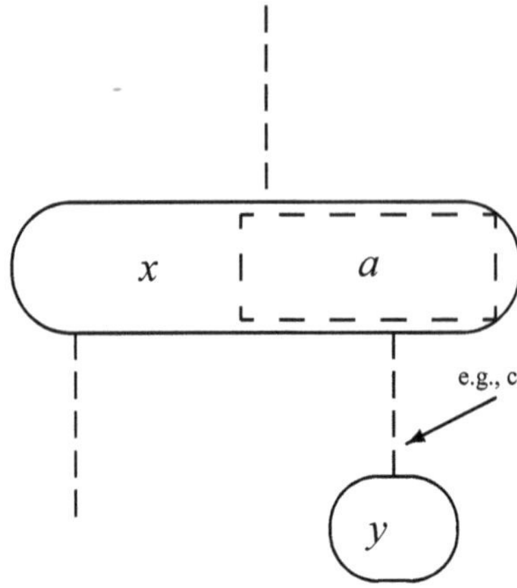

Definition 4.1 An **extensive decision problem** with player set I is a pair (T, C), where $T = (N, \supseteq)$ is a game tree with set of plays W and $C = (C_i)_{i \in I}$ is a system of collections C_i (the sets of players' choices) of nonempty unions of nodes (hence, sets of plays) for all $i \in I$, such that the four conditions below hold. Let $A_i(x) = \{c \in C_i | x \in P(c)\}$ denote the choices available to i at x for all $i \in I$, and $J(x) = \{i \in I | A_i(x) \neq \emptyset\}$ the set of decision makers at x which is required to be nonempty for all $x \in X$.

(EDP1) If $P(c) \cap P(c') \neq \emptyset$ and $c \neq c'$ then $P(c) = P(c')$ and $c \cap c' = \emptyset$, for all $c, c' \in C_i$ and all $i \in I$;
(EDP2) $x \cap \left[\cap_{i \in J(x)} c_i\right] \neq \emptyset$ for all $(c_i)_{i \in J(x)} \in \times_{i \in J(x)} A_i(x)$ and all $x \in X$;
(EDP3) if $y, y' \in N$ satisfy $y \cap y' = \emptyset$ then there are $i \in I$ and $c, c' \in C_i$ such that $y \subseteq c$, $y' \subseteq c'$, and $c \cap c' = \emptyset$;
(EDP4) if $x \supset y \in N$ then for every $i \in J(x)$ there is $c \in A_i(x)$ such that $y \subseteq c$, for all $x \in X$.

The set of players I may contain a distinguished player $i = 0$, called "chance," that models events not under the control of personal players. The behavior of "chance" can then be determined by fixing a (possibly randomized) strategy for this particular player.

What is added to the tree to obtain an extensive decision problem are collections of choices $c \in C_i$ (i.e. collections of sets of plays) for all players $i \in I$. With the choices in place, Definition 4.1 captures the following quasi-operational specification of an extensive decision problem. At every move $x \in X$ each player $i \in I$ is told (by an "umpire") which choices $c \in C_i$ she has available (in the sense that $x \in P(c)$, $c \in C_i$) and asked to select one of those. No other information is released to players. Given the decisions by all personal players and the possible chance moves, taking the intersection gives a nonempty set of plays, which determines how the game will continue.

We now briefly comment on the four conditions (EDP1-4). First, (EDP1) stands in for information sets. If two distinct choices are available at a common move, then their immediate predecessors are identical and the choices are disjoint. Thus, the player cannot infer from the available menu of choices at which move (in the common set of predecessors, i.e. information set) she chooses. And, two choices that are simultaneously available cannot overlap.[3] Section 4.2.1 below discusses information sets in further detail.

Second, (EDP2) ensures that any combination of available choices yields something nonempty. If a combination of choices (one for each decision maker) is available at a common move, then the combination has a nonempty intersection. Note that, in general, several players might be active at the same move (see Sect. 4.2.2 below).

Third, (EDP3) deals with the "residual" that remains after personal players have "made their choices." If two nodes differ, then at some point someone (possibly chance) takes a decision that separates them. It will be shown below (Theorem 4.1) that this implies that the intersection of all choices, belonging to personal players or to chance, that contain a particular play, yields precisely this play. Hence, "in the end" whatever is not decided by personal players will be decided by chance.

Fourth, (EDP4) implies the traditional exclusion of absent-mindedness (Kuhn 1953, p. 195), as discussed in Isbell (1957). In the absence of such a condition a play may cross an information set more than once (Piccione and Rubinstein 1997) or, in the present formalism, the same choice may be available more than once along the same play. It will be shown below (Proposition 4.1) that (EDP4) implies "no-absent-mindedness." That is, unlike in Kuhn's original formalism, the exclusion of absent-mindedness need not be assumed, but is an implication of the model. Section 4.2.3 below discusses absent-mindedness in more detail.

The concept of extensive decision problem is crucial for the remainder of the book. Hence, we will devote the following subsections to clarify some subtle points which might otherwise escape awareness, and which in part concern the differences with other approaches.

4.2.1 Information Sets

In graphical representations of extensive form games, it has become customary to mark the sets of nodes where a certain menu of choices is available and call them "information sets,". The interpretation is as sets of nodes that cannot be distinguished by a certain player. In our framework information sets are not primitive objects, but rather derived ones (if they can be derived at all). They

[3] Property (EDP1) is not necessary for the behavior of the distinguished player "chance."

correspond to the sets $P(c)$ for choices $c \in C_i$, that is, information sets are the sets of immediate predecessors of available choices (hence they are sets of nodes). Accordingly let $X_i = \{x \in X | \exists c \in C_i : x \in P(c)\}$ denote player i's moves or *decision points*, for each $i \in I$.

Consider the extensive decision problem depicted on the left-hand side of Fig. 4.2 (which is defined on the left-hand-side tree of Fig. 2.1). Player 1 moves first, choosing among Out, A, and B. If he chooses Out, the game ends. If he chooses A or B, player 2 has to choose L or R without knowing whether A or B has been chosen. The tree is defined on the set $W = \{w_1, \ldots, w_5\}$ and has three moves, $r = W$, $y_1 = \{w_2, w_3\}$, and $y_2 = \{w_4, w_5\}$ (and five terminal nodes, which are singletons). In the graphical representation, two conventions are used. First, a move is labeled with the name of the player it "belongs to," that is, r is labeled with 1 and y_1, y_2 are labeled with 2 indicating that $I = \{1,2\}$ with $X_1 = \{r\}$ and $X_2 = \{y_1, y_2\}$. Second, the choices are indicated by labeling the edges from decision points to the (immediate) successors that are covered by the choice. That is, the choices of player 1 are Out $= \{w_1\}$, which is hence noted on the edge leading to the node $\{w_1\}$, A $= \{w_2, w_3\}$, depicted on the edge leading to the node y_1, and B $= \{w_4, w_5\}$, depicted on the edge leading to the node y_2. The case of player 2 is slightly more involved. Both choices L and R are available at both nodes y_1 and y_2, because this player does not know whether A or B has been chosen. Hence, L $= \{w_2, w_4\}$ has to be depicted both on the edge leading to the terminal node $\{w_2\}$ and on the one leading to $\{w_4\}$. Analogously, R $= \{w_3, w_5\}$ is depicted both on the edge leading to the terminal node $\{w_3\}$ and on the one leading to $\{w_5\}$.

The right-hand side of Fig. 4.2 might be more user-friendly. First, the names of the nodes have been dropped, since the set structure (both for nodes and choices) can be readily deduced from the graph. Second, the information set $P(\text{L}) = P(\text{R}) = \{y_1, y_2\}$ has been represented by a dashed line joining the two moves of player 2. Of course, the latter convention is redundant, but might help ease the transition from typical textbook representations of finite games.

An important feature of Definition 4.1, however, is that information sets need *not* exist. If they do, they are given by the set of immediate predecessors of available choices. But at this level of generality nothing ensures that all choices have immediate predecessors. Still, in the following version of the differential game information sets do exist.

Example 4.1 (Differential EDP) Turn the tree of the differential game (Sect. 2.2.2.6) into an extensive decision problem, say, with two personal players, as follows. Let the set of actions A be a product set $A = A_1 \times A_2$. Given any function $f \in V$, denote $f = (f_1, f_2)$. The interpretation is as follows. At any point in time the two players $i = 1,2$ simultaneously decide on an action $a_i \in A_i$ for $i = 1,2$. Up to that moment, they know the entire history, but cannot anticipate the decision taken by the other player at t. Choices are of the form

$$c = c_{it}(f) = \{g \in V | g \in x_t(f), g_i(t) = f_i(t)\}$$

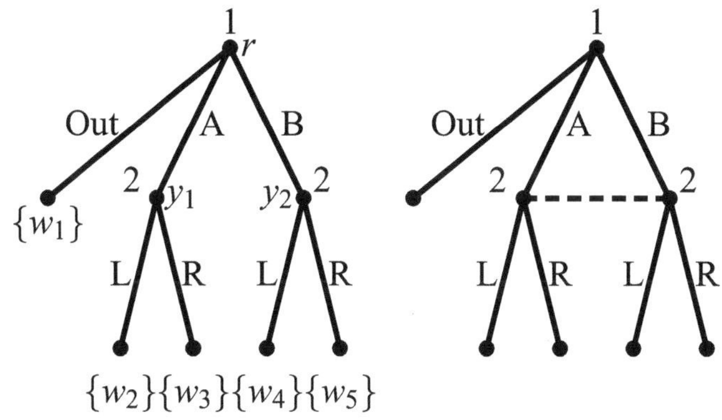

Fig. 4.2 Representation of an extensive decision problem. The *left-hand-side* version contains the names of the nodes for concreteness, but those can be omitted. The *dashed line* representing the information set of player 2 on the *right-hand-side* version is redundant. On the *left-hand-side* version, the fact that player 2 cannot distinguish between his two moves is captured by the fact that the same choices (L and R) are available

for some $f \in V$, some $t \in \mathbb{R}_+$, and $i = 1,2$. That choices are unions of nodes follows from $c_{it}(f) = \cup_{\tau > t} \cup_{g \in c_{it}(f)} x_\tau(g)$. Information sets are then given by $P(c_{it}(f)) = \{x_t(f)\}$. When two players decide their choices at t by picking, say, $c_{it}(f^i) \in C_i$ for $i = 1,2$, their intersection

$$c_{1t}(f^1) \cap c_{2t}(f^2) = \{g \in V \,|\, (g_1(\tau), g_2(\tau)) = (f_1^1(\tau), f_2^2(\tau)), \forall \tau \in [0,t]\}$$

keeps track of both decisions while leaving all possibilities open for the future.

If player i were not to observe previous decisions by the other player, but still recall his own, choices would be defined only by the property that $g_i(\tau) = f_i(\tau)$ for all $\tau \in [0,t]$, rather than $g \in x_t(f)$. In this case, information sets would be given by $P(c_{it}(f)) = \{x_t(g) \mid g_i(\tau) = f_i(\tau) \; \forall \, \tau \in [0,t)\}$.

4.2.2 *Simultaneous Decisions*

Whenever several players take decisions simultaneously, this may (but need not) be represented by a single move, in contrast to the usual approach of specifying several nodes and using information sets to preserve the informational simultaneity.[4]

[4]Friedman (1990, p.109) coined the term "semi-extensive form" for such a representation of true simultaneity. Another alternative formalization that allows several players to choose at the same move was introduced by Selten (1998).

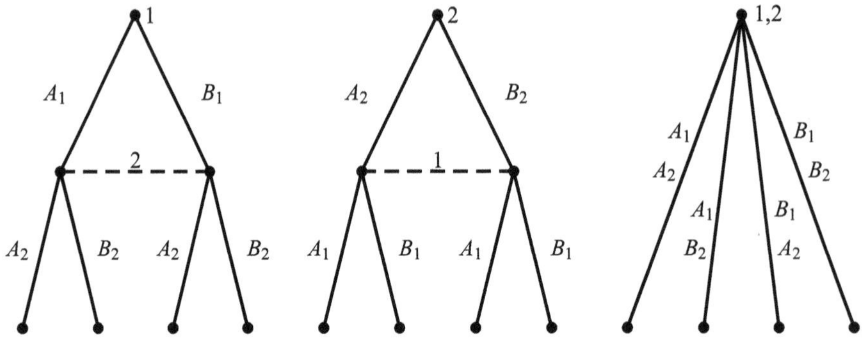

Fig. 4.3 Representations of an extensive decision problem with simultaneous decisions. The *left-hand-side* and *center* versions rely on cascading information sets and artificial order of decisions. The *right-hand-side* version uses a simpler tree with straightforward choices and no artificial intermediate moves, hence capturing true simultaneous decisions

To drive this point home, consider a game where two players, 1 and 2, make simultaneous binary choices. Player 1 chooses between A_1 and B_1 and, simultaneously, player 2 chooses between A_2 and B_2. The traditional depiction of such a normal form game in extensive form, following a convention due to Reinhard Selten (Selten 1975), leads to misrepresent the game as one where player 1 plays first, but player 2 does not observe the choice, leading to the graph on the left-hand side of Fig. 4.3. That is, an artificial information set is used to capture simultaneous decisions, declaring them indistinguishable from non-simultaneous but unobserved decisions. Of course, the choice of player 2 to be the one not observing a previous decision is arbitrary, and the game can be equally represented by the central depiction in Fig. 4.3, creating an uncomfortable multiplicity. In our approach, there is a more natural representation, given on the right-hand side of Fig. 4.3. Simultaneous decisions are captured by several players being active at the same node, and the specification of choices follows naturally. Condition (EDP2) requires that every combination of choices yields an outcome.

Under our approach, all three representations in Fig. 4.3 are feasible. All of them are based on trees with a four-element set of plays, say $W = \{w_1, w_2, w_3, w_4\}$. The right-hand-side representation merely contains the root $r = W$ and the singleton sets as terminal nodes, with choices being given by $A_1 = \{w_1, w_2\}$, $B_1 = \{w_3, w_4\}$, $A_2 = \{w_1, w_3\}$, and $B_2 = \{w_2, w_4\}$. The other two representations require additional (and, in our view, artificial) nodes, e.g., $y_1 = \{w_1, w_2\}$ and $y_2 = \{w_3, w_4\}$ for the case on the left-hand side. In our approach, one can reserve the simpler representation for true simultaneous decisions, while the other representations capture non-simultaneous but unobserved decisions.

Imagine that not two but three players would move simultaneously. The traditional approach following Selten's convention then would require two levels of artificial information sets. Such "cascading information sets" very quickly grow out of proportion as the number of players moving simultaneously increases. The next example illustrates that allowing a (very) large number of players to decide at the same move produces much simpler representations.

Example 4.2 (Nonatomic One-Shot Games) Let I be an arbitrary player set, e.g. the interval $[0,1]$ or an abstract measure space.[5] For every i, let A_i be an arbitrary action set. All players $i \in I$ decide simultaneously. Let $A = \times_{i \in I} A_i$ be the set of possible action profiles. With $N = \{A\} \cup \{\{a\} \,|a \in A\}$ a simple (but possibly large) rooted tree $(N, \supseteq)$ is obtained that is already in set-theoretic form: The set A is the root, where all outcomes are still possible; then every player makes a choice and those determine an action profile $a \in A$ that leads to the corresponding terminal node $\{a\} \in N$. Plays are trivial: each is of the form $w = \{A, \{a\}\}$. Choices available at A for player $i \in I$ are given by $c_i(a_i) = \{a' \in A \,|a'_i = a_i\}$ for all $a_i \in A_i$.

Suppose that in this example the traditional convention is used, where every player controls a separate information set even if she decides simultaneously with the others. Then the "infinite width" of the tree would directly (but artificially) translate into "infinite height," while under our approach the tree has just two levels, the root and the terminal nodes. This is a general property: under our approach, every normal form game has an extensive form of "height one."

Remark 4.1 For games with a continuum of players (see e.g. Schmeidler 1973, Mas-Colell 1984, or Carmona 2009), of course, other problems remain. For instance, straightforward aggregation of a continuum of independent mixed strategies is prevented by the failure of the law of large numbers for such a framework (Alós-Ferrer 2002). This problem can be bypassed in the very definition of the mixed extension of the game, by e.g. changing the concept of aggregation as in Uhlig (1996). Khan and Sun (1999) proposed abandoning the standard continuum framework in favor of a hyperfinite agent set. More recently, Sun (2006) and Podczeck (2010) showed that the law of large numbers holds in appropriately extended probability spaces, without the recourse to constructions from nonstandard analysis. A closely related problem is the rigorous mathematical foundation of models involving random matching of a continuum of agents (see Alós-Ferrer 1999, or Podczeck and Puzzello 2012).

4.2.3 *Absent Mindedness*

The current definition of an extensive decision problem encompasses all traditional ones inclusive of a few exotic cases—but *not* cases of absent-mindedness (see Piccione and Rubinstein 1997), where a play crosses an information set more than once. A decision theory for games with absent-mindedness would require that the node that is realized within a given information set depends on the player's choice at

[5]Strictly speaking, the term "nonatomic" refers to the measure which is not specified in this example, because it is immaterial to the argument.

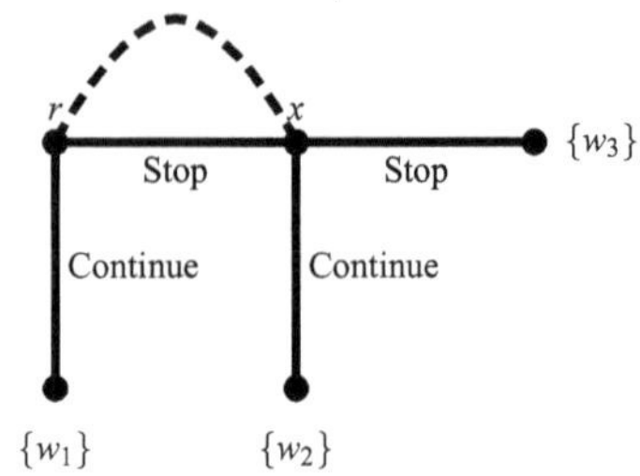

Fig. 4.4 (Absent-minded driver) This example is impossible in the refined-partitions approach (violation of (EDP1))

that information set. (In decision theoretic terms the state that obtains would depend on the decision maker's choice.) Since we do not know how to handle that, we find it worthwhile to follow Kuhn (1953) by excluding absent-mindedness in the definition of an EDP. However, this is only partially a modeling choice. While the general exclusion of absent-mindedness follows from condition (EDP4), the most basic examples of games with absent-mindedness can be seen to be simple misspecifications (violations of (EDP1)).

Consider the absent-minded driver example postulated by Piccione and Rubinstein (1997) as an example of the difficulties with traditional textbook formulations of extensive form games. The (with apologies) *naïve* graphical representation of this example is given in Fig. 4.4. This representation, however, makes no sense, and it just shows that a casual translation of the refined-partitions approach into graphs can lead to trouble.

The story behind the absent-minded driver is as follows. A single player can choose Stop (in which case the game ends) or Continue, but if he continues he will forget that he has done so and face the decision again. Apparently, this can be represented by a tree on three outcomes $W = \{w_1, w_2, w_3\}$ and an intermediate move $x = \{w_2, w_3\}$ (see Fig. 4.4). This deceivingly simple example wreaks havoc with textbook formulations of games, because, for instance, there is no sequence of choices (later to be called a "strategy") which allows the outcome w_2 to be reached.

Actually, this example is a non-issue in our framework or any properly specified version of the refined-partitions approach. It is simply a misspecified game. For, how should the choices be defined? The graph approach would specify the "information set" $\{r,x\}$, but this is not a primitive and should be derived from the choices. The choice to stop appears to be unproblematic, Stop $= \{w_1, w_2\}$. But a choice to continue in this formulation has to include the plays w_2 and w_3, because both are still possible after Continue has been chosen the first time. This leads to a violation of (EDP1) (simultaneously available choices cannot intersect), showing that the example cannot be formulated as a extensive decision problem. This is unrelated to condition (EDP4). Rather, the traditional absent-minded driver is simply an impossible example when extensive decision problems are properly defined (rather than intuitively interpreted out of graphical structures).

It is possible, however, to build more sophisticated examples reflecting the idea of absent-mindedness without violating (EDP1), as the following example shows. The story is again simple: a driver can exit on the right or exit on the left, but there

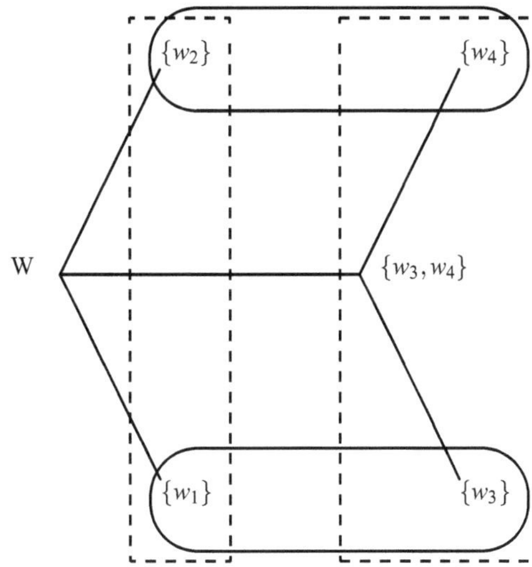

Fig. 4.5 (Two-sided absent-minded driver) *Rounded boxes* represent personal player's choices, *dashed boxes* chance's choices. As choices are sets of plays, the choice $\{w_3, w_4\}$ includes the node $\{w_3, w_4\}$

is a chance event that he does not notice the point at which he should have stopped and faces the same decision at the following intersection. The event where he does or does not notice is modeled as chance's choice, but the player is fully unaware of it (Fig. 4.5).

Example 4.3 (Two-sided absent-minded driver) Let $W = \{w_1, \ldots, w_4\}$ and $N = \{W, \{w_3, w_4\}, (\{w\})_{w \in W}\}$, $I = \{0, 1\}$, $C_0 = \{\{w_1, w_2\}, \{w_3, w_4\}\}$, and $C_1 = \{\{w_1, w_3\}, \{w_2, w_4\}\}$, see Fig. 4.5. That

$$P(\{w_1, w_3\}) = \{W, \{w_3, w_4\}\} = P(\{w_2, w_4\})$$

verifies (EDP1). Because $J(W) = \{0, 1\}$ and $J(\{w_3, w_4\}) = \{1\}$ and $c_0 \cap c_1 \neq \emptyset$ for all $(c_0, c_1) \in C_0 \times C_1$, (EDP2) also holds. Since for two nodes to be disjoint requires at least one of them not to be a move and none of them to be the root, the hypothesis of (EDP3) applies only if either $y = \{w_3, w_4\}$ and $y' = \{w\}$ for some $w \in \{w_1, w_2\}$ or both y and y' are singletons. For pairs of singletons $y, y' \in N$ there clearly is always a disjoint pair of choices of the same decision maker that separates them. Furthermore, the singletons $\{w_1\}$ and $\{w_2\}$ are separated from $\{w_3, w_4\}$ by the two choices of chance. Hence, (EDP3) also holds true. But (EDP4) fails, as $1 \in J(W)$ and $x = W \supset y = \{w_3, w_4\}$, but there is no $c \in A_1(W) = C_1$ such that $\{w_3, w_4\} \subseteq c$.

That is, this example fulfills (EDP1-3), but fails (EDP4). The key observation is that choices are sets of plays, not sets of nodes. Chance's choice to "continue" must include both plays w_3 and w_4, corresponding to the possibilities of stopping on the right or the left at the second point. But this implies that the node $\{w_3, w_4\}$ is a subset of that choice of chance. This makes chance's choice to continue available at the root (as it should), but there is no choice of the personal player leading to that node.

The point made in the last example is a general one. Under (EDP4), absent-mindedness is excluded from the framework. The reason is that, because in the present framework players choose among sets of ultimate outcomes, they cannot "choose not to choose," that is, pick a choice that will become available once more, later on. The formal argument relies on the definition of immediate predecessors, (4.1), and conditions (EDP1) and (EDP4). It demonstrates that an extensive decision problem satisfies "no-absent-mindedness," i.e., if a choice is available at two distinct moves, then these moves cannot be ordered.

Proposition 4.1 *Let (T,C) be an extensive decision problem with player set I as in Definition 4.1. Then, for all $x,y \in X$,*

$$\text{if } A_i(x) \cap A_i(y) \neq \emptyset \text{ and } y \subseteq x \text{ then } y = x, \text{ for all } i \in I \tag{4.2}$$

Proof Suppose for some $i \in I$ there are $c \in C_i$ and $x,y \in N$ such that $x,y \in P(c)$, i.e. $c \in A_i(x) \cap A_i(y)$, and $y \subseteq x$. By $y \in P(c)$ there is $y' \in \downarrow c$ such that $\uparrow y = \uparrow y' \setminus \downarrow c$. Hence, $y' \subset y \subseteq x$ and $y \setminus c \neq \emptyset$. If $y \subset x$ would hold, then by (EDP4) there would be $c' \in A_i(x)$ such that $y \subseteq c'$, implying that $c \neq c'$ from $y \setminus c \neq \emptyset$. But then $x \in P(c) \cap P(c')$ would imply that $c \cap c' = \emptyset$ by (EDP1), in contradiction to $y' \subseteq c \cap c'$. Hence, $y \subseteq x$ must imply $y = x$, as desired. ∎

Given (EDP1), condition (EDP4) implies the following "node partition" property:

(NP) For all $x \in X$ the collection $\{x \cap c \,|\, c \in A_i(x)\}$ is a partition of x, for all $i \in J(x)$.

For, if two choices are available at x, then by (EDP1) they must be disjoint. But, for any play $w \in x$, because $x \in X$, there is a node $y \in N$ such that $w \in y \subset x$. By (EDP4) y (and hence w) must be contained in a choice $c \in A_i(x)$ available at x.

If (EDP4) were weakened to (NP), though, some examples of absent-mindedness would be feasible, as Example 4.3 shows. There, conditions (EDP1)–(EDP3) hold, while (EDP4) fails. But (NP) holds, as the choices of the personal player partition W. This illustrates the implications of (EDP4): by imposing that all successors of a given node (where a player has choices available) be contained in some available choice, players cannot "jump ahead" in the tree and select a node, skipping an intermediate step. As it turns out, this simple intuition rules out absent-mindedness.

4.2.4 Independence of the Conditions

Example 4.3 shows that (EDP4) is independent of (EDP1-3), that is, there are examples where the latter conditions hold but (EDP4) fails. The following examples complete the verification that conditions (EDP1)–(EDP4) are independent from each other.

Example 4.4 Let $W=\{w_1,w_2,w_3\}$ and $N=\{W,(\{w\})_{w\in W}\}$, $I=\{1\}$, and $C_1=\{\{w_1,w_2\},\{w_3\}\}$. As $P(\{w_1,w_2\})=P(\{w_3\})=W$, (EDP1) holds true. As $X=\{W\}$ and $J(W)=\{1\}$, (EDP2) is also satisfied. Since for every singleton $\{w\}\in N\backslash X$ there is $c\in C_1$ such that $\{w\}\subseteq c$, (EDP4) also holds. But (EDP3) fails, because $\{w_1\}\cap\{w_2\}=\emptyset$, but either $\{w_1,w_2\}\subseteq c$ or $\{w_1,w_2\}\cap c=\emptyset$ for all $c\in C_1$.

Example 4.5 Let $W=\{w_1,w_2,w_3\}$, $N=\{W,(\{w\})_{w\in W}\}$, $I=\{1,2\}$, $C_1=\{\{w_1,w_2\},\{w_3\}\}$, and $C_2=\{\{w_1\},\{w_2,w_3\}\}$. Then $P(c)=\{W\}$ for all $c\in C_1\cup C_2$ verifies (EDP1). If $y,y'\in N$ are such that $y\cap y'=\emptyset$, both y and y' must be singletons. But for every pair of singletons there exists a disjoint pair of choices of the same personal player that separates the two singletons. Hence, (EDP3) holds true. Since $X=\{W\}$ and $J(W)=\{1,2\}$, (EDP4) is also fulfilled. But (EDP2) fails, as $W\cap\{w_1\}\cap\{w_3\}=\emptyset$.

Example 4.6 Let $W=\{w_1,\ldots,w_5\}$, $N=\{W,(\{w\})_{w\in W},\{w_1,w_2,w_3\},\{w_4,w_5\}\}$, $I=\{0,1\}$, $C_0=\{\{w_1,w_2,w_3\},\{w_4,w_5\}\}$, and $C_1=\{\{w_1,w_4\},\{w_2\},\{w_3,w_5\}\}$. Then $J(W)=\{0\}$, $J(\{w_1,w_2,w_3\})=J(\{w_4,w_5\})=\{1\}$, $A_1(\{w_1,w_2,w_3\})=C_1$, and $A_1(\{w_4,w_5\})=C_1\setminus\{\{w_2\}\}$ show that (EDP2) holds. Verification of (EDP3) and (EDP4) is direct (but tedious). Yet, the first part of (EDP1) fails, as $\{w_1,w_2,w_3\}\in P(\{w_2\})\cap P(\{w_1,w_4\})$, but $\{w_4,w_5\}\in P(\{w_1,w_4\})\setminus P(\{w_2\})$.

Example 4.7 Let $W=\{w_1,\ldots,w_4\}$, $N=\{W,(\{w\})_{w\in W}\}$, $I=\{1\}$, and $C_1=\{\{w_1,w_2\},\{w_3,w_4\},\{w_2,w_3\},\{w_1,w_4\}\}$. Then $X=\{W\}$, $J(W)=\{1\}$, and $A_1(W)=C_1$. Properties (EDP2), (EDP3), and (EDP4) are trivially satisfied. But the second part of (EDP1) fails, as, say, $\{w_1,w_2\}\cap\{w_2,w_3\}\neq\emptyset$.

4.2.5 EDP2 When Chains Have Lower Bounds

If every chain in N has a lower bound in N, then condition (EDP2) can be reformulated as follows: For all $(c_i)_{i\in J(x)}\in\times_{i\in J(x)}A_i(x)$,

$$\exists y\in\downarrow x\setminus\{x\}:y\subseteq x\cap\left[\cap_{i\in J(x)}c_i\right], \tag{4.3}$$

For, let $w\in x\cap\left[\cap_{i\in J(x)}c_i\right]$. As choices are unions of nodes, there are $z_i\in N$ such that $w\in z_i\in\downarrow c_i$ for all $i\in J(x)$. Because T is a game tree, Trivial Intersection (2.4) implies that the collection $\{z_i\,|i\in J(x)\}$ is a chain. By hypothesis this chain has a lower bound $y\in\downarrow x$. If $y=x$ would hold, then $y=x\subseteq c_i$ would contradict $x\in P(c_i)$ for all $i\in J(x)$. Therefore, $x\supset y\in N$, as desired.

The condition that every chain in N has a lower bound in N is, in fact, without loss of generality. Since every game tree can be completed by adding the singletons without changing the tree (by Proposition 2.11), and in a complete game tree every chain has a lower bound (the singleton in its intersection), condition (EDP2) is equivalent to (4.3) for all practical purposes.

4.3 Choices and Strategies

4.3.1 Plays and Choices

The next result shows that every play is realized when all the choices compatible with the play are actually taken. That is, the players' decisions ultimately do lead to the realization of a play resp. outcome, and there are no "unreachable outcomes."

Theorem 4.1 *Let (T, C) be an extensive decision problem with player set I. Then,*

$$\cap\{c \in \cup_{i\in I} C_i \,|w \in c\} = \{w\}$$

for all plays $w \in W$.

Proof If $N = \{W\}$, i.e. the tree is trivial, there is nothing to prove. In nontrivial cases all plays pass through at least two nodes.

Let $C(w) = \{c \in \cup_{i\in I} C_i \,|w \in c\}$ and note that this set is nonempty by (EDP3). For, let $x, y \in N$ be such that $w \in x \cap y$ and $x \supset y$. Since a game tree satisfies Separability, there is a third node $z \in N$ such that $x \supset z$ and $y \cap z = \emptyset$. By (EDP3) there is $i \in I$ and disjoint choices $c, c' \in C_i$ such that $y \subseteq c$ and $z \subseteq c'$. Since $w \in y \subseteq c$, $c \in C(w)$ verifies that $C(w)$ is nonempty.

That $w \in \cap_{c\in C(w)} c$ follows from the definition. Suppose there is $w' \in W \setminus \{w\}$ such that $w' \in \cap_{c\in C(w)} c$. Because T is a game tree, there are $x, x' \in N$ such that $w \in x$, $w' \in x'$, and $x \cap x' = \emptyset$ (by Irreducibility). By (EDP3) there are $i \in I$ and $c, c' \in C_i$ such that $x \subseteq c$, $x' \subseteq c'$, and $c \cap c' = \emptyset$. Since $w \in x \subseteq c$, the choice c belongs to $C(w)$, so that by hypothesis $w' \in c$. But $w' \in x' \subseteq c'$ contradicts $c \cap c' = \emptyset$. ■

This shows that plays "build up" from consecutive decisions by players (and/or chance) on sets of plays. Hence, the framework achieves what it was designed for. Theorem 4.1 does *not* say, however, that the players' decisions *as described by strategies* (defined in the next subsection) will always induce outcomes. In this sense, representability as an extensive decision problem is only a weak consistency check.

4.3.2 Strategies

A key question about an EDP (Definition 4.1) is whether *strategies* can be defined on it. This is a much more modest question than to ask whether or not strategies induce outcomes. Accordingly, the former can be answered affirmatively in the present framework. The latter will have to await treatment in Chap. 5. To see that strategies can be defined on EDPs, let $X_i = \{x \in N \mid A_i(x) \neq \emptyset\}$ denote player i's decision points and define a *pure strategy* for player $i \in I$ as a function $s_i : X_i \to C_i$ such that

$$s_i^{-1}(c) = P(c) \text{ for all } c \in s_i(X_i) \equiv \cup_{x \in X_i} s_i(x). \tag{4.4}$$

Since $s_i^{-1}(c) = \{x \in X_i | s_i(x) = c\}$, that $s_i^{-1}(c) \supseteq P(c)$ says that, if a choice $c \in C_i$ is selected at all (i.e. if $c \in s_i(X_i)$), then at every move $x \in X_i$, where c is available, strategy s_i picks this choice c. Hence, s_i picks the *same* choice at all moves, where this choice is available. The reverse inclusion, $s_i^{-1}(c) \subseteq P(c)$, states that, if $c \in C_i$ is chosen by s_i, then it is chosen only where it is available. In other words, $s_i^{-1}(c) \supseteq P(c)$ guarantees that a pure strategy does not use more information than the player has, i.e., it is measurable with respect to the player's information partition. And $s_i^{-1}(c) \subseteq P(c)$ guarantees that choices are only taken where they are *available*, i.e. where $x \in P(c)$.

This specification illustrates that the familiar notion of a pure strategy can be defined naturally for EDPs. Hence, representing a strategic situation in extensive form does not necessarily require the finiteness and discreteness assumptions implicit in textbook definitions.

4.4 Game Trees Revisited

With the concept of an extensive decision problem (EDP, Definition 4.1) at hand we now turn back to game trees. The purpose is to identify the structures in game trees that are responsible for the next step: Identifying game trees such that all strategies in fact induce outcomes and do so uniquely.

We begin with three properties of a game tree that are critical. To introduce these properties takes some terminology: For a game tree $T = (N, \supseteq)$ a *filter* is a chain h in N such that $\uparrow x \subseteq h$ for all $x \in h$. A *history* is a nonempty filter that is not maximal in T, i.e. that is not a play. For a history h in T a *continuation* is the complement of h in a play that contains h.

Definition 4.2 A game tree $T = (N, \supseteq)$ is

(a) **weakly up-discrete** if all maximal chains in $\downarrow x \setminus \{x\}$ have maxima, for all nodes $x \in N$ for which $\downarrow x \setminus \{x\} \neq \emptyset$;
(b) **coherent** if every history without minimum has at least one continuation with a maximum;
(c) **regular** if every history of the form $\uparrow x \setminus \{x\}$ for $x \in N$ has an infimum.

Intuitively, the first property allows inductive arguments to proceed from any non-terminal node by guaranteeing existence of all immediate successors. The second requires that, after an infinite chain of decisions, (transfinite) induction is still able to proceed. This would be prevented by a "hole" in the tree, which amounts to a history without minimum whose continuations have no maxima. The third property makes sure that any node can be "singled out" by (possibly transfinite) induction.

These properties are "local." That is, to verify them one considers a history and looks at what happens "after" it. If the history is the up-set of a node, then weak up-discreteness requires that all of its continuations have a maximum. If the history has no minimum, then coherence asks for at least one continuation with a maximum. If the history consists of the proper predecessors of a node, then regularity requires that it has an infimum.

What may go wrong, if the properties from Definition 4.2 fail, is illustrated next by a series of examples. The examples in the following subsection illustrate the properties, show that they are independent, and foreshadow the impact that they have on playability.

Remark 4.2 The following condition is equivalent to regularity and coherence:

$$\cap \{x | x \in h\} \in N, \text{ for every history } h \subseteq N \tag{4.5}$$

(If "history" is strengthened to "nonempty filter," the equivalence still holds in the class of complete game trees.) For, if this condition holds, then $\uparrow x \setminus \{x\}$ has an infimum, namely $\cap \{y | y \in \uparrow x \setminus \{x\}\} \in N$, for every $x \in N$ (recall Lemma 4.2(a)), so that the tree is regular. Furthermore, if h is a history without minimum, then $y = \cap \{x | x \in h\} \in N$ provides a maximum for the continuation $\uparrow \{w\} \setminus h$ (with $w \in y$) of h, so that the tree is coherent. Conversely, if for a regular and coherent game tree a history h has a minimum $y \in h$, then $y = \cap \{x | x \in h\} \in N$ verifies (4.5). If not, then by coherence there is a continuation g of h with a maximum $y \in g$. In that case $h = \uparrow y \setminus \{y\}$ and this up-set has an infimum by regularity; by Lemma 4.2(a) and the definition of an infimum $y = \inf h = \cap \{x | x \in h\} \in N$ and this verifies (4.5). (If (4.5) holds for every nonempty filter, h may be a play; then $\cap \{x | x \in h\} = \{w\} \in N$ for some $w \in W$ by Theorem 2.3(c), provided the game tree is complete.)

4.4.1 Examples

As pointed out before, game trees and EDPs as defined above encompass all classical examples of games, from finite to infinitely repeated and stochastic games, and even "Long Cheap Talk" (Aumann and Hart 2003), where the length of the plays is $\omega + 1$, with ω being the first infinite ordinal. Many effects that arise in the abstract can be illustrated with instances from the following family that encompasses the, in many respects, simplest trees of all.

Example 4.8 (General Centipedes) Let W be any completely ordered set, i.e. there is an order relation $\geq$ defined on W such that either $w \geq w'$ or $w' \geq w$ for all $w, w' \in W$. Define $x_t = \{\tau \in W | \tau \geq t\}$ for all $t \in W$, and let $N = \left\{ (\{t\})_{t \in W}, (x_t)_{t \in W} \right\}$.

The tree $(N, \supseteq)$ is irreducible. Since the play $v_\infty = \{x_t\}_{t \in W}$ may not have a lower bound (e.g. if W is the set of natural numbers), the tree may not be bounded. This is, however, a case in which the tree can be completed by the addition of an "infinite" element which does not affect the order-theoretic structure (see Proposition 2.8). Formally, if W has a maximum, $w_\infty \geq w$ for all $w \in W$, it follows that $w_\infty \in x$ for all $x \in v_\infty$. Then, $(N, \supseteq)$ is bounded, i.e. it is a game tree. Since all singleton sets are nodes, it is even a complete game tree. We refer to this tree as the *W-centipede*.[6]

Decision problems in continuous time are often formalized as "differential games." Those tend to be implicitly defined as normal form games, where strategies and payoffs are taken as the primitives, and a differential equation is used to compute the payoffs from strategy combinations (Friedman 1994, p. 784). Thus, the set of strategies is taken small enough so that payoffs can be computed, e.g. a certain set of Lipschitz continuous functions. The general concept of a game tree allows us to treat decision problems in continuous time as non-cooperative games in extensive form.

Example 4.9 (Differential game: coherence and regularity) As in Sect. 2.2.2.6 let W be the set of functions $f : \mathbb{R}_+ \to A$, where A is some fixed set of "actions," with at least two elements, and let $N = \{x_t(g) \,|\, g \in W,\ t \in \mathbb{R}_+\}$, where $x_t(g) = \{f \in W \,|\, f(\tau) = g(\tau),\ \forall \tau \in [0, t)\}$, for any $g \in W$ and $t \in \mathbb{R}_+$. Intuitively, at each point in time $t \in \mathbb{R}_+$ a decision $a_t \in A$ is taken. The "history" of all decisions in the past (up to, but exclusive of, time t) is a function $f : [0, t) \to A$, i.e. $f(\tau) = a_\tau$ for all $\tau \in [0, t)$. A node at time t is the set of functions that coincide with f on $[0, t)$, all possibilities still open for their values thereafter.

It was shown in Examples 2.12 and 2.14 that $(N, \supseteq)$ is a game tree. At each node $x_t(f)$, the decision that an agent has to take is merely her action at time t. Ultimately, a function $f \in W$ becomes a complete description of all decisions taken from the beginning to the end.

This tree is coherent. For, if h is a history without a minimum, let $f \in \bigcap_{x \in h} x$ and $\tau = \sup\{t \in \mathbb{R}_+ \,|\, x_t(f) \in h\}$. By hypothesis $x_\tau(f) \notin h$ and, thus, $x_\tau(f)$ provides a maximum for the continuation $\{x_t(f) \in N \,|\, t \geq \tau\}$.

The tree is also regular, because $x_t(f) = \inf \uparrow x_t(f) \setminus \{x_t(f)\}$ for all nodes. But weak up-discreteness fails. For, if $x_\tau(g) \in \downarrow x_t(f) \setminus \{x_t(f)\}$ were a maximum for a continuation of $\uparrow x_t(f)$, then $g|_{[0,t)} = f|_{[0,t)}$ and $\tau > t$ would imply the contradiction $x_t(f) \supset x_{(t+\tau)/2}(g) \supset x_\tau(g)$.

The formalization offered in Example 4.9 differs from Simon and Stinchcombe (1989), who view continuous time as "discrete time, but with a grid that is "infinitely fine" (p. 1171). Stinchcombe (1992) determines strategies and outcomes

[6] This family of examples shows that there are (complete) game trees with plays of arbitrary cardinality. Consider a set W' of the appropriate cardinality, endow it with a well order (hence, a total order) by applying Zermelo's well-order theorem (see e.g. Hewitt and Stromberg 1965), and adjoin a "top" (maximum) to it. If the resulting set is called W, the corresponding W-centipede proves the claim.

simultaneously to find the maximal set of strategies that give well defined outcomes for such problems. By contrast, we start with an unrestricted set of outcomes and will define strategies from the associated extensive form. Therefore, we do not employ any additional restrictions on strategies, like measurability or continuity (on the formalization of games in continuous time, see also Sect. 5.7).

The problems which arise from pushing the limits of extensive form games beyond the confines of finite games are often difficult to see in the differential game, but are easy to identify in the following examples. In fact, the following three examples also show that the three properties from Definition 4.2 are independent.

Example 4.10 (Hole in the Middle) Let $x_t = [(t-1)/(4t), (3t+1)/(4t)]$, $y_t = [1/4, (2t-1)/(4t)]$, and $y'_t = [(2t+1)/(4t), 3/4]$ for all $t = 1, 2, \dots$ and $W = [0,1]$, and let $N = \{(x_t)_{t=1}^{\infty}, (y_t)_{t=1}^{\infty}, (y'_t)_{t=1}^{\infty}, [1/4, 3/4], (\{w\})_{w\in W}\}$. So, $\{x_t\}_{t=1}^{\infty}$ is a decreasing sequence of intervals starting with $W = [0,1]$ and converging to $[1/4, 3/4] \in N$; $\{y_t\}_{t=1}^{\infty}$ resp. $\{y'_t\}_{t=1}^{\infty}$ are increasing sequences of intervals that start at $\{1/4\} \in N$ resp. $\{3/4\} \in N$ and converge to $[1/4, 1/2)$ resp. $(1/2, 3/4]$ (which are not nodes). See the left-hand side of Fig. 4.6 for a qualitative graphical representation.

This is a complete game tree. Since $\{1/2\} \in N$ provides a maximum for a continuation of the history $\{x_t\}_{t=1}^{\infty}$, this tree is coherent. It is regular, because $[1/4, 3/4] \in N$ provides a minimum for $\{x_t\}_{t=1}^{\infty}$ and $\{1/4\} \in N$ resp. $\{3/4\} \in N$ provide minima for $\{y_t\}_{t=1}^{\infty}$ resp. $\{y'_t\}_{t=1}^{\infty}$. But it is not weakly up-discrete, since the chains $\{y_t\}_{t=1}^{\infty}$ and $\{y_t\}_{t=1}^{\infty}$ in $\downarrow[1/4, 3/4] \setminus \{[1/4, 3/4]\}$ have no maxima.[7]

Consider the following single-player perfect information extensive form. Assign all nodes except the root as choices to the personal player and add $[1/4, 1/2) \cup (1/2, 3/4] = (\cup_{t=1}^{\infty} y_t) \cup (\cup_{t=1}^{\infty} y'_t)$ as a further choice (but not as a node), so that at $[1/4, 3/4]$ the player chooses between $[1/4, 1/2) \cup (1/2, 3/4]$ and (the node) $\{1/2\}$. There exists a strategy that assigns to any x_t the non-singleton choice x_{t+1}, to every y_t resp. y'_t the corresponding singleton choice $\{(2t-1)/(4t)\}$ resp. $\{(2t+1)/(4t)\}$, and to $[1/4, 3/4]$ the choice $[1/4, 1/2) \cup (1/2, 3/4]$. This strategy "continues" along $\{x_t\}_{t=1}^{\infty}$, discards $\{1/2\}$, and "stops" at every y_t resp. y'_t. And it induces no outcome/play!

This example has a coherent and regular, but *not* weakly up-discrete tree, and a strategy that induces no outcome at all. A modification illustrates that with a regular and weakly up-discrete tree, that is *not* coherent, the same problem may appear.

Example 4.11 (Modified Hole in the Middle) Reconsider Example 4.10. Remove the nodes $[1/4, 3/4]$ and $\{1/2\}$ from N and the element $1/2$ from the underlying set W. See Fig. 4.6 (center) for a representation. The resulting tree is regular and now weakly up-discrete (because $[1/4, 3/4] \notin N$), but not coherent anymore. Consider a strategy that assigns to each x_t the choice x_{t+1}, to each y_t resp. y'_t the (singleton) choice $\{(2t-1)/(4t)\}$ resp. $\{(2t+1)/(4t)\}$, i.e., a strategy that "continues" at all

[7] This game tree also satisfies that $\cap\{x \mid x \in h\} \in N$ for every history h. This condition is equivalent to regularity and coherence (see Remark 4.2 above).

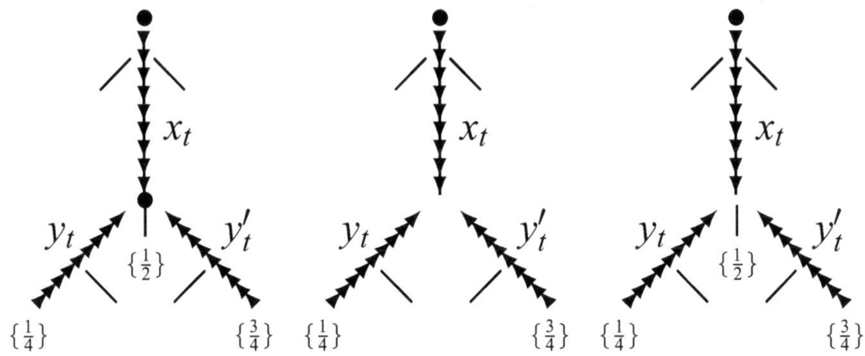

Fig. 4.6 (Hole in the middle example and variants) Qualitative graphical representations of Examples 4.10 (*left-hand side*), 4.11 (*center*), and 4.12 (*right-hand side*)

x_t, but "stops" at all y_t and y'_t, for all t. Once again, this strategy does not induce any outcome at all.

Reinserting the node $\{1/2\}$ to N and the element $1/2$ to W in the last example yields a tree that is weakly up-discrete (because still $[1/4, 3/4] \notin N$) and coherent (because $\{1/2\} \in N$ provides a maximum for a continuation of the history $\{x_t\}_{t=1}^{\infty}$), but not regular anymore, since $\uparrow \{1/2\} \setminus \{1/2\}$ has no infimum.[8] As a result, a strategy can be constructed that induces multiple outcomes.

Example 4.12 (Hole in the Middle, modified again) Reconsider Example 4.10. Let

$$N = \left\{ (x_t)_{t=1}^{\infty}, (y_t)_{t=1}^{\infty}, \left(y'_t\right)_{t=1}^{\infty}, (\{w\})_{w \in W} \right\}$$

(see the right-hand side of Fig. 4.6 for a qualitative graphical representation). Consider again the single-player extensive form with perfect information. The strategy that assigns to each x_t the choice x_{t+1}, to each y_{t+1} the choice y_t, and to each y'_{t+1} the choice y'_t for all $t = 1, 2, \ldots$ always "continues." It is intuitively clear that this strategy selects the outcomes/plays $1/4, 1/2, 3/4 \in W$. The reason for this multiplicity is that there is no move where a decision between $\cup_{t=1}^{\infty} y_t = [1/4, 1/2)$ and $\cup_{t=1}^{\infty} y'_t = (1/2, 3/4]$ is taken, because a node of the form $[1/4, 3/4]$ is missing.

Examples 4.10 and 4.11 illustrate that a weakly up-discrete and coherent game tree is necessary for an arbitrary strategy to induce an outcome. Example 4.12 shows that a regular tree is necessary for a strategy to induce a unique outcome. Sufficiency is a different matter, though, and will be considered later. For, to make it precise

[8] The node $\{1/2\}$ is a lower bound for the chain $(x_t)_{t=1}^{\infty}$, as is any of the nodes y_t and y'_t. Since $y_t \cap y'_t = \emptyset$, the chain $\uparrow \{1/2\} \setminus \{\{1/2\}\} = (x_t)_{t=1}^{\infty}$ has no infimum.

when a strategy induces an outcome/play requires further ingredients. A key one, that purely concerns the tree, is introduced next.

4.4.2 A Classification of Nodes

Recall that nodes that are properly followed by other nodes are called *moves* and the set of moves is denoted by $X = \{x \in N \mid \downarrow x \setminus \{x\} \neq \emptyset\}$, while nodes that are not properly followed by other nodes are called *terminal* and $E = N \setminus X$ is the set of terminal nodes. The following preliminary result gives a useful characterization.

Lemma 4.1 *Let $T = (N, \supseteq)$ be a game tree.*[9]

(a) If $x \in N$ contains at least two distinct elements, then for each $w \in x$ there is a node $x' \in N$ such that $w \in x' \subset x$;
(b) A node $x \in N$ is terminal if and only if there is $w \in W$ such that $x = \{w\}$.

Proof (a) Let $w \in x \in N$, and let $w' \in x$ such that $w' \neq w$. By (GT2) there is $x' \in N$ such that $w \in x'$, but $w' \notin x'$. As $w \in x \cap x'$, but $w' \in x \setminus x'$, (GT1) implies that $x' \subset x$.
(b) The "if"-part is trivial. The "only if"-part follows from (a). ■

Thus, when $(N, \supseteq)$ is a game tree, $X = N \setminus \{\{w\}\}_{w \in W}$ can be taken as an alternative definition of (the set of) moves. Yet, this result does not imply that $\{w\} \in N$ for all $w \in W$, unless $(N, \supseteq)$ is a complete game tree. If all chains are finite, then $(N, \supseteq)$ is a game tree *if and only if* it satisfies Trivial Intersection and all singleton sets are nodes (see Proposition 2.13). In particular, this always holds if W is finite. Yet, even if W is infinite, certain nodes may have properties analogous to the finite case.

Definition 4.3 Let $(N, \supseteq)$ be a game tree and $x \in N \setminus \{W\}$. The node x is

(a) **finite** if $\uparrow x \setminus \{x\}$ has a minimum,
(b) **infinite** if $x = \inf \uparrow x \setminus \{x\}$, and
(c) **strange** if $\uparrow x \setminus \{x\}$ has no infimum.

Denote by $F(N)$ the set of finite nodes of N, by $I(N)$ the set of infinite nodes, and by $S(N)$ the set of strange nodes.

The three possibilities in this definition are exhaustive, i.e., all nodes (other than the root) are either finite, infinite, or strange. For, if $\uparrow x \setminus \{x\}$ has an infimum z, it is either a minimum (and then x is finite), or $z \notin \uparrow x \setminus \{x\}$. In the latter case, it follows from the definition of an infimum that $z = x$.

Lemma 4.2 *For a game tree $(N, \supseteq)$ a node $x \in N \setminus \{W\}$ is*

(a) the infimum of a chain $h \in 2^N$ if and only if $x = \bigcap_{y \in h} y$;

[9] This result also holds for all irreducible W-set trees, where irreducibility is defined in (2.12).

(b) *infinite if and only if* $x = \bigcap_{y \in \uparrow x \setminus \{x\}} y$;
(c) *strange if and only if* $\uparrow x \setminus \{x\}$ *has no minimum and* $x \subset \bigcap_{y \in \uparrow x \setminus \{x\}} y$.
(d) *Further, x is not infinite (that is, it is either finite or strange) if and only if, for every chain* $h \in 2^N$,

$$\text{if } x \supseteq \bigcap_{y \in h} y \text{ then } x \supseteq y \text{ for some } y \in h. \tag{4.6}$$

Proof (a) The "if"-part is trivial. For the "only if" part, let $x = \inf h$. Obviously, $x \subseteq \bigcap_{y \in h} y$. Fix $w \in x$, and suppose there exists $w' \in \bigcap_{y \in h} y$ such that $w' \notin x$. By Irreducibility there are $z, z' \in N$ such that $w \in z \setminus z'$ and $w' \in z' \setminus z$.

Consider any $y \in h$. Since $w' \in z' \setminus z$ and $w' \in y$, it follows from Trivial Intersection (2.4) that either $y \subseteq z'$ or $z' \subseteq y$. In the first case, it would follow that $x \subseteq z'$, in contradiction to $w \in z \setminus z'$. Hence, $z' \subseteq y$ for all $y \in h$, i.e., z' is a lower bound for the chain h. Since x is its infimum, $z' \subseteq x$, a contradiction to $w' \notin x$. This shows that $x = \bigcap_{y \in h} y$.

(b) follows from (a).

(c) "if": If $z = \inf \uparrow x \setminus \{x\}$, it follows from (a) that $z = \bigcap_{y \in \uparrow x \setminus \{x\}} y$. By hypothesis, $x \subset z$ and thus $z = \min \uparrow x \setminus \{x\}$, a contradiction.

"only if": If $x \in N$ is strange, then $\uparrow x \setminus \{x\}$ has no infimum, in particular, no minimum. If $x = \bigcap_{y \in \uparrow x \setminus \{x\}} y$ would hold, then by (a) $x = \inf \uparrow x \setminus \{x\}$ and x could not be strange.

(d) For the "if"-part of (d) let x be an infinite node. Consider the chain $h = \uparrow x \setminus \{x\}$. By (b), $x = \bigcap_{y \in h} y$, but $x \subset y$ for all $y \in h$, so that (4.6) cannot hold. For the "only if"-part suppose x fails (4.6). There exists a chain h in N such that $x \supseteq \bigcap_{y \in h} y$, but there is no $y \in h$ such that $x \supseteq y$. Since $x \supseteq \bigcap_{y \in h} y$ implies $x \cap y \neq \emptyset$ (because $\bigcap_{y \in h} y \neq \emptyset$, by Boundedness (2.16)) for all $y \in h$, it follows from Trivial Intersection (2.4) that $x \subset y$ for all $y \in h$, i.e. $y \in \uparrow x \setminus \{x\}$ for all $y \in h$. Thus, $x \supseteq \bigcap_{y \in h} y \supseteq \bigcap_{y \in \uparrow x \setminus \{x\}} y \supseteq x$. Hence, x is infinite. ∎

By Lemma 4.2(b) infinite nodes can be reconstructed from the other nodes. Therefore, an infinite node has infinitely many predecessors, justifying its name. For, if $\uparrow x \setminus \{x\}$ were finite, it would have a minimum $z \in \uparrow x \setminus \{x\}$, implying that x was finite.

Example 4.13 (Infinite and Continuous Centipedes) If in Example 4.8 W is the set of natural numbers together with "infinity" ∞, $W = \{1, 2, \ldots, \infty\}$, an "infinite centipede" as in Example 2.13 emerges. Terminal nodes are singletons $\{t\}$ (including $\{\infty\}$), moves are of the form $x_t = \{t, t+1, \ldots, \infty\}$, except for $x_\infty = \{\infty\}$ which is terminal. All nodes, except the terminal node $\{\infty\}$, are finite. Only $\{\infty\}$ is infinite, since

$$\bigcap \{x_t \,|\, 1 \leq t < \infty\} = \{\infty\}$$

yet $\{\infty\} \notin \{x_t \in N \,|\, t = 1, 2, \ldots\} = \uparrow \{\infty\} \setminus \{\{\infty\}\}$. This represents a never-ending chain of decisions.

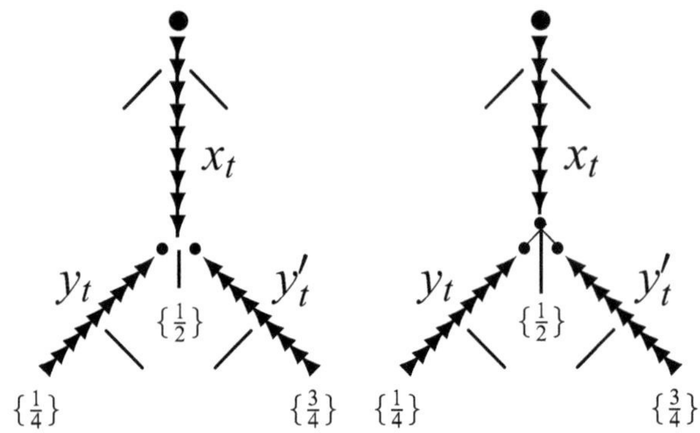

Fig. 4.7 (More variants of the hole in the middle example) The *left-hand-side* example has three strange nodes, the *right-hand-side* one has none

If in Example 4.8 $W = [0,1]$, a "continuous centipede" emerges. This is again a game tree, as $[0,1]$ has a maximum. Nodes are either singletons $\{t\}$ or of the form $x_t = [t,1]$. All singletons $\{t\}$ are finite except the "last" node, $\{1\}$. For, $\cap\{y \,|\, y \in \uparrow\{t\} \setminus \{\{t\}\}\} = \cap\{x_\tau \,|\, \tau < t\} = \cap_{\tau<t}[\tau,1] = [t,1] = x_t$, for any $t \leq 1$. The latter is only identical to $\{t\}$ if $t = 1$. All moves, though, are infinite. For, $\cap\{y \,|\, y \in \uparrow x_t \setminus \{x_t\}\} = \cap\{x_\tau \,|\, \tau < t\} = [t,1] = x_t$, provided $t > 0$ (otherwise the intersection is empty).

Trivially, a game tree is regular if and only if there are no strange nodes. All centipedes are regular. An example with strange nodes is the "hole in the middle" (Example 4.12 with $\{1/2\} \in N$), where $\{1/2\} \in S(N)$. If nodes $y = [1/4, 1/2)$ and $y' = (1/2, 3/4]$ were added, resulting in the example depicted on the left-hand side of Fig. 4.7, these would also be strange.[10] If, however, a node $x = [1/4, 3/4]$ were also added, resulting in the example on the right-hand side of Fig. 4.7, then none of the previous would be strange, but $[1/4, 3/4]$ would be infinite. But this later addition changes the tree by adding a decision.

Finally, note that a node $x \in N$ in a game tree has an immediate predecessor, $P(x) \neq \emptyset$, if and only if x is finite; in this case, $P(x) = \{\min \uparrow x \setminus \{x\}\}$. For, if x is finite, $P(x) = \{\min \uparrow x \setminus \{x\}\}$ by uniqueness of the minimum. Conversely, if $x' \in P(x)$ for $x \in N$, then there is $y \in \downarrow x$ such that $\uparrow x' = \uparrow y \setminus \downarrow x$; since $\uparrow y \setminus \downarrow x = \uparrow x \setminus \{x\}$ for all $y \in \downarrow x$, that $\uparrow x' = \uparrow x \setminus \{x\}$ implies $x' = \min \uparrow x \setminus \{x\}$, and x is finite.

[10] If a node x is strange, then any maximal lower bound of $\uparrow x \setminus \{x\}$ is also strange.

4.4.3 Removing Singletons

In a game tree the set of plays can be identified with the set of underlying outcomes, thus making it the natural domain for preferences or payoff functions. Of course, for finite trees plays are one-to-one with the set of terminal nodes E, so specifying payoffs on plays or terminal nodes is equivalent. For e.g. infinite horizon games, though, this equivalence may break down, since it is not guaranteed that all plays correspond to terminal nodes. For example, in the infinite centipede (Example 2.13) the terminal infinite node $\{\infty\}$ could be removed without affecting the structure of the tree. The corresponding payoffs have to be associated to the play $\uparrow\{\infty\}$. A similar comment applies to infinitely repeated games.

Recall that by Lemma 4.1 a node is terminal if and only if it is a singleton. This does not mean, however, that $\{w\} \in N$ for all $w \in W$. The intersection of the elements of an infinite chain of nodes may contain only a single element $w \in W$, even though the singleton $\{w\}$ does not belong to N. It was shown in Proposition 2.11 that the singletons from the underlying set W can be *added* (e.g. for repeated games) to the set of nodes, as "terminal nodes," without affecting the structure of the tree. This is clearly the case in the infinite centipede where, if node $\{\infty\}$ were absent, it could be added without changing the set of plays. Hence, a *complete* game tree (Definition 2.6) has been defined as one where $\{w\} \in N$ for all $w \in W$. That is, any game tree can be "completed" by adding singletons.

In Examples 2.13 the infinite terminal node $\{\infty\}$ could be *removed* without affecting the structure of the tree. Example 2.15, though, features terminal nodes, $\{0\}$ and $\{1\}$, that cannot be removed without changing the tree. This is so, because the terminal nodes $\{0\}$ and $\{1\}$ in Example 2.15 are not infinite nodes. That is, while some terminal nodes seem to be irrelevant for the specification of the game (meaning they can be added or removed at will), others are clearly relevant. This raises the issue which terminal nodes can be removed from the tree.

Any node $x \in N \setminus W$ that is either finite or strange satisfies $x \subset \bigcap_{y \in \uparrow x \setminus \{x\}} y$, by definition if it is finite or by Lemma 4.2(c) if it is strange. Hence, there are $w \in x$ and $w' \in \left(\bigcap_{y \in \uparrow x \setminus \{x\}} y\right) \setminus x$. Therefore, if a node $x \in N \setminus W$, which is terminal and either finite or strange, were removed, then in the resulting tree $w \in y$ would imply $w' \in y$ (because $y \in \uparrow x \setminus \{x\}$ in the original tree) for all $y \in N \setminus \{x\}$ in contradiction to Irreducibility (2.12).[11] Hence, if at all, only infinite terminal nodes may be removed without harm.

Indeed, suppose that a singleton $\{w\} \subseteq W$, that is not already a node (i.e. $\{w\} \notin N$), can be added without affecting the set of plays (i.e. such that the set of plays of the resulting tree is bijective to the set of plays of the original tree; see Proposition 2.11). If $\{w\}$ were finite or strange in $(N \cup \{\{w\}\}, \supseteq)$, then $\{w\} \subset \bigcap_{x \in \uparrow\{w\} \setminus \{\{w\}\}} x$ by definition or by Lemma 4.2(c). Thus, there would be some

[11] In order to recover Irreducibility (2.12), one would need to remove also the element w from the underlying set W of plays. But this, of course, changes the structure of the tree by changing the set of plays.

element $w' \neq w$ in W such that $w' \in x$ for all $x \in\uparrow \{w\} \setminus \{\{w\}\}$. Since $(N, \supseteq)$ is a game tree, $\uparrow \{w\}$ and $\uparrow \{w'\}$ would be two distinct plays for $(N, \supseteq)$ by Theorem 2.3(c). But it has just been found that $\uparrow \{w\} \subseteq\uparrow \{w'\}$—a contradiction. It follows that $\{w\}$ must be infinite in $(N \cup \{\{w\}\}, \supseteq)$. In other words, any singleton $\{w\} \notin N$ that can properly be added to a game tree must be infinite in the resulting tree.

The following result states the converse that infinite terminal nodes can be *removed* from a game tree without affecting its structure. The resulting tree has the same order-theoretic structure as the original tree and essentially the same plays. Furthermore, if the resulting set tree is completed by adding the now-missing singletons, the original game tree reemerges. In other words, for game trees infinite terminal nodes are *precisely* those, the presence or absence of which is immaterial to the structure of the tree. (Hence, the following complements Proposition 2.11.)

Proposition 4.2 *If $(N, \supseteq)$ is a complete game tree with set of plays W, then for every $Y \subseteq E \cap I(N) \equiv E_I$ the partially ordered set $(N \setminus Y, \supseteq)$ is a game tree with set of plays W', the mapping $\Upsilon : W \to W'$ given by $\Upsilon(w) = w \setminus Y$ is bijective, and if $(N \setminus Y, \supseteq)$ is completed by adding all singletons (as in Proposition 2.11), the result is the complete game tree $(N, \supseteq)$.*

Proof Denote by E_I the set of infinite terminal nodes. Next, recall that, since $(N, \supseteq)$ is a (complete) game tree, one can take the set W both as the set of plays and as the underlying set (Theorem 2.3). We now verify that $(N \setminus Y, \supseteq)$ is a game tree (Definition 2.4). The if-part of (GT1) (viz. Trivial Intersection (2.4)) for $(N, \supseteq)$ is trivially inherited by $(N \setminus Y, \supseteq)$. If h is a chain in $N \setminus Y$, then it is a chain in N and, therefore, has a lower bound in W; that is, there is $w \in \bigcap_{x \in h} x$, because $(N, \supseteq)$ is complete. Thus $(N \setminus Y, \supseteq)$ satisfies also the only if-part of (GT1) (viz. Boundedness (2.16)).

To verify (GT2) (viz. Irreducibility (2.12)), let $v, v' \in W$ be such that $v \neq v'$. By Irreducibility (2.12) for $(N, \supseteq)$ there are $x, x' \in N$ such that $v \in x \setminus x'$ and $v' \in x' \setminus x$. If $x, x' \in N \setminus Y$, we are done. If $x \in Y \subseteq E_I$ then by Lemma 4.1(b) $x = \{v\}$. Since x is infinite, $\{v\} = \bigcap_{y \in \uparrow x \setminus \{x\}} y$ by Lemma 4.2(b). Thus, there is $z \in\uparrow x \setminus \{x\}$ such that $v \in z$ and $v' \notin z$. If $x' \in N \setminus Y$, then $v \in z \setminus x'$ and $v' \in x' \setminus z$, as required. If $x' \in Y \subseteq E_I$ then, analogously, $x' = \{v'\}$ and there is $z' \in\uparrow x' \setminus \{x'\}$ such that $v' \in z'$ and $v \notin z'$. Hence, $(N \setminus Y, \supseteq)$ also satisfies Irreducibility (2.12) and is a game tree.[12]

Let $w' \in W'$. If $w' \in W$, there is a unique play for $(N, \supseteq)$, namely w' itself, such that $\Upsilon(w') = w'$. If $w' \notin W$ then there is $y \in Y$ such that $w' \cup \{y\}$ is a chain for $(N, \supseteq)$. Because $y \in Y \subseteq E_I$, by Lemma 4.1(b) $y = \{v\}$ for some $v \in W$. Therefore, $v \in x$ for all $x \in w'$, since $w' \cup \{y\}$ is a chain. Since $(N \setminus Y, \supseteq)$ is a game tree, Theorem 2.3(c) implies that $\uparrow \{v\} \setminus \{\{v\}\} \in W'$ is a play (because $\{v\} \notin N \setminus Y$), so that $w' \subseteq\uparrow y \setminus \{y\}$ and maximality imply $w' =\uparrow y \setminus \{y\}$. Because y is infinite, $y = \bigcap_{x \in w'} x$ by Lemma 4.2(b), and there can be no other $y' \in Y \setminus \{y\}$ such that

[12]Since the underlying set has not changed, Theorem 2.3 implies that the sets of plays in the new and the former trees are bijective. The present proof, however, is constructive.

$w' \cup \{y'\}$ is a chain for $(N, \supseteq)$. Therefore, $w = w' \cup \{y\}$ is the unique play for $(N, \supseteq)$ such that $\Upsilon(w) = w'$. Since this shows that every element of W' has a unique preimage under Υ, this mapping is bijective.

Finally, since all terminal nodes are singletons by Lemma 4.1(b), adding to $(N \setminus Y, \supseteq)$ all singletons must yield the (complete) game tree $(N, \supseteq)$ with all singletons present. ■

4.5 Available Choices

Whenever it is possible to define an EDP on a tree, it should be possible to define a single-player perfect information game on the same tree, where the single player takes all possible decisions under the best possible information. Also, the single-player perfect information game should be determined by the tree *alone*, in the sense that there should exist a unique form of defining it.

4.5.1 Perfect Information Choices

If the tree were discrete, choices available to the single player at $x \in X$ under perfect information would be the immediate successors of x. In the abstract framework used here a more general construction is required. For any move $x \in X$ and a play $w \in x$ define the *perfect information choice* $\gamma(x, w) \subseteq W$ as the set of plays

$$\gamma(x, w) = \cup \{z \,|\, w \in z \in \downarrow x \setminus \{x\}\}, \tag{4.7}$$

i.e. the union of all proper successors of x that contain w. Note that by Trivial Intersection (2.4) these successors form a chain.

Example 4.14 (Differential EDP, revisited) A typical example where this construction goes beyond immediate successors is the differential game (Example 4.9). Given $g \in x_t(f) \in X$, notice that $x_t(g) = x_t(f)$. Then,

$$\gamma(x_t(f), g) = \left\{h \in x_t(f) \,|\, \exists \tau > t : h|_{[0,\tau)} = g|_{[0,\tau)}\right\}.$$

As a consequence, for the tree of the differential game two distinct single-player perfect information problems can be defined. In the first perfect information choices at $x_t(f) \in X$ are $c_t(f, g) = \gamma(x_t(f), g)$ for any $g \in x_t(f)$. In the second

$$c_t(f, a) = \{g \in x_t(f) \,|\, g(t) = a\}$$

for any $g \in x_t(f)$ with $g(t) = a \in A$ defines choices under perfect information (recall Example 4.1). A single-player game with perfect information choices $\gamma(x_t(f), g)$ is easily seen to be an EDP. Yet, using the choices $c_t(f,a)$ at $x_t(f) \in X$ also defines a single-player EDP, though a slightly different one. We will refer to the latter version as the “differential game” and to the former as $\Pi(T)$ for the tree of the differential game.

Note that both specifications differ from, e.g., Stinchcombe (1992), who allows the players to change their action only at a set of time-points that is order isomorphic to an initial segment of the countable ordinals (see also Sect. 5.7).

The properties of perfect information choices determine when they can indeed be considered choices in a well-defined EDP. To demonstrate this (Proposition 4.3 below) requires an auxiliary result.

Lemma 4.3 *Let $(N, \supseteq)$ be a game tree with set of plays W. If $w, w' \in x \in X$, then there are $z, z' \in \downarrow x \setminus \{x\}$ such that $w \in z$ and $w' \in z'$, where $w \neq w'$ implies $z \cap z' = \emptyset$.*

Proof First, $x \in X$ implies $\downarrow x \setminus \{x\} \neq \emptyset$. Suppose $w \neq w'$. By Irreducibility ((2.12) resp. (GT2)) and Trivial Intersection (2.4) there are $z, z' \in N$ such that $w \in z$, $w' \in z'$, and $z \cap z' = \emptyset$. Since $w \in x \cap z$ ($w' \in x \cap z'$), by Trivial Intersection either $x \subseteq z$ ($x \subseteq z'$) or $z \subset x$ ($z' \subset x$). Since $x \subseteq z$ ($x \subseteq z'$) would contradict $w' \notin z$ ($w \notin z'$), it follows that $z \subset x$ ($z' \subset x$). If $w = w'$, the conclusion follows taking $w'' \neq w$, $w'' \in x$, which is possible because $x \in X$ (recall Lemma 4.1(b)). ■

Proposition 4.3 *Let $(N, \supseteq)$ be a game tree, let $x \in X$, and $w \in x$. Then:*

(a) if $w' \in x$ and $\gamma(x,w) \cap \gamma(x,w') \neq \emptyset$, then $\gamma(x,w) = \gamma(x,w')$;
(b) if $\gamma(x,w) \subset x$, then $P(\gamma(x,w)) = \{x\}$ and there exists at least one $w' \in x$ such that $\gamma(x,w) \cap \gamma(x,w') = \emptyset$;
(c) if $\gamma(x,w) = x$, then the chain $\{y \in \downarrow x \setminus \{x\} | w \in y\}$ has no maximum, and no choice can be available at x, hence, no EDP can be defined on $(N, \supseteq)$.

Proof (a) Let $w'' \in \gamma(x,w) \cap \gamma(x,w')$. Then there are nodes $z, z' \in \downarrow x \setminus \{x\}$ such that $w, w'' \in z$ and $w', w'' \in z'$. Since $w'' \in z \cap z'$, Trivial Intersection (2.4) implies either $z \subset z'$ or $z' \subseteq z$. If $z \subset z'$, then $w \in z'$ implies $\gamma(x,w') = \cup_{y \in \uparrow z' \setminus \uparrow x} y = \gamma(x,w)$. If $z' \subseteq z$, then $w' \in z$ implies $\gamma(x,w) = \cup_{y \in \uparrow z \setminus \uparrow x} y = \gamma(x,w')$.

(b) Denote $c = \gamma(x,w) \subset x$. Since $x \in X$, there is some $y \in \downarrow x \setminus \{x\}$ such that $w \in y$ by Lemma 4.3 and, therefore, $y \in \downarrow c$. If $z \in \uparrow x$, then $y \subseteq c \subset x \subseteq z$ implies $z \in \uparrow y$ and $z \notin \downarrow c$, so $z \in \uparrow y \setminus \downarrow c$. Since $z \in \uparrow x$ was arbitrary, $\uparrow x \subseteq \uparrow y \setminus \downarrow c$. If $z \in \uparrow y \setminus \downarrow c$, i.e. $y \subseteq z$ and $z \setminus c \neq \emptyset$, then $x \subseteq z$. For, $y \subseteq z \cap x$ implies $z \subset x$ or $x \subseteq z$ by Trivial Intersection (2.4); but $z \subset x$ would imply $z \subseteq c$, because $w \in y \subseteq z \in \downarrow x \setminus \{x\}$, in contradiction to $z \setminus c \neq \emptyset$, so that $x \subseteq z$ must obtain. Since $z \in \uparrow y \setminus \downarrow c$ was arbitrary, $\uparrow x \supseteq \uparrow y \setminus \downarrow c$. Hence, $x \in P(c)$.

Let $x' \in P(c)$. Then there is $y' \in \downarrow c$ such that $\uparrow x' = \uparrow y' \setminus \downarrow c$. Since $y' \in \downarrow c$, there exists some $z \in \downarrow x \setminus \{x\}$ such that $w \in z$ and $y' \subseteq z$. Thus $z \in \downarrow c$. Since $y' \subseteq x'$ but $x' \notin \downarrow c$, Trivial Intersection (2.4) implies $z \subseteq x'$, in particular $w \in x'$. Then, $w \in x' \cap x$ implies, again by Trivial Intersection (2.4), that $x \subset x'$ or $x' \subseteq x$. If $x \subset x'$, then $x \in \uparrow y' \setminus \downarrow c$ but $x \notin \uparrow x'$, a contradiction. If $x' \subset x$, that $w \in x'$ implies $x' \in \downarrow c$, a contradiction. Thus $x = x'$. It follows that $P(c) = \{x\}$.

To prove the second statement, choose $w' \in x \setminus c$ and let $c' = \gamma(x, w')$. If $c \cap c' \neq \emptyset$ would hold, part (a) would imply that $c = c'$, yielding the contradiction $w' \in c$. Therefore, $c \cap c' = \emptyset$. Since $c' \subseteq x \setminus c$, it also follows that $x \in P(c')$.

(c) Suppose that $\gamma(x,w) = x$. If the chain $h = \{z \in \downarrow x \setminus \{x\} \,|\, w \in z\}$ had a maximum $y \in \downarrow x \setminus \{x\}$, then $\cup_{w \in z \in \downarrow x \setminus \{x\}} z = \gamma(x,w) = y \subset x$, in contradiction to the hypothesis. To see the second part of the claim, suppose that an EDP is defined on this tree. Because $x \in X$ and $w \in x$, there is $z \in N$ such that $w \in z \subset x$ by Lemma 4.3, hence, $\gamma(x,w) = \cup_{y \in \uparrow z \setminus \uparrow x} y$. Since $z \subset x$, there is $c_i \in A_i(x)$ such that $z \subseteq c_i$ for all $i \in J(x) \neq \emptyset$ by (EDP4). By $x \in P(c_i)$ there is $y_i \in \downarrow c_i$ such that $\uparrow x = \uparrow y_i \setminus \downarrow c_i$ for all $i \in J(x)$. Since $x \notin \downarrow c_i$, there is $w_i \in x \setminus c_i$ for all $i \in J(x)$. Yet, by hypothesis there is $x_i \in \uparrow z \setminus \uparrow x$ such that $\{w_i\} \cup y_i \subseteq x_i$ for all $i \in J(x)$. Because $x_i \in \uparrow y_i$, it follows from $\uparrow x = \uparrow y_i \setminus \downarrow c_i$ that $x_i \in \downarrow c_i$ (i.e. $x_i \subseteq c_i$), in contradiction to $w_i \notin c_i$, for all $i \in J(x)$. ■

The last impossibility suggests that, in order to have an EDP defined on a tree, an extra condition is needed.

Definition 4.4 A game tree $(N, \supseteq)$ has **available** (perfect information) **choices** if $\gamma(x,w) \subset x$ for all $w \in x$ and all $x \in X$.

The terminology is motivated by the fact that, by Proposition 4.3(c), if $\gamma(x,w) = x$ then there can be no choice available at x, i.e. $J(x) = \emptyset$. If, by contrast, $\gamma(x,w) \subset x$, then the sets $\gamma(x,w)$ can serve as choices—at least in a perfect information game. This may not always be the case.

Example 4.15 (Inverse Infinite Centipede) Let $W = \{\ldots, -2, -1\}$ be the set of negative integers with the natural order and consider the corresponding centipede. This is a game tree by Example 4.8, as -1 provides a maximum (see Fig. 4.8). The set of plays for $(N, \supseteq)$ consists of sets of the form $\{(x_{-\tau})_{\tau=t}^{\infty}, \{-t\}\}$ for all $t = 1, 2, \ldots$ (since $x_{-1} = \{-1\}$, the play $\{(x_{-\tau})_{\tau=1}^{\infty}\}$ is included). Every play $\{(x_{-\tau})_{\tau=t}^{\infty}, \{-t\}\}$ can be represented by the negative integer $-t$, for all $t = 1, 2, \ldots$ But none of the chains $\{x_{-\tau}\}_{\tau=t}^{\infty}$ has a maximum. Therefore, the perfect information choice $\gamma(W, -1)$ coincides with the root itself, i.e. $\gamma(W, -1) = \cup_{t=1}^{\infty} x_{-t} = W$. By Proposition 4.3(c) it is impossible to define an EDP on this tree.[13]

Proposition 4.3(c) gives a sufficient condition for available choices: that all maximal chains in $\downarrow x \setminus \{x\}$ have a maximum, for all $x \in X$, viz. weak up-discreteness.

Corollary 4.1 *If a game tree is weakly up-discrete, then it has available choices.*

[13]A similar example is presented by Schwarz (1974) to illustrate when different ways of randomizing may not be equivalent.

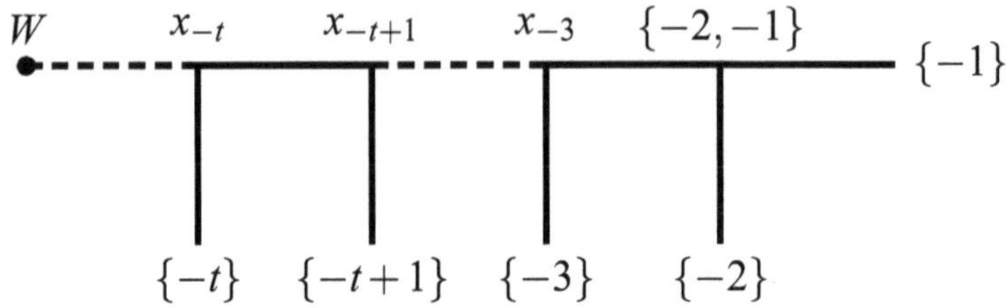

Fig. 4.8 The inverse infinite centipede. $W=\{\ldots,-2,-1\}, x_{-t}=\{-t,-t+1,\ldots,-1\}$

4.5.2 Existence of EDPs

Denote the set of perfect information choices for a game tree $T=(N,\supseteq)$ by $\Gamma(T)=\{\gamma(x,w)\,|w\in x\in X\}$ and recall that $S(N)$ denotes the set of strange nodes. By Proposition 4.3(b), if a game tree has available choices, then all choices in $\Gamma(T)$ are available at some node. The following theorem shows that available choices indeed *characterize* game trees on which EDPs can be defined.

Theorem 4.2 *Let $T=(N,\supseteq)$ be a (rooted) game tree. The following are equivalent:*

(a) Some EDP (T,C) can be defined on T;
(b) T has available choices;
(c) $\Pi(T)=(T,C_1)$ is a well-defined single-player EDP, where $C_1=\Gamma(T)\cup S(N)$.

Proof (c) trivially implies (a), and (a) implies (b) by Proposition 4.3(c). To show that (b) implies (c), suppose that $T=(N,\supseteq)$ has available choices. Let $I=\{1\}$ and $C_1=\Gamma(T)\cup S(N)$. Then choices are unions of nodes by construction. Proposition 4.3(b) implies that $A_1(x)\neq\emptyset$, hence, $J(x)=\{1\}\neq\emptyset$ for all $x\in X$.

It remains to verify (EDP1) to (EDP4) from Definition 4.1. Property (EDP2) follows trivially from $J(x)=\{1\}$ for all $x\in X$. (EDP4) is also simple: Consider $x\in X$ and $x\supset y\in N$. Choose $w\in y$ and $c=\gamma(x,w)$. Then $y\subseteq c$ and $c\in A_1(x)$ by Proposition 4.3(b).

Consider now (EDP1). If $P(c)\cap P(c')\neq\emptyset$, then Proposition 4.3(b) implies that $P(c)=\{x\}=P(c')$ for some $x\in X$, where $c=\gamma(x,w)$ and $c'=\gamma(x,w')$ for some $w,w'\in x$. That $c\cap c'=\emptyset$ then follows from Proposition 4.3(a).

Turn to (EDP3). Strange nodes are choices in C_1 by construction. We claim that finite nodes are also in C_1. Let y be a finite node and $x=\min\uparrow y\setminus\{y\}$. Let $c=\gamma(x,w)$ for some $w\in y$. By definition $y\subseteq c$. If $y\subset c\subset x$ (the latter by (b)), then there exists $z\in\downarrow x\setminus\{x\}$ such that $w\in y\subset z$, in contradiction to $x=\min\uparrow y\setminus\{y\}$. Therefore, $y=c$. Hence, (EDP3) holds trivially for pairs of non-infinite nodes. Consider now an infinite node y. By definition, $y=\inf\uparrow y\setminus\{y\}$. Consider any other node $y'\in N$ (infinite or not) such that $y\cap y'=\emptyset$. It follows that $y'\notin\uparrow y$. Then, there exists $z\in\uparrow y\setminus\{y\}$ such that $z\cap y'=\emptyset$. For, if not, Trivial Intersection (2.4) would imply that $y'\subseteq z$ for all $z\in\uparrow y\setminus\{y\}$; by the definition of the infimum, it would follow that $y \supseteq y'$, a contradiction. Take now $c = \gamma(z,w)$ for some $w \in y$. It follows that $y \subseteq c$ and $c \cap y' = \emptyset$. Hence, (EDP3) is verified if y is infinite and y' is not, since then $y' \in C_1$. If y, y' are infinite disjoint nodes, then, as above, there exist a choice $c = \gamma(z,w)$ with $z \in \uparrow y \setminus \{y\}$, $z \cap y' = \emptyset$, $y \subseteq c \subseteq z$, and $c \cap y' = \emptyset$. Repeating the argument with the (disjoint) nodes z, y' we obtain a choice c' such that $y' \subseteq c'$ and $c' \cap z = \emptyset$, implying that $c \cap c' = \emptyset$. This completes the verification of (EDP3). ■

The strange nodes that are choices in $C_1 = \Gamma(T) \cup S(N)$ are never available. This makes the interpretation of Theorem 4.2 odd. Obviously, if the tree is regular, the single-player game only requires the choices $C_1 = \Gamma(T)$ that are always available.

Say that an EDP (T,C) has *quasi-perfect information* if $P(c) \neq \emptyset$ implies that $P(c) = \{x\}$ for some $x \in X$, for all $c \in C_i$ and all $i \in I$. Quasi-perfect information differs from the traditional notion of "perfect information" in an extensive form in two respects. First, there may be choices that are never available at any move; second, if there are several players, they may choose at the same node. That is, unavailable choices aside, quasi-perfect information corresponds to the standard concept of perfect information, except that several players may decide at the same move.

Theorem 4.2(c) (and Proposition 4.3(b)) reveals that if an EDP (T,C) can be defined on the game tree T at all, then there exists one with quasi-perfect information defined on T, which is unique up to an assignment of decision points to players. In particular, it is unique if a single player is assumed. The associated single player decision problem with perfect information is henceforth denoted $\Pi(T) = (T, C_1)$, as in Theorem 4.2(c).

4.6 Summary

This chapter introduces the second milestone on the way towards a general representation theory for extensive form games: Extensive Decision Problems (EDPs, Definition 4.1). This adds the second, "subjective," layer to the formalization that brings in the players and their decisions. Due to the generality of the underlying game trees an EDP is capable of capturing all instances of extensive form games in the literature. This generality has its costs, though. For instance, the familiar notion of an *information set* may not make sense in a general EDP. Therefore, *choices* need to capture who can when do what and under which informational constraints.

Still, in an EDP every play is realized when all the players' choices that are compatible with the play are actually taken (Theorem 4.1). The familiar notion of a pure strategy is also well defined for EDPs. Whether or not an EDP can be defined on an arbitrary game tree is a different matter, though. To clarify that takes a closer look at game trees.

A classification of nodes in the game tree (Definition 4.3) enables a complementary result to Proposition 2.11: Infinite terminal nodes are precisely those nodes that may be added or removed from the tree without affecting its structure (Proposition 4.2).

Yet, this insight is not enough to identify those trees on which EDPs can be defined. In particular, if the tree is such that at some move no choices can be available (Definition 4.4), then no EDP can be defined on this tree (Proposition 4.3(c)). In fact, the condition that the game tree has available (perfect information) choices is not only necessary, but also sufficient to define an EDP on the tree (Theorem 4.2). Hence, available choices *characterize* game trees on which EDPs can be defined. A sufficient condition for the tree to have available choices is that every maximal chain in the set of (proper) successors of a move has a maximum—that the tree is "weakly up-discrete" (Corollary 4.1). This condition will play an even more important role in the next chapter.

References

Alós-Ferrer C (1999) Dynamical systems with a continuum of randomly matched agents. J Econ Theory 86:245–267
Alós-Ferrer C (2002) Individual randomness with a continuum of agents. Mimeo, Vienna
Alós-Ferrer C, Ritzberger K (2005a) Trees and decisions. Econ Theory 25(4):763–798
Alós-Ferrer C, Ritzberger K (2008) Trees and extensive forms. J Econ Theory 43(1):216–250
Aumann RJ, Hart S (2003) Long cheap talk. Econometrica 71:1619–1660
Carmona G (2009) A remark on the measurability of large games. Econ Theory 39:491–494
Friedman JW (1990) Game theory with applications to economics, 2nd edn. Oxford University Press, Oxford
Friedman A (1994) Differential games. In: Aumann RJ, Hart S (eds) Handbook of game theory, vol 2. Elsevier, Amsterdam/Princeton, pp 781–799
Hewitt E, Stromberg K (1965) Real and abstract analysis. Graduate texts in mathematics, vol 25. Springer, Berlin/Heidelberg/New York
Isbell JR (1957) Finitary games. In: Dresher D, Tucker AW, Wolfe P (eds) Contributions to the theory of games, volume III. Annals of mathematics studies, vol 39. Princeton University Press, Princeton, pp 79–96
Khan MA, Sun Y (1999) Non-cooperative games on hyperfinite Loeb spaces. J Math Econ 31:455–492
Kuhn H (1953) Extensive games and the problem of information. In: Kuhn H, Tucker A (eds) Contributions to the theory of games, volume II. Annals of mathematics study, vol 28. Princeton University Press, Princeton, pp 217–243
Mas-Colell A (1984) On a theorem of Schmeidler. J Math Econ 13:201–206
Piccione M, Rubinstein A (1997) On the interpretation of decision problems with imperfect recall. Games Econ Behav 20:3–24
Podczeck K (2010) On the existence of rich Fubini extensions. Econ Theory 45:1–22
Podczeck K, Puzzello D (2012) Independent random matching. Econ Theory 50:1–29
Schmeidler D (1973) Equilibrium points of nonatomic games. J Stat Phys 7:295–300
Schwarz G (1974) Randomizing when time is not well-ordered. Isr J Math 19:241–245
Selten R (1975) Reexamination of the perfectness concept for equilibrium points in extensive games. Int J Game Theory 4:25–55
Selten R (1998) Multistage game models and delay supergames. Theory Decis 44:1–36
Simon L, Stinchcombe MB (1989) Extensive form games in continuous time, Part I: pure strategies. Econometrica 57:1171–1214

Stinchcombe MB (1992) Maximal strategy sets for continuous-time game theory. J Econ Theory 56:235–265
Sun Y (2006) The exact law of large numbers via Fubini extension and characterization of insurable risk. J Econ Theory 126:31–69
Uhlig H (1996) A law of large numbers for large economies. Econ Theory 8:41–50

Chapter 5
Extensive Forms

Now that it has been unveiled what it takes to define an EDP on a game tree, it is time to ask when players can indeed play the extensive form game.[1] After all, non-cooperative game theory is the theory of games with complete rules. Its hallmark is a thought experiment that leaves all decisions exclusively to the players. Unlike cooperative game theory, where axioms like efficiency or symmetry restrict the solutions, or competitive equilibrium, where an auctioneer provides the agents with prices, a non-cooperative game provides an idealized 'world of its own,' where the effects of individual decisions can be studied without external intervention—a purely individualistic "interactive decision theory" (Aumann 1987, p. 460).

The formal device to verify complete rules is the specification of an EDP. It employs a game tree to model the order of decisions by players—who can do what when—and how those generate an outcome. The previous chapters have shown that the traditional finiteness assumptions[2] are not needed. Theorem 4.2 provides necessary and sufficient conditions on the game tree for an EDP to be definable on it. These conditions thus characterize when the rules are complete.

Are these conditions enough for players to be able to plan their strategies in some "solution" of the game? Somewhat surprisingly this is not the case. The concept of a game tree with available choices is so general that it encompasses all examples from the literature, including exotic cases like "differential games" (decision problems in continuous time), repeated games, stochastic games (Shapley 1953), infinite bilateral bargaining (Rubinstein 1982), and long cheap-talk (Aumann and Hart 2003) (see Sect. 2.2.2). In this Chapter it will be shown that there is a sense in which game trees are "too general."

---

[1]This chapter is based on the second part of an expanded version of Alós-Ferrer and Ritzberger (2008), modified as discussed in Alós-Ferrer et al. (2011).

[2]A notable early exception is Aumann (1964), who generalizes Kuhn's theorem to infinite extensive form games.

C. Alós-Ferrer, K. Ritzberger, *The Theory of Extensive Form Games*,
Springer Series in Game Theory, DOI 10.1007/978-3-662-49944-3_5

It was shown in Chap. 4, for instance, that differential games can be rigorously defined as extensive form games with the aid of the general concept of a set-tree. But are they such that players can indeed always decide on the basis of a well-defined outcome being associated with every strategy? This is important, because if a non-cooperative game is meant to leave all decisions to players, then players ought to be able to evaluate their strategies. Solution concepts like Nash equilibrium are also defined in terms of strategy combinations.[3] Hence, it is essential that they yield an object in the domain of the players' preferences—a play or an outcome. If there are strategy combinations that "evaporate" (induce no outcome at all) or yield multiple outcomes, they cannot be evaluated by the decision makers.

Identifying when players can actually play the game is the purpose of this Chapter. It will be show that this leads to further restrictions, beyond available choices. In fact, these considerations will also lead to a constraint on choices in an EDP, one that gives rise to the ultimate concept of an *extensive form*.

5.1 Preview

In this chapter we characterize extensive forms that satisfy the following three desiderata: (A0) Every play can be induced by some strategy combination. (A1) Every strategy combination induces some outcome/play. (A2) The outcome/play induced by a given strategy combination is unique.

These desiderata take a global perspective. Completeness of rules is a local criterion in the sense that at each "when" it is clear "who can do what." The differential game, for instance, has an extensive form representation and, thereby, complete rules. But, as shown in this chapter, its rules are such that they allow for decisions that do not combine to a definite outcome.

The results have two virtues. First, because they are characterizations, they make explicit which assumptions have been implicitly used—and which have to be used necessarily—in applications that push the limits of extensive form analysis beyond finite games. Second, the main characterization is in terms of easily verifiable properties of the tree. This makes it more tractable to ascertain whether or not a given extensive form game can be analyzed by classical methods. This is illustrated by the insights about differential games.

The three desiderata are also necessary and sufficient for an extensive form game to have a normal form representation. Many non-cooperative solution concepts are defined directly in the normal form—on the implicit assumption that strategies always yield well-defined outcomes. Such an approach is meaningful if and only if the desiderata hold. Therefore, our results provide the first *characterization* of the class of trees for which the traditional techniques of game theory are appropriate.

[3]This is in contrast to cooperative solution concepts that are often defined in terms of utility allocations or "imputations."

A further motivation of this work is that extensive form games are often solved by backwards induction procedures. Those require that strategies induce outcomes after every history (roughly, in every subgame) even if those are unreached (counterfactuals). It will be shown that this is guaranteed for the class of games described by the main characterization. Thus, indeed the trees identified in this chapter provide an adequate domain for sequential decision theory.

In the following section the three desiderata (A0), (A1), and (A2) are defined. Then, as a first step, it is shown that (A0) is always true without further restrictions (Theorem 5.1). Thereafter, a slightly stronger criterion than (A1) is characterized: that every strategy induces an outcome after every history (Theorem 5.2). Since subgames correspond to histories, this forms the basis on which backwards induction concepts can be built. For the class of regular game trees this stronger criterion is, in fact, equivalent to (A1), as shown in Theorem 5.3.

The latter parts of this chapter deal with the uniqueness criterion (A2). Since this also requires a restriction on the choice system, it takes us to the third milestone of the development: Extensive forms (EFs). These guarantee that (A2) holds by Theorem 5.4. Finally, we characterize the class of trees on which extensive forms satisfy all three desiderata (Theorem 5.5 and Corollary 5.4).

These results combine to a framework that is appropriate for the analysis of non-cooperative games. When (A0), (A1), and (A2) hold, then there is a surjection $\phi : S \to W$ that uniquely assigns to each strategy combination the outcome that it induces. This enables players to evaluate their strategies. Furthermore, it is equivalent to the existence of a normal form representation of the game.

5.2 Strategies and the Desiderata

Recall that a *pure strategy* for a player $i \in I$, as derived from an EDP (T, C), is a function $s_i : X_i = \{x \in X \,|\, i \in J(x)\} \to C_i$ such that $s_i^{-1}(c) = P(c)$ for all $c \in s_i(X_i)$, where $s_i(X_i) \equiv \cup_{x \in X_i} s_i(x)$ (see (4.4)). That is, s_i assigns to every move $x \in X_i$ a choice $c \in C_i$ such that (a) choice c is *available* at x, i.e. $s_i(x) = c \Rightarrow x \in P(c)$ or $s_i^{-1}(c) \subseteq P(c)$, and (b) to every move x in an *information set* $\xi = P(c)$ the same choice gets assigned, i.e. $x \in P(c) \Rightarrow s_i(x) = c$ or $P(c) \subseteq s_i^{-1}(c)$, for all $c \in C_i$ that are chosen somewhere, viz. $c \in s_i(X_i)$. Let S_i denote the set of all pure strategies for player $i \in I$. A *pure strategy combination* is an element $s = (s_i)_{i \in I} \in S \equiv \times_{i \in I} S_i$.

If an EDP captures complete and consistent rules, then there are a few desiderata that need to be satisfied, like that every strategy combination ought to induce an outcome. First, of course, it has to be clarified when a strategy combination "induces" a play. To this end define, for every $s \in S$, the correspondence $R_s : W \twoheadrightarrow W$ by

$$R_s(w) = \cap \{s_i(x) \,|\, w \in x \in X, i \in J(x)\}. \tag{5.1}$$

Say that the strategy combination s *induces* the play w if $w \in R_s(w)$, i.e., if w is a fixed point of R_s. The following are the three key desiderata on the mapping R_s:

(A0) For every play $w \in W$ there is some $s \in S$ such that $w \in R_s(w)$.
(A1) For every $s \in S$ there is some $w \in W$ such that $w \in R_s(w)$.
(A2) If for $s \in S$ there is $w \in W$ such that $w \in R_s(w)$, then R_s has no other fixed point and $R_s(w) = \{w\}$.

The desiderata (A1) and (A2) together with preference profiles for the players are precisely what is needed for a normal form representation of the game. In *finite* games there is no problem with the transition from the extensive to the normal form: Pure strategies are derived from the extensive form and every pure strategy combination is associated with precisely one play, so that a utility representation of the players' preferences over plays generates a utility profile for each pure strategy combination. In general (A1) defines a nonempty-valued correspondence ϕ from strategies to plays. Under (A2) the map $\phi : S \to W$ is a function that, in turn, defines the normal form.

5.2.1 *Randomized Strategies*

This also enables a treatment of *mixed* strategies, that is, probability distributions on pure strategies. For the function ϕ defined by (A1) and (A2) the preimage of a set $V \subseteq W$ of plays, $\phi^{-1}(V)$, is unambiguously defined. A mixed strategy combination gives a (product) probability distribution σ on pure strategy combinations $s \in S$. Assuming that ϕ is measurable, the measure that σ associates to the preimage $\phi^{-1}(V)$ of a set V of plays gives the probability $\sigma\left(\phi^{-1}(V)\right)$ of the set V of plays under the mixed strategy combination. In this sense (A1) and (A2) for pure strategies implicitly also cover mixed strategies.

A similar comment applies to behavioral strategies. This is because by splitting players into agents—one agent per information set—behavioral strategies become the mixed strategies of the agents. Since the present framework imposes no restrictions on the player set I, the above construction works for behavioral strategies, too: Given a probability measure ρ on the pure strategy combinations (pure choices) of the agents, the probability mass assigned to a set V of plays is $\rho\left(\phi_a^{-1}(V)\right)$, where ϕ_a is the function defined by (A1) and (A2) for the game played by the agents.

The caveat to this argument is that it ignores all measurability issues.[4] Consider a Stackelberg-duopoly with continuum action spaces. What the follower chooses is a function of the leader's choice. When both players randomize, the follower's

[4]For instance, let a continuum of players decide simultaneously between two actions, 0 and 1. If all select i.i.d. randomizations, it is known that the relevant derived sets that will be needed in order to properly define payoffs, e.g. the set of plays where half the players choose 0 ex post, are not measurable in the standard Kolmogorov product measure space; see e.g. Feldman and Gilles (1985) or Alós-Ferrer (2002).

strategy becomes a measure on a function space. As Aumann (1961) observed, it is not clear which measurable structure on the function space is required so that a distribution on the functions and one on their inputs generates a distribution on the outputs.[5] Since all plays consist of three nodes only, this problem is not related to "long plays," but haunts any sequential game with large action spaces. Therefore, we view this as beyond the scope of the present treatment.

5.3 Plays Reached by Strategies

The first issue is what it takes to fulfill (A0), that is, whether every play is reachable by some pure strategy combination (a condition also imposed e.g. by Stinchcombe 1992, p. 236). This would fail, for instance, if the game had absent-mindedness (Piccione and Rubinstein 1997). Since the current definition of an EDP rules out absent-mindedness, see Proposition 4.1, every play is reachable by some strategy combination, without further assumptions on the EDP. When (A1) and (A2) are fulfilled, this implies that the function $\phi : S \to W$ defined by the desiderata is onto (surjective).

Theorem 5.1 *If (T, C) is an EDP, then for every play $w \in W$ there is a pure strategy combination $s \in S$ such that $w \in R_s(w)$.*

Proof Given $w \in W$, we construct $s \in S$ in three steps. First, consider the moves $x \in X$ with $w \in x$ and let $i \in J(x)$. Let $y \in \downarrow x \setminus \{x\}$ with $w \in y$. By (EDP4), there exists a choice $c_i \in A_i(x)$ such that $y \subseteq c_i$, so $w \in c_i$. Define $s_i(x) = c_i$.

Second, consider the moves $x \in X$ with $w \notin x$. Suppose there exists $x' \in X$ with $w \in x'$ and $A_i(x) \cap A_i(x') \neq \emptyset$. Then, necessarily $A_i(x) = A_i(x')$. (Let $c^* \in A_i(x) \cap A_i(x')$, so that $x, x' \in P(c^*)$. Now let $c \in A_i(x)$. Then $x \in P(c^*) \cap P(c)$ and (EDP1) imply that $x' \in P(c) = P(c^*)$ and $c \in A_i(x')$.) Hence, by (EDP4) and (EDP1) there exists a unique $c \in A_i(x) = A_i(x')$ with $w \in c$. Set $s_i(x) = c$. To see that this definition is consistent, suppose there exist two different nodes x', x'' with $w \in x', x''$, $A_i(x) \cap A_i(x') \neq \emptyset$ and $A_i(x) \cap A_i(x'') \neq \emptyset$. The latter imply $A_i(x') = A_i(x'')$. By Trivial Intersection (2.4) x' and x'' are ordered, and then Proposition 4.1 implies that $x' = x''$, a contradiction.

Third, consider the moves $x \in X$ with $w \notin x$ such that for all $x' \in X$ with $w \in x'$ it follows that $A_i(x) \cap A_i(x') = \emptyset$. In that case choose $s_i(x)$ arbitrarily, for every $i \in J(x)$. Then, $w \in R_s(w)$ holds by construction. ■

Theorem 5.1 shows, somewhat surprisingly, that (A0) is true for any EDP without further assumptions on the underlying game tree. Unfortunately, this is not so for the two remaining desiderata.

---

[5]Aumann (1964) proposes to circumvent (but not resolve) this difficulty by treating mixed strategies as jointly measurable mappings (into pure strategies) with respect to an extraneous probability space and decision points in X_i, rather than as distributions.

5.4 When Do Strategies Induce Outcomes?

This section is devoted to (A1). The focus is on properties of the tree. This is because the crucial condition on choices is already part of the definition of an EDP (Definition 4.1): Clearly, property (EDP2) is necessary for (A1) to hold. For, if (EDP2) were not true, players could choose such that the game cannot continue from some move.

But the restriction on choices incorporated in (EDP2) is not enough to fulfill (A1). Below its is shown that weak up-discreteness and coherence characterize the class of trees for which every strategy induces outcomes *after every history*—a slightly stronger criterion than (A1). This, then, turns into a characterization of (A1) for the class of *regular* game trees.

5.4.1 Examples for Non-existence

The following example provides a transparent illustration for what can go wrong with existence of outcomes.

Example 5.1 (Augmented Inverse Infinite Centipede) No EDP can be defined on the inverse infinite centipede from Example 4.15. Construct, though, an *augmented* inverse centipede by adding a new element $-\infty$ to the underlying set W (which was previously just the negative integers), such that $-\infty < -t$ for any $t = 1,2,\ldots$, and consider the corresponding W-centipede (see Fig. 5.1). Now the root has an "immediate successor," $\{-\infty\}$, and $\gamma(W,-1) \subset W$. This (regular) game tree has available choices and, hence, admits an EDP. The corresponding single-player problem $\Pi(T)$ with perfect information is easy to construct: $\gamma(x_{-t},-\tau) = x_{1-t} \subset x_{-t}$ if $\tau < t$, and $\gamma(x_{-t},-\tau) = \{-t\} \subset x_{-t}$ if $\tau = t$, for all $t = 1,2,\ldots$ The interpretation of these choices as "continue" or "stop" is obvious.

Consider a strategy s which prescribes to continue at the beginning and to stop at every other move, i.e. $s(W) = \gamma(W,1) = \{-1,-2,\ldots\}$ and $s(x_{-t}) = \gamma(x_{-t},-t) = \{-t\}$ for all $t = 1,2,\ldots$ There is no play that is consistent with this strategy. For, $R_s(-\infty) = s(W) = \{-1,-2,\ldots\}$ so that $-\infty \notin R_s(-\infty)$, and for any $t = 1,2,\ldots$ one obtains $R_s(-t) = s(W) \cap [\cap_{\tau=t,t+1,\ldots} s(x_{-\tau})] = \cap_{\tau=t,t+1,\ldots} \{-\tau\} = \emptyset$. Thus, the strategy s induces no outcome at all.

A similar point can be made with the continuous centipede (Example 4.13), when in Example 4.8 the set W of plays is the unit interval, $W = [0,1]$. Even though this problem is not peculiar to continuous time, it may well plague continuous-time decision problems. It is known from Simon and Stinchcombe (1989) and Stinchcombe (1992) that for decision problems in continuous time the relation between outcomes/plays and strategies is subtle. The following example shows that in the differential game some strategies may not induce an outcome at all.

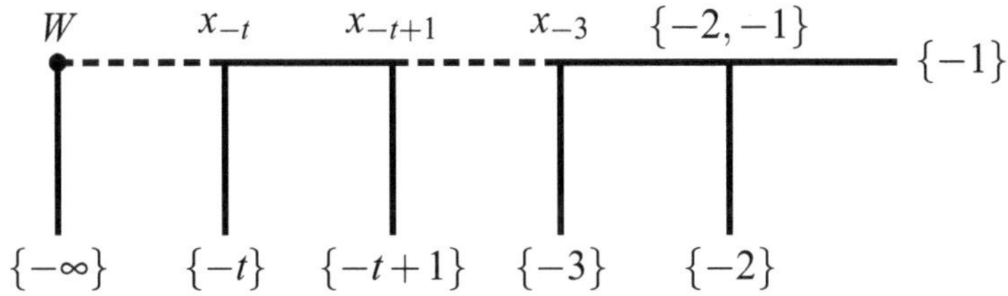

Fig. 5.1 The augmented inverse infinite centipede. $W = \{-\infty,\ldots,-2,-1\}$, $x_{-t} = \{-t,-t+1,\ldots,-1\}$

Example 5.2 (Differential game: outcome nonexistence) Consider the differential game with a single player, perfect information, and $A = \{0,1\}$ as in Examples 4.9 and 4.14. Specify a strategy $s \in S$ by $s(W) = c_0(h,1)$, $s(x_t(f)) = c_t(f,0)$ if $f(r) = 1$ for all $r < t$, and $s(x_t(f)) = c_t(f,1)$ otherwise, for any $t > 0$ and any $f \in W$. Clearly, the constant function $\mathbf{1}$ (viz. $\mathbf{1}(t) = 1\ \forall t$) is not a fixed point of R_s, as $\mathbf{1}(s) = 1$ for all $s < t$ for any $t > 0$, so that by the construction of s it would follow that $\mathbf{1}(t) = 0$, a contradiction.

Suppose R_s has a fixed point f. It follows that $f(0) = 1$ but, since $f \neq \mathbf{1}$, there exists $t > 0$ such that $f(t) = 0$. Thus, the set of real numbers $\{t \geq 0 | f(t) = 0\}$ is nonempty and bounded below by 0. By the Supremum Axiom, this set has an infimum t^*. If $t^* > 0$, consider $t' = t^*/2$. Then, $f(t') = 1$, but also $f(r) = 1$ for all $r < t'$. By the definition of s we should have $f(t') = 0$, a contradiction.

It follows that $t^* = 0$. But then, consider any $t > 0$. By the definition of an infimum, there exists $0 < r < t$ such that $f(r) = 0$. By the definition of s we have that $f(t) = 1$. Since $t > 0$ was arbitrary, it follows that f must be identically 1, a contradiction.

5.4.2 Undiscarded Nodes

While desideratum (A1), that strategies induce outcomes, is clearly appealing, it is not always sufficient. It may well be sufficient for pure one-shot decisions among strategies. But for truly sequential decision making it is necessary to evaluate counterfactuals, that is, 'continuation' strategies after arbitrary histories. This requires that strategies not only induce outcomes, but do so after every history. This is a prerequisite for any "backwards induction" solution concept, like subgame perfection. But in the present framework such a criterion is pushed beyond subgames, as arbitrary (possibly infinite) histories need to be accounted for.

Define, for any history h in N, $W(h) = \bigcap_{x \in h} x$ as the set of outcomes that have still not been discarded after h. Clearly, $W(\{W\}) = W$ corresponds to the null history that consists only of the root. Let (T,C) be an EDP and s a pure strategy combination. Say that (a) s *induces outcomes* if there exists $w \in W$ such that

$w \in R_s(w)$, where R_s is defined in (5.1), and that (b) s *induces outcomes after* history h if there exists $w \in W(h)$ such that $w \in R_s^h(w)$, where

$$R_s^h(w) = \cap \{s_i(x) \,|\, w \in x \subseteq W(h), x \in X, i \in J(x)\}. \tag{5.2}$$

An EDP is *playable* if every strategy combination induces outcomes. It is *playable everywhere* if every strategy combination induces outcomes after every history.

That is, an EDP is playable (everywhere) if the mapping R_s (R_s^h) has a fixed point for every $s \in S$ (and every history h). Histories take the role of subgames in a general EDP, but only correspond to subgames (under quasi-perfect information) when they have infima. This has a useful characterization in terms of the sets $W(h) = \cap_{x \in h} x$.

Lemma 5.1 *Let $T = (N, \supseteq)$ be a game tree and h a history in T. Then:*

(a) $\emptyset \neq W(h) = \{w \in W \,|\, \uparrow\{w\} = h \cup g$ for some continuation g of $h\}$, and
(b) $W(h) \in N$ if and only if h has an infimum.

Proof (a) Let $w \in W(h)$. Then, $w \in x$ and hence $x \in \uparrow\{w\}$ for all $x \in h$. Hence, $\uparrow\{w\}$ is a play containing h and $\uparrow\{w\} = h \cup g$ for the continuation $g = \uparrow\{w\} \setminus h$. Conversely, let $w \in W$ be such that $\uparrow\{w\} = h \cup g$ for some continuation g of h. Then, $x \in \uparrow\{w\}$ and hence $w \in x$ for all $x \in h$. Hence, $w \in W(h)$. This proves the equality. $W(h)$ is nonempty by definition of a game tree.

(b) Is an immediate consequence of Lemma 4.2(a). ■

Fix a history h and a strategy combination $s \in S$. Define the set of *discarded nodes at h*, denoted $D^h(s)$, as the set of nodes $y \in N$ that are properly contained in $W(h)$ and for which there are $x \in \uparrow y \setminus \{y\}$, $i \in J(x)$, and $c \in A_i(x)$ such that $x \subseteq W(h)$ and $y \subseteq c \neq s_i(x)$. The set of *undiscarded nodes at h*, denoted $U^h(s)$, is the set of nodes contained in $W(h)$ that are not discarded. The sets of *discarded* and *undiscarded nodes* are defined as $D(s) = D^{\{W\}}(s) \subseteq N \setminus \{W\}$ and $U(s) = U^{\{W\}}(s) = N \setminus D(s)$ respectively. Clearly, $W \in U(s)$ by construction. Let $\downarrow W(h) = \{x \in N \,|\, x \subseteq W(h)\}$ be the set of nodes contained in $W(h)$.

Proposition 5.1 *Consider an EDP, a history h, and a strategy combination $s \in S$. Then, there exists $w \in W$ such that $w \in R_s^h(w)$ if and only if $U^h(s)$ contains a maximal chain in $\downarrow W(h)$.*

Proof "if:" If $u \subseteq U^h(s)$ is a maximal chain in $\downarrow W(h)$, then $u \cup h$ is a maximal chain in N. By Boundedness (2.16), there is $w \in W$ such that $w \in \cap_{x \in u \cup h} x$. By Theorem 2.3(c), $\uparrow\{w\} = u \cup h$. Hence, $w \in R_s^h(w) = \cap_{x \in u} \cap_{i \in J(x)} s_i(x)$.

"only if:" Let $w \in W$ be such that $w \in R_s^h(w)$. By construction, $u = \uparrow\{w\} \setminus h \subseteq U^h(s)$. Since $\uparrow\{w\}$ is a play and h is a history, it follows that u is maximal in $\downarrow W(h)$. ■

This result shows that existence of outcomes is equivalent to existence of plays consisting of undiscarded nodes. This is, of course, almost a tautology. But it illustrates what goes wrong in the "augmented inverse infinite centipede,"

Example 5.1. There, a strategy that "continues" at the root, but "stops" everywhere else, generates a set of undiscarded nodes that consists only of the root.

5.4.3 Perfect Information and Playability

It will now be shown that for playability it suffices to consider the perfect information case. The following concerns an arbitrary EDP (T,C) and the associated single-player perfect information problem $\Pi(T) = (T, C_1)$. Let $s \in S$ denote the strategy combinations in (T,C) and $s' \in S'$ the strategies in $\Pi(T)$. Whether or not every strategy combination induces outcomes in a given EDP is purely a matter of the tree and, therefore, independent of the choice (information) structure (granted (EDP2) holds). This is the essence of Proposition 5.2 below.

The following lemma is needed for the proof of Proposition 5.2, and also for Propositions 5.5 and 5.8.

Lemma 5.2 *For an EDP (T,C): If $x \in P(c)$, with $c \in C_i$ for some $i \in I$, and $w \in x \cap c$, then $\gamma(x,w) \subseteq x \cap c$.*

Proof If $x \in P(c)$, for $c \in C_i$ and $i \in I$, and $x \supset y \in N$, then by (EDP4) there is $c_i \in A_i(x)$ such that $y \subseteq c_i$ for all $i \in J(x)$. Therefore, if $c \in A_i(x)$ and $w \in x \cap c$, then $w \in y \in \downarrow x \setminus \{x\}$ implies $y \subseteq c$, because $c \neq c_i$ would imply $c \cap c_i = \emptyset$ by (EDP1) for all $i \in J(x)$. It follows that $\gamma(x,w) \subseteq x \cap c$, as desired. ■

Proposition 5.2 *Fix a history h. If every strategy $s' \in S'$ for $\Pi(T)$ induces outcomes after h, then for any EDP (T,C) with the same tree every strategy combination $s \in S$ induces outcomes after h.*

Proof Pick any $s \in S$ and construct $s' \in S'$ as follows: By (EDP2) and the Axiom of Choice there is $w_x \in x \cap \left[\cap_{i \in J(x)} s_i(x)\right]$ for every $x \in X$; set $s'(x) = \gamma(x, w_x)$ for all $x \in X$. Then $s'(x) \subseteq x \cap \left[\cap_{i \in J(x)} s_i(x)\right]$ for all $x \in X$ by Lemma 5.2.

Next, we claim that if $s \in S$ and $s' \in S'$ are such that $s'(x) \subseteq s_i(x) \cap x$ for all $i \in J(x)$ for all $x \in X$, then $U^h(s') \subseteq U^h(s)$, for any history h. To see this, let $y \in D^h(s)$, i.e., there are $x \in \uparrow y \setminus \{y\}$ with $x \subseteq W(h)$, $i \in J(x)$, and $c \in A_i(x)$ such that $y \subseteq c \neq s_i(x)$. Then $y \cap s_i(x) = \emptyset$. Because $s'(x) \subseteq x \cap \left[\cap_{i \in J(x)} s_i(x)\right]$ by hypothesis, $s'(x) \cap y = \emptyset$. Hence, $y \in D^h(s')$. Since $y \in D^h(s)$ was arbitrary, $D^h(s) \subseteq D^h(s')$ or, equivalently, $U^h(s') \subseteq U^h(s)$, verifying the claim.

This implies that $U^h(s') \subseteq U^h(s)$. By hypothesis and the "only if"-part of Proposition 5.1, the set $U^h(s')$ of undiscarded nodes contains a maximal chain u that is also maximal in $\{x \in N \,|\, x \subseteq W(h)\}$. By $U^h(s') \subseteq U^h(s)$ this chain u is also contained in $U^h(s)$. Hence, the "if"-part of Proposition 5.1 implies the statement. ■

Choosing $h = \{W\}$, for a fixed game tree, Proposition 5.2 implies the following.

Corollary 5.1 *If $\Pi(T)$ is playable (resp. playable everywhere), then any EDP (T,C) with the same tree is playable (resp. playable everywhere).*

Even though playability is a matter of the tree, it remains a surprisingly subtle problem. To clarify it, two issues need to be addressed. First, a history may or may not have a minimum, e.g., the chain of proper predecessors of an infinite or strange node does not. Second, a continuation of a history may or may not have a maximum. In "classical" games, all histories have minima and all continuations have maxima. Large games, e.g. in continuous time, provide examples, where this is not the case.

Recall that T is *weakly up-discrete* if for every move $x \in X$ all maximal chains in $\downarrow x \setminus \{x\}$ have a maximum (Definition 4.2). The next result gives two characterizations.

Lemma 5.3 *For a game tree $T = (N, \supseteq)$ the following statements are equivalent:*

(a) T is weakly up-discrete;
(b) for every history with a minimum, every continuation has a maximum;
(c) $x \supset \gamma(x,w) \in N$ for all $w \in x$ and all $x \in X$.

Proof "(a) implies (b):" If h is a history with a minimum x, then by Lemma 5.1, $W(h) = x$ and the continuations of h are maximal chains in $\downarrow x \setminus \{x\}$.

"(b) implies (c):" Let $w \in x \in X$. As $\uparrow \{w\}$ is a play,

$$u = \{z \in N \,|\, w \in z \in \downarrow x \setminus \{x\}\}$$

is a maximal chain in $\downarrow x \setminus \{x\}$. Obviously, $h = \uparrow x$ is a history with minimum x and u is a continuation of h. Hence, u has a maximum by (b), $y = \max u$, and $y \subset x$. It follows that $\gamma(x,w) = \cup \{z | z \in u\} = y \in N$ and $\gamma(x,w) \subset x$.

"(c) implies (a):" Suppose that $\gamma(x,w) \in N$ for all $w \in x$ and all $x \in X$. Consider $x \in X$ and let u be a maximal chain in $\downarrow x \setminus \{x\}$. Since T is a game tree, u is contained in a maximal chain in N (play), i.e. there exists $w \in W$ such that $w \in y$ for all $y \in u$. By hypothesis $w \in \gamma(x,w) \in \downarrow x \setminus \{x\}$ and thus $\gamma(x,w) \in u$. By the construction of γ also $y \subseteq \gamma(x,w)$ for any $y \in u$. Therefore, $\gamma(x,w) = \max u$. ■

Weak up-discreteness implies available choices (Corollary 4.1), but Lemma 5.3 (c) additionally states that perfect information choices are, in fact, nodes. This is not the case, for instance, in the differential game (Example 4.9), whose tree is not weakly up-discrete.

5.4.4 Everywhere Playable EDPs

For sequential decision theory it is essential that decision makers can evaluate their 'continuation' strategies after arbitrary histories, in particular, if a game is to be solved by backwards induction. Therefore, a domain appropriate for sequential decision theory has to be playable everywhere. Hence, we turn now to a characterization of the class of trees, where this is fulfilled, by two of the properties defined in Definition 4.2.

Theorem 5.2 *Let $T = (N, \supseteq)$ be a game tree with available choices. Then, every EDP (T, C) is playable everywhere if and only if T is coherent and weakly up-discrete.*

Proof "if:" By Proposition 5.2 it suffices to consider $\Pi(T)$. Fix a history h and a strategy s. We first show: (a) $D^h(s)$ is an ideal, i.e. $x \in D^h(s) \Rightarrow \downarrow x \subseteq D^h(s)$; (b) every maximal chain in $U^h(s)$ is a filter in $\downarrow W(h)$, i.e. $x \in U^h(s) \Rightarrow \uparrow x \cap \downarrow W(h) \subseteq U^h(s)$. Part (a) follows by definition. Then, every chain in $U^h(s)$ is a filter. For, if $x \in U^h(s)$ and $y \in D^h(s)$ with $y \supseteq x$, then $x \in D^h(s)$, a contradiction.

Next, $U^h(s)$ is nonempty: If h has an infimum, then $W(h) \in N$ by Lemma 5.1 and, by definition, $W(h) \in U^h(s)$. If h has no infimum (and hence no minimum), by coherence there exists a continuation g of h which has a maximum, $z = \max g$. By definition, $z \in U^h(s)$, because there exists no node $x \in \uparrow z \setminus \{z\}$ such that $x \subseteq W(h)$.

Second, suppose that there is no $w \in W$ such that $w \in R_s^h(w)$. Since $U^h(s) \neq \emptyset$, there exists a maximal chain u in $U^h(s)$ by the Hausdorff Maximality Principle. Let $w \in W$ be such that $u \subseteq \uparrow \{w\}$. If $u = \uparrow \{w\} \setminus h$, then $\uparrow \{w\} = u \cup h$ and, by construction, $w \in R_s^h(w)$, a contradiction. Thus, $u \subset \uparrow \{w\} \setminus h$ and $u \cup h$ is a history.

Third, u has no minimum. If it had, say, $x = \min u$, then $x = \min\ u \cup h$. By weak up-discreteness and Lemma 5.3, every continuation of $u \cup h$ would then have a maximum. Let $w \in s(x)$, and let z be the maximum of the continuation $\uparrow \{w\} \setminus (u \cup h)$. Hence, $P(z) = \{x\}$. Since $z \subseteq s(x)$ by (EDP1) and (EDP4), and $x \in U^h(s)$, it follows from the fact that $U^h(s)$ is a filter that $z \in U^h(s)$, a contradiction to maximality of u.

Since u has no minimum, coherence implies that there exists a continuation g of $u \cup h$ which has a maximum, $z' = \max g$. Let $w \in W$ be such that $\uparrow \{w\} = u \cup h \cup g$. Since $w \in x$ for all $x \in u$, it follows that $w \in s(x)$ for all $x \in u$. For, since u has no minimum, for any $x \in u$ there is $x' \in u$ such that $x' \subseteq x$. Since $x' \in u \subseteq U^h(s)$, it follows that $w \in x' \subset s(x)$, using (EDP1) and (EDP4) again. But then that $w \in z'$ implies $z' \subseteq s(x)$ for all $x \in u$. Since $\uparrow z' \setminus \{z'\} \subseteq u \cup h$, it follows that $z' \in U^h(s)$, a contradiction.

"only if:" It has to be shown that if either weak up-discreteness or coherence fail, then some strategy induces no outcome after some history. The EDP used for this purpose is again $\Pi(T)$.

Suppose, first, that weak up-discreteness fails. Then, by Lemma 5.3, there exists a history h which has a minimum, $z = \min h = W(h)$ (by Lemma 5.1), and a continuation g of h which has no maximum. Let $w^* \in W$ be such that $\uparrow \{w^*\} = h \cup g$. Define a strategy s as follows. For every $x \in h$ (which includes $z = \min h$) set $s(x) = \gamma(x, w^*)$. For every $x \in g$, choose $s(x) \neq \gamma(x, w^*)$. Choose arbitrary choices at all other nodes. Obviously, $w^* \notin R_s^h(w^*)$. If $w \in z \setminus \gamma(z, w^*)$, also $w \notin R_s^h(w)$ because $s(z) = \gamma(z, w^*)$.

Let $w \in \gamma(z, w^*)$ be such that $w \neq w^*$, and consider the choice $\gamma(z, w)$. Since $w \in \gamma(z, w^*) = \bigcup \{x \in \downarrow z \setminus \{z\} \,|\, w^* \in x\}$, there exists $x \in N$ such that $x \subset z = W(h)$

and $w, w^* \in x$. Since $x \subset z$ and $w^* \in x$, hence, $x \in g$ (which has no maximum), there exists $y \in g$ such that $x \subset y \subset z$. Hence, $\gamma(y,w) \cap \gamma(y,w^*) \neq \emptyset$ and it follows from Proposition 4.3(a) that $\gamma(y,w) = \gamma(y,w^*)$. Thus, $s(y) \neq \gamma(y,w)$, implying that x is discarded at y and hence $w \notin R_s^h(w)$. Therefore, s does not induce an outcome after the history h.[6]

Suppose, second, that coherence fails. Then, there exists a history h without minimum, such that no continuation g of h has a maximum. Define a relation on $W(h)$ as follows. Given $w, w' \in W(h)$, say that wRw' if there exists $x \in N$ such that $x \subseteq W(h)$ and $w, w' \in x$. This is an equivalence (it is clearly reflexive and symmetric; transitivity follows from Trivial Intersection) and, hence, its quotient set $W(h)/R$ induces a partition of the set $W(h)$. Thus, given any $x \in N$ such that $x \subseteq W(h)$, there exists a unique $z \in W(h)/R$ such that $x \subseteq z$.

For each $z \in W(h)/R$, choose an element $w(z) \in z$ and let $g(z) = (\uparrow \{w(z)\}) \setminus h$. Define a strategy s as follows. For every $x \in h$, set $s(x) = \gamma(x, w(z))$ for any $w(z)$. For every $z \in W(h)/R$ and every $x \in g(z)$, choose $s(x) \neq \gamma(x, w^*)$.

Consider an arbitrary $z \in W(h)/R$. Clearly, $w(z) \notin R_s^h(w(z))$. Now consider any $w \in z$ such that $w \neq w(z)$. By definition of R, there exists a node $x \in N$ such that $x \subseteq W(h)$ and $w, w' \in x$. Since $x \in g(z)$ (which has no maximum), there exists $y \in g(z)$ such that $x \subset y \subset z$. Hence, $\gamma(y,w) \cap \gamma(y,w(z)) \neq \emptyset$ and it follows from Proposition 4.3(a) that $\gamma(y,w) = \gamma(y,w(z))$. Thus, $s(y) \neq \gamma(y,w)$, implying that x is discarded at y and, hence, $w \notin R_s^h(w)$. Therefore, s indeed does not induce an outcome after the history h.[7] ■

The Theorem fails if coherence is replaced by regularity. In Example 4.11, with $\{1/2\} \notin N$ and $1/2 \notin W$, the strategy that "continues" at all x_t and "stops" otherwise fails to induce a play, but the tree is regular and weakly up-discrete. On the other hand, the Theorem applies to some non-regular trees. Add in Example 4.12, where $\{1/2\}$ is a strange node and $1/2 \in W$, the (also strange) nodes $y = [1/4, 1/2)$ and $y' = (1/2, 3/4]$. An EDP can be defined, but the three strange nodes are unavailable choices. Yet, the tree is weakly up-discrete and coherent. Thus, any EDP defined on it is everywhere playable, despite the presence of strange nodes.

The distinction between "everywhere playable" and "playable" may first appear puzzling. It may seem that if every strategy induces outcomes, then every strategy should induce outcomes after every history. An intuitive (but false) supportive argument could be as follows. Suppose every strategy induces an outcome, but there is a strategy that induces no outcome after history h. Then, construct a new strategy that coincides with the first after history h, but "selects" h before. Since the new strategy must induce an outcome, and this outcome must necessarily "come after" h, the original strategy combination must induce an outcome after history h.

---

[6]This does not mean that it does not induce an outcome in the whole game. The strategy could have selected a play which bifurcates from h before z, e.g. if there is a proper subhistory of h which has a continuation that is simply a terminal node and no other continuation has a maximum.

[7]A similar comment as in the previous footnote applies.

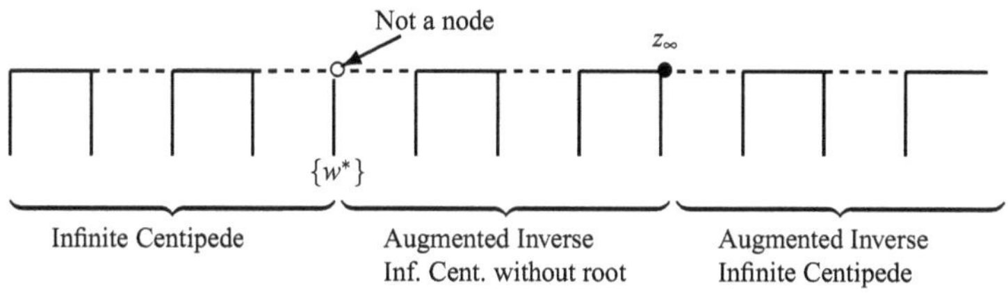

Fig. 5.2 Lexicographic centipede

This argument is an instance of an intuition that is guided by the finite case, but fails in the general case. The following example presents a *playable* EDP, where not every strategy induces an outcome after every history (Fig. 5.2).

Example 5.3 (Lexicographic Centipede) The idea is to start with an infinite centipede, replace its infinite terminal node by an augmented inverse infinite centipede (Example 5.1) without its root, and add another augmented inverse infinite centipede, now with root. Then every strategy induces an outcome, but not after every history. For, roughly, if it does not end during the first centipede, it is forced to select the first augmented inverse infinite centipede. That the latter "has no beginning" generates an "unavoidable" outcome, but it does not impose anything on the rest of the tree. In particular, in the rest things can be arranged so as to violate playability. Formally, let $W = \{((-1)^\tau t, \tau)\,|t = 1,2,\ldots,\, \tau = 0,1,3\} \cup \{(\infty,0)\}$, endow W with the natural lexicographic order, consider the associated W-centipede, and remove the node $\{(-t,\tau)\,|\tau = 1,3,\, t = 1,2,\ldots\} \cup \{(\infty,0)\}$ viz. the root of the first augmented inverse infinite centipede.[8] Denote $w^* = (\infty,0)$. Note that the ordering of the nodes for $\tau = 1,3$ is the reverse than for $\tau = 0$, and that w^* does not belong to any one of the moves $y_t = \{(-k,\tau) \in W\,|\tau = 1,3,\ \tau = 1 \Rightarrow k = 1,\ldots,t\}$ or $z_t = \{(-k,3) \in W\,|k = 1,\ldots,t\}$, but to all moves $x_t = \{(k,\tau) \in W\,|\tau = 0 \Rightarrow k \geq t\}$.

At each move two choices (for $\Pi(T)$) are available, a singleton and the remainder of the plays in the move. That is, choices at x_t are $\{(t,0)\}$ and x_{t+1} for $t = 1,2,\ldots$, at y_t they are $\{(-t,1)\}$ and y_{t-1} for $t = 2,3,\ldots$, at y_1 they are $\{(-1,1)\}$ and $\{(-t,3) \in W\,|t = 1,2,\ldots\} = \gamma(y_1,(-1,3))$, at z_t they are $\{(-t,3)\}$ and z_{t-1} for $t = 2,3,\ldots$, and at z_1 they are $\{(-2,3)\}$ and $\{(-1,3)\}$.

The strategy s^* given by $s^*(x_t) = x_{t+1}$ for $t = 1,2,\ldots$, $s^*(y_t) = y_{t-1}$ for $t = 2,3,\ldots$, $s^*(y_1) = \gamma(y_1,(-1,3))$, and $s^*(z_t) = \{(-t,3)\}$ does not induce an outcome after the history $\{x_1,x_2,\ldots,y_3,y_2\}$. This is so, because the remainder of the tree is an augmented inverse infinite centipede with a strategy like in Example 5.1: "continue" at the root, but "stop" everywhere else. So, this EDP is *not* everywhere playable.

Let, on the other hand, s be any strategy. If there is $t = 1,2,\ldots$ such that $s(x_t) = \{(t,0)\}$, let t^* be the smallest such t. Then, s induces the outcome $(t^*,0)$. If $s(x_t) = x_{t+1}$ for all $t = 1,2,\ldots$, the strategy s induces the outcome w^*, because $\uparrow\{w^*\} \setminus \{w^*\} = \{x_t\,|t = 1,2,\ldots\}$ and $w^* \in x_t$ for all $t = 1,2,\ldots$

[8]The resulting tree is technically not a centipede, because it not regular anymore.

The crux of the matter is that the node $\{w^*\}$ is strange. For a finite node x, a strategy that continues towards x along $\uparrow x \setminus \{x\}$ will reach the immediate predecessor of x and may or may not choose x. For an infinite node x, a strategy that continues towards x along $\uparrow x \setminus \{x\}$ must select x, but cannot select anything else. For a strange node, a strategy that continues towards x along $\uparrow x \setminus \{x\}$ must select x, but also selects anything else "after" (contained in the union over) $\uparrow x \setminus \{x\}$. If x were infinite and terminal, the strategy would select it and the game would end. If x is strange and terminal, as in the example, the strategy selects x, but the game goes on.

5.4.5 Up-Discrete Trees

If there were no strange nodes, the two conditions in Theorem 5.2 would simplify. As commented in Chap. 3, in order theory (Koppelberg and Monk 1992) and theoretical computer science it is often assumed that for trees the sets $\uparrow x$ are (dually) well-ordered: all their subsets have a first element (a maximum; see Koppelberg 1989, chap. 6). This is equivalent to the following:

Definition 5.1 A game tree is **up-discrete** if all (nonempty) chains have a maximum.

For, if $(N, \supseteq)$ is an up-discrete tree, all subsets $g \subseteq \uparrow x$ for $x \in N$ have maxima. Conversely, let $(N, \supseteq)$ be a tree, where all subsets of $\uparrow x$ have a maximum, for all x. Let g be a chain in N and $x \in g$. The chain $g \cap \uparrow x$ must have a maximum, z. But, if $y \in g$ and $y \notin \uparrow x$, then (as g is a chain), $z \supseteq x \supseteq y$, which shows that $z = \max g$. This proves the equivalence between up-discreteness and (dually) well-ordered sets $\uparrow x$. A further characterization strengthens Lemma 5.3(b).

Lemma 5.4 *A game tree $T = (N, \supseteq)$ is up-discrete if and only if for every history, every continuation has a maximum.*

Proof The "only if"-part is obvious, as every continuation is a chain. To see the "if"-part, suppose every continuation of every history has a maximum. Let g be an arbitrary (nonempty) chain in T. If the root W is in g, then $W = \max g$. Thus, suppose $W \notin g$. Let w be a play such that $g \subseteq \uparrow \{w\}$. Let $g^* = \{x \in \uparrow \{w\} \,|\, x \subseteq x' \text{ for some } x' \in g\}$. Then, $h = \uparrow \{w\} \setminus g^*$ is a history and g^* is a continuation of h. (To see that h is a history, let $x \in h$ and $y \in \uparrow x \subseteq \uparrow \{w\}$. If $y \notin h$, then $y \in g^*$, hence there exists $y' \in g$ with $x \subseteq y \subseteq y'$ and thus $x \in g$, a contradiction.) By hypothesis, g^* has a maximum, $z = \max g^*$. Then, $z \in g^*$ and there exists $z' \in g$ such that $z \subseteq z'$. But, since $g \subseteq g^*$, it follows that $z = z'$, i.e. z is also the maximum of g. ■

Theorem 5.2 works even in the presence of strange nodes. If regularity, $S(N) = \emptyset$, were assumed, the previous lemmata imply that weak up-discreteness and coherence would collapse to up-discreteness.

Corollary 5.2 *A regular game tree $T = (N, \supseteq)$ is up-discrete if and only if it is weakly up-discrete and coherent.*

The "only if"-part of Corollary 5.2 holds even in the presence of strange nodes. But its "if"-part is false without regularity: In Example 4.12 the chains $\{y_t\}_{t=1}^{\infty}$ and $\{y'_t\}_{t=1}^{\infty}$ have no maxima, even though the tree is weakly up-discrete and coherent.

Recall that in Example 5.3 the strange node $\{w^*\}$ generated an outcome even though the EDP was not everywhere playable. This is suggestive: If there are no strange nodes and a strategy induces no outcome after history h, then there exists another strategy which prescribes to go along the history h and *nothing else.* Hence, this new strategy does not induce an outcome. In other words, playability may indeed be equivalent to playability everywhere—but only for regular trees. Regularity is necessary for this, as Example 5.3 demonstrates.

That playability and everywhere playability are equivalent for regular game trees will now be formally demonstrated. The key technical step before Theorem 5.3 shows how to construct a strategy that discards all the nodes of a continuation without maximum. In the augmented inverse infinite centipede (Example 5.1) this is accomplished by "continuing" towards a continuation without maximum (at the root) and "stopping" everywhere else. In the differential game (Example 5.2) we construct a strategy such that for every node (after the root) there is a previous node, where the former gets discarded. The Supremum Axiom allows us to do this consistently for all nodes in the tree (except the root). Analogous constructions yield non-existence in Examples 4.10 and 4.11. Generalizing this to arbitrary game trees requires the Axiom of Choice (an application of Zorn's Lemma in the proof of Proposition 5.3; see e.g. Hewitt and Stromberg 1965).

Formally, for a game tree with available choices $T = (N, \supseteq)$ and its associated problem $\Pi(T)$, define a *partial strategy* on a set of nodes $Y \subseteq N$ to be the restriction of a strategy to $Y \cap X$, i.e. a mapping that assigns to every $y \in Y \cap X$ a choice available at y. The set of partial strategies on Y is denoted S_Y.

Proposition 5.3 *Let $T = (N, \supseteq)$ be a game tree with available choices and consider the associated problem $\Pi(T)$. Let h be a history and g a continuation that has no maximum. Then, there exists a partial strategy s on $N(h,g) = \{x \in N \mid x \subset \cup_{y \in g} y\}$ such that all nodes in $N(h,g)$*[9] *are discarded under s, i.e. for all $x \in N(h,g)$ there is $y \in N(h,g)$ such that $x \subset y$ and $x \cap s(y) = \emptyset$.*

Proof The proof proceeds in several steps. First, observe that for every $x \in N(h,g)$, there exists $y \in N(h,g)$ such that $x \subset y$. That is, every node in $N(h,g)$ has a proper predecessor in $N(h,g)$. To see this, suppose $x \in N(h,g)$ has no immediate predecessor in $N(h,g)$. Let $w \in x$. Then, $g' = \uparrow \{w\} \cap N(h,g)$ is a continuation of h with $\max g' = x$. Clearly, $\cup \{x \in N | x \in g\} = \cup \{x \in N | x \in g'\} = x$. It follows that $x \in g$ (because $x \notin h$) and $\max g = x$, a contradiction.

[9]Note that the defining inclusion in $N(h,g)$ is strict. If h had a minimum, then the minimum would not be in this set. Also, if h had an infimum but not a minimum, then there could not exist a continuation without maximum.

Second, define the set

$$A=\left\{(Y,s)\;\middle|\;\begin{array}{l}Y\subseteq N(h,g),Y\cap X\neq\emptyset,\text{ and }s\in S_Y\text{ such that}\\ \forall x\in Y\;\exists y\in Y\text{ with }x\subset y\text{ and }x\cap s(y)=\emptyset\end{array}\right\}$$

and the partial order on A given by $(Y,s)\geq(Y',s')\Leftrightarrow Y\supseteq Y'$ and $s|_{Y'}=s'$ (reflexivity and transitivity are obvious; antisymmetry follows by construction).

Third, for every $x\in N(h,g)$, there exists $(Y,s)\in A$ such that $x\in Y\subseteq\uparrow x$. To see this, let $x\in N(h,g)$. By the first step, there exists $x_1\in N(h,g)$ such that $x\subset x_1$. By (EDP4), there exists a choice c, available at x_1, such that $x\subseteq c$. By available choices, there exists an available choice at x_1 which is disjoint from c. Thus one can define $s(x_1)\neq c$. Analogously, there is $x_2\in N(h,g)$ such that $x_1\subset x_2$ and an available choice at x_2 that discards x_1. The conclusion follows from an induction argument.[10]

Fourth, apply Zorn's Lemma to $(A,\geq)$. Observe that A is nonempty by the third step. We have to show that every chain in $(A,\geq)$ has an upper bound in A. Let C be a chain in A. That is, for every $(Y,s),(Y',s')\in C$, either $(Y,s)\geq(Y',s')$ or $(Y',s')\geq(Y,s)$. Define $Z=\cup\{Y|(Y,s)\in C\}$ and construct $\overline{s}$ as follows. Given $z\in Z$, define $\overline{s}(z)=s(z)$ for any $(Y,s)\in C$ such that $z\in Y$. Such a (Y,s) exists by construction of Z, and $\overline{s}$ is well-defined because C is a chain. Further, given $z\in Z$, taking $(Y,s)\in C$ such that $z\in Y$ shows that there exists $y\in Y$ with $x\subset y$ and $x\cap s(y)=\emptyset$, and hence $x\cap\overline{s}(y)=\emptyset$. Thus $(Z,\overline{s})$ is an upper bound for C in A.

Zorn's Lemma implies that $(A,\geq)$ has a maximal element (Z^*,s^*). Then, for any $x\in N(h)$, there exists $z\in Z^*$ such that $x\cap s^*(z)=\emptyset$. That is, any node in $N(h,g)$ (and not only in Z^*) is discarded under the partial strategy given by s^*.

If $\uparrow x\cap Z^*\neq\emptyset$, then let $z\in\uparrow x\cap Z^*$. Since $(Z^*,s^*)\in A$, there exists $z'\in Z^*$ such that $z\cap s^*(z')=\emptyset$, and thus $x\cap s^*(z')=\emptyset$ (because $x\subseteq z$). Suppose, then, $\uparrow x\cap Z^*=\emptyset$. By the third step above, there exists $(Y,s)\in A$ such that $x\in Y\subseteq\uparrow x$. Define now $Z_1=Z^*\cup Y$. Clearly, $Z^*\subset Z_1$. Define also $s_1\in S_{Z_1}$ as follows. For every $y\in Y$, let $s_1(y)=s(y)$. For every $z\in Z^*$, define $s_1(z)=s^*(z)$. Since $Y\cap Z^*=\emptyset$, s_1 is well defined. It follows that $(Z_1,s_1)\in A$ and $(Z_1,s_1)\geq(Z^*,s^*)$ but $Z\subset Z_1$, a contradiction with the maximality of (Z^*,s^*).

The conclusion now follows by specifying $s^*(x)$ arbitrarily for any node $x\in N(h,g)\setminus Z^*$. ■

Theorem 5.3 *For a regular (rooted) game tree $T=(N,\supseteq)$ with available choices the following statements are equivalent:*

(a) every EDP (T,C) is playable everywhere;
(b) every EDP (T,C) is playable;
(c) T is up-discrete.

Proof (a) implies (b) is trivial. Furthermore, (c) implies (a) by Theorem 5.2 and Corollary 5.2. It remains to show that (b) implies (c). By Corollary 5.2, it suffices

[10] More precisely, this step takes the Axiom of Dependent Choices, a consequence of the Axiom of Choice.

to establish weak up-discreteness and coherence. Three observations are useful for that:

Claim A The sets $W(h,g) = \cup\{x \in N \,|x \in g\}$, where g is a continuation of h, form a partition of $W(h)$.

To see this, let g be a continuation of h. Since $x \subseteq y$ for any $x \in g$ and $y \in h$, it follows that $W(h,g) \subseteq W(h)$. That $\cup\{W(h,g)\,|g$ is a continuation of $h\} = W(h)$ follows since, for any $w \in W(h)$, $\uparrow\{w\} \setminus h$ is a continuation of h. To see that the union is disjoint, let g, g' be continuations of h such that $W(h,g) \cap W(h,g') \neq \emptyset$ and $w \in W(h,g) \cap W(h,g')$. Then, there are $x \in g$ and $x' \in g'$ such that $w \in x \cap x'$, hence $x \cap x' \neq \emptyset$. By Trivial Intersection (2.4) either $x \subseteq x'$ or $x' \subseteq x$. In the first case (the second is analogous), since $h \cup g$ is a play, it follows that $x' \in g$. This implies that $g \cap \uparrow x' = g' \cap \uparrow x'$ and thus $W(h,g) = W(h,g')$.

Claim B If h has no minimum and no continuation of h has a maximum, then h has no infimum and the set $W(h) = \cap\{x \in N\,|x \in h\}$ is not a node in N.

Let g be a continuation of h. If h had an infimum z, then $z \in g$, because $h \cup g$ is a play and h has no minimum. Thus, h has no infimum and, by Lemma 5.1, $W(h)$ is not a node.

Claim C Let $(N, \supseteq)$ be a regular game tree and h a history without minimum. Then, for any subhistory $h' \subset h$ which also has no minimum, no alternative continuation of h' (i.e. g' with $g' \cap (h \setminus h') = \emptyset$) has a maximum.

To see this, suppose h' is a subhistory of h such that one continuation g' with $g' \cap (h \setminus h') = \emptyset$ has a maximum, $x = \max g'$. Then, by construction $\uparrow x \setminus \{x\} = h'$ and, since $g' \cap (h \setminus h')$ is empty, h' has no infimum. Thus, x is strange, in contradiction to regularity.

Coherence. Start now with verifying coherence. Suppose, by contradiction, that coherence is violated. We construct a strategy that induces no outcome.

Let h be a history without minimum such that no continuation has a maximum. Define a strategy s as follows. Fix $w \in W(h)$ and, for any $x \in h$, let $s(x) = \gamma(x,w)$. Thus, $h \subseteq U(s)$. For the nodes in $N(h) = \{x \in N\,|x \subset W(h)\}$, proceed as follows. Partition $W(h)$ into the sets $W(h,g)$ (apply Claim A) where g are the possible continuations of h. Clearly, every node in $N(h)$ belongs to exactly one $N(h,g)$. For every $W(h,g)$, define s on $N(h,g)$ as the partial strategy identified by Proposition 5.3. Thus, $N(h) \subseteq D(s)$.

Consider any subhistory h' of h having no minimum. By Claim C no continuation has a maximum. For all $W(h',g)$ such that g is a continuation of h' and such that $W(h',g) \neq W(h',\uparrow\{w\} \setminus h')$, define s on $N(h',g)$ as the partial strategy identified by Proposition 5.3. Thus $N(h',g) \subseteq D(h)$ for all such g. At all other nodes s is arbitrary.

Let $x \notin h$ be such that $x \notin N(h)$. Choose any $w' \in x$. Define $g = \uparrow\{w'\} \setminus h$ and let $h' = \uparrow\{w'\} \cap h$ (which is nonempty, because h is a history). Then, h' is a subhistory of h and g is a continuation of h'. If h' has no minimum, it follows by construction that $x \in D(s)$. Suppose, then, that there exists $z = \min h'$. Since $z \in h$, we have $s(z) = \gamma(z,w)$. Then $\gamma(z,w') \cap \gamma(z,w) = \emptyset$. For, if not, $\gamma(z,w') = \gamma(z,w)$ by

Proposition 4.3(a) and there would exist $y \in \uparrow\{w'\}$ with $y \subset z$ such that $w \in y$. It follows that $w \in y \in g = \uparrow\{w'\} \setminus h$. If $y \in h$, then $y \in \uparrow\{w'\} \cap h = h'$, in contradiction to $y \subset z = \min h'$. Thus $y \notin h$. Then $w \in y$ implies $y \in \uparrow\{w\} \setminus h$, thus $w' \in W(h)$, in contradiction to $x \notin N(h)$.

In conclusion, $s(z) = \gamma(z,w)$ and $x \subseteq \gamma(z,w')$ with $\gamma(z,w') \cap \gamma(z,w) = \emptyset$, i.e. $x \in D(s)$. It has been shown that $U(s) = h$ and, hence, $U(s)$ contains no play, i.e. s induces no outcome by Proposition 5.1.

Weak Up-Discreteness. Next, turn to weak up-discreteness. The proof again proceeds by contradiction. Let h be a history with a minimum such that one continuation g of h has no maximum (recall Lemma 5.3). By Claim B, $W(h,g)$ is not a node. The construction is similar to the one in the coherence part above. Let $z^* = \min h$, and define a strategy s as follows. Fix $w \in W(h,g)$. For any $x \in h$, including z^*, let $s(x) = \gamma(x,w)$. Thus all nodes in h are undiscarded, $h \subseteq U(s)$.

For the nodes in $N(h) = \{x \in N \,|\, x \subset W(h)\}$, proceed as follows. Define s on $N(h,g)$ as the partial strategy given by Proposition 5.3. For every $x \in N(h)$ not in $W(h,g)$, specify s arbitrarily. Since $s(z^*) = \gamma(z^*,w)$, it follows that $N(h) \subseteq D(s)$.

For subhistories h' of h having no minimum, proceed as in the proof of coherence to obtain $N(h',g') \subseteq D(h)$ for all alternative continuations g' of h'. At all other nodes s is specified arbitrarily. As in the coherence part above, for any $x \notin h$ such that $x \notin N(h)$, we obtain $x \in D(s)$. This shows that $U(s) = h$ and, hence, $U(s)$ contains no play, i.e. s induces no outcome by Proposition 5.1. ■

That not every playable EDP is necessarily everywhere playable may be regarded as an argument in favor of focusing on the class of regular games trees, banning strange nodes. Thus, (A1) would lead to up-discrete game trees.

5.5 Uniqueness

This section is devoted to the class of EDPs for which the uniqueness criterion (A2) holds: "extensive forms." Those satisfy both an extra condition on the choice (information) structure and one on the tree. The condition on choices becomes redundant, though, in the class of EDPs that satisfy (A1).

Thus in this section we finally arrive at a formulation that is appropriately called an *extensive form*, rather than EDP. As will be seen shortly, the key point is that one of the conditions defining an EDP has to be strengthened in order to satisfy (A2). Without this strengthening strategies may induce multiple outcomes, which then players would not be able to evaluate.

5.5.1 *Examples with Multiple Outcomes*

That (A2) imposes further restrictions on the tree is illustrated by the third "hole in the middle" (Example 4.12), where a strategy induces multiple outcomes. This holds even in Example 4.11, where the tree is turned regular by removing $1/2$ from W and $\{1/2\}$ from N. The absence of a node $[1/4,3/4]$ still dictates that the strategy described in Example 4.12 induces multiple plays, $\{1/4\} = R_s(1/4)$ and $\{3/4\} = R_s(3/4)$.

The same argument applies in a variant of Example 4.11, when the (strange) nodes $[1/4,1/2)$ and $(1/2,3/4]$ are added. This tree is then weakly up-discrete and coherent, therefore, (everywhere) playable by Theorem 5.2. But a strategy that always "continues" still induces multiple plays. Thus, (A1) holds but (A2) fails.

The defect in these examples may appear to be the absence of a minimum for the chain $\uparrow y_t \cap \uparrow y'_t$, like in the "Twins" example (Example 2.15).[11] The problem is deeper, though. In the following example *every* chain of the form $\uparrow x \cap \uparrow y$ with $x,y \in N$ has a minimum,[12] and still (A2) fails. (Anderson 1984, Simon and Stinchcombe 1989, p. 1172, and Stinchcombe 1992, p. 239, provide similar examples.)

Example 5.4 (Differential game: outcome non-uniqueness) Reconsider the differential game (Example 4.9), when perfect information choices are $c_t(f,a) = \{g \in x_t(f) \,|\, g(t) = a\}$ for any $g \in x_t(f)$ with $g(t) = a \in A$ (see Example 4.14).

In this tree every chain of the form $\uparrow x \cap \uparrow y$ with $x,y \in N$ has a minimum. To see this, let $x_r(g)$ and $x_s(h)$ be any two nodes. If they are not disjoint, the conclusion is obvious (by Trivial Intersection). Hence, they can be assumed disjoint. The infimum $t^* = \{t \geq 0 | g(t) \neq h(t)\}$ exists, because the defining set of real numbers is bounded from below. It follows that $g(t) = h(t)$ for all $t < t^*$. Moreover, since the two nodes are disjoint, it follows that $t^* < \min\{r,s\}$. Define $x = x_{t^*}(g) = x_{t^*}(h)$. Obviously, $x \supseteq x_r(g)$ and $x \supseteq x_s(h)$. Suppose $x \supset x' \supseteq x_r(g) \cup x_s(h)$. It follows that $x' = x_t(g)$ for some $t \in (t^*, \min\{r,s\})$. But then there is $t' \in (t^*,t)$ such that $g(t') \neq h(t')$. It follows that $h \notin x'$, a contradiction which establishes that $x = \min \uparrow x_r(g) \cap \uparrow x_s(h)$.

But the tree of the differential game still allows strategies that induce multiple outcomes. More concretely, let $A = \{0,1\}$ and consider $\Pi(T)$. Let $\mathbf{1} \in W$ be constant 1. Note that $g \in x_t(f) \Leftrightarrow f \in x_t(g)$, for all t and all $f,g \in W$. Define $s \in S$ by $s(x_t(f)) = c_t(f,1)$ if $f \in x_t(\mathbf{1})$ and $s(x_t(f)) = c_t(f,0)$ otherwise. Consider the function $f_r \in W$ defined by $f_r(t) = 1$ for all $t \in [0,r]$ and $f_r(t) = 0$ for all $t > r$, for any $r > 0$. Then,

$$\begin{aligned} R_s(f_r) &= \cap_{g \in x_t(f_r)} s(x_t(g)) = [\cap_{t \leq r} c_t(\mathbf{1},1)] \cap [\cap_{r<t} \{h \in x_t(f_r) \,|\, h(t) = 0\}] \\ &= [\cap_{t \leq r} c_t(\mathbf{1},1)] \cap [\cap_{r<t} \{h \in x_t(f_r) \,|\, h(\tau) = 0, \forall \tau \in (r,t]\}] = \{f_r\} \end{aligned}$$

In other words, for every $r > 0$ the play $f_r \in W$ is a fixed point of R_s, so $s \in S$ induces a continuum of outcomes![13]

---

[11] There, $\uparrow \{0\} \cap \uparrow \{1\}$ has no minimum. A strategy (for $\Pi(T)$) that assigns to each move its immediate successor among the moves (and not an available singleton) induces two distinct plays.

[12] If the chain $\uparrow x \cap \uparrow y$ has a minimum for all $x,y \in N$, the tree is "well-joined," the dual of the concept of a "well-met" (pseudo)tree in Baur and Heindorf (1997). If the tree is well-joined, then $N \cup \{\emptyset\}$ is a *lattice* (see Chap. 3).

[13] We are grateful to Nicolas Vieille for suggesting this example.

That strategies induce unique outcomes (if they do at all) is not a property of the tree alone. This is illustrated by the next example, which uses precisely the same tree as the previous one.

Example 5.5 (Continuum of players, one at a time) Let again T be the tree of the differential game. But now consider an EDP (T,C) based on this tree as follows. There is a continuum of players, $I = \mathbb{R}_+$. Each player chooses an action $a \in A$. Player t is the only player, who plays at time t. All nodes at period t belong to the same information set, i.e., no player ever learns any previous decision. That is, the choices of player t are of the form $c_t(a) = \{f \in W | f(t) = a\}$. The set of nodes where player t is active, is the "slice" $X_t = \{x_t(f) | f \in W\}$. Further, each such slice is the only information set of the corresponding player, where all choices of the form $c_t(a)$ are available, $P(c_t(a)) = X_t$ for all $a \in A$.

This game is just the "cascading information sets" version of the one-shot game, where each player in $I = \mathbb{R}_+$ chooses an action $a \in A$. A strategy of player t is simply an action $a \in A$ and the outcomes (plays) of the game are simply functions $f : \mathbb{R}_+ \to A$. Hence, (A1) and (A2) follow immediately.

5.5.2 Extensive Forms

In order to understand the origin of multiplicity of outcomes, the following introduces a subclass of EDPs. The only difference between the definition of an EDP and the one of an *extensive form* below is the stronger property (EDP3′). For the reader's convenience we replicate the full definition.

Definition 5.2 An **extensive form** (EF) with player set I is a pair (T,C), where $T = (N, \supseteq)$ is a game tree with set of plays W and $C = (C_i)_{i \in I}$ is a system of collections C_i (the sets of players' choices) of nonempty unions of nodes (hence, sets of plays) for all $i \in I$, such that the four conditions below hold. Let $A_i(x) = \{c \in C_i | x \in P(c)\}$ denote the choices available to i at x for all $i \in I$, and $J(x) = \{i \in I | A_i(x) \neq \emptyset\}$ the set of decision makers at x which is required to be nonempty for all $x \in X$.

(EDP1) If $P(c) \cap P(c') \neq \emptyset$ and $c \neq c'$ then $P(c) = P(c')$ and $c \cap c' = \emptyset$, for all $c, c' \in C_i$ and all $i \in I$;

(EDP2) $x \cap \left[\bigcap_{i \in J(x)} c_i\right] \neq \emptyset$ for all $(c_i)_{i \in J(x)} \in \times_{i \in J(x)} A_i(x)$ and all $x \in X$;

(EDP3′) for all $y, y' \in N$, if $y \cap y' = \emptyset$, then there are $x \in X$, $i \in J(x)$, and $c, c' \in C_i$ such that $x \in P(c) \cap P(c')$, $y \subseteq x \cap c$, $y' \subseteq x \cap c'$, and $c \cap c' = \emptyset$;

(EDP4) if $x \supset y \in N$ then for every $i \in J(x)$ there is $c \in A_i(x)$ such that $y \subseteq c$, for all $x \in X$.

(EDP3′) is almost identical to (EDP3) except that the two choices $c, c' \in C_i$ have to be available simultaneously at a common move, i.e. $x \in P(c) \cap P(c')$. This stronger condition is sufficient for strategy combinations to induce unique outcomes, provided they induce outcomes at all (Theorem 5.4 below).

To begin with, a property slightly short of (A2) will be considered. Say that an EDP satisfies *outcome uniqueness* if, for all strategy combinations $s \in S$, $w \in R_s(w)$ and $w' \in R_s(w')$ imply $w = w'$. (Unlike (A2), outcome uniqueness does not require that $R_s(w) = \{w\}$.) In other words, outcome uniqueness fails if there are $s \in S$ and $w, w' \in W$ with $w \neq w'$ such that $w \in R_s(w)$ and $w' \in R_s(w')$.

Whether or not an EDP is everywhere playable is a matter of the underlying game tree. Outcome uniqueness is different, though. Condition (EDP3′) combines a restriction on the tree with one on choices. To identify the restriction on the tree the following property is introduced:

Definition 5.3 A game tree $T = (N, \supseteq)$ is **selective** if, for all $w, w' \in W$, that $w \neq w'$ implies $\exists x \in X$ such that $w, w' \in x$ and $\gamma(x, w) \neq \gamma(x, w')$.

Selectiveness is related to properties of game trees as introduced earlier, in particular, regularity. To see this (Proposition 5.4 below), an auxiliary step is needed.

Lemma 5.5 *For a game tree $T = (N, \supseteq)$, a history h has a continuation with a maximum if and only if there is $z \in N$ such that $h = \uparrow z \setminus \{z\}$.*

Proof To see the sufficiency part of this statement, assume that $h = \uparrow z \setminus \{z\}$ for some $z \in N$ and choose $y \in \downarrow z$. By the Hausdorff Maximality Principle there is a maximal chain g in $\downarrow z$ that contains $\uparrow y \cap \downarrow z$, and $z = \max g$ by construction. It remains to show that $h \cup g$ is a play. If not, there is $x \in N \setminus (h \cup g)$ such that $h \cup g \cup \{x\}$ is a chain. Then, by $z \in g$ either $z \subset x$ or $x \subseteq z$. The latter would imply $x \in g$ by $x \in \downarrow z$ and maximality of g in $\downarrow z$; therefore, $z \subset x$. But that means $x \in \uparrow z \setminus \{z\} = h$, again a contradiction. Thus, $h \cup g$ is maximal in N, hence, a play.

To see the necessity part, let g be a continuation with maximum $z \in g$ of a history h, so that $z \notin h$. If $x \in h$, then $z \subset x$, because $h \cup g$ is a chain and $x \subseteq z$ would imply $z \in h$, since h is a history. Hence, $h \subseteq \uparrow z \setminus \{z\}$. If $x \in \uparrow z \setminus \{z\}$, then $x \notin g$, because $z = \max g$. On the other hand, $z \subset x \in N$ implies $x \in h \cup g$, because $h \cup g$ is a maximal chain in N, so that $x \in h$ follows. Hence, $\uparrow z \setminus \{z\} \subseteq h$ combines with the previous inclusion to $h = \uparrow z \setminus \{z\}$. ■

Proposition 5.4 *(a) If a game tree $T = (N, \supseteq)$ is selective, then it is regular.*

(b) If a game tree $T = (N, \supseteq)$ is regular and every history h has a continuation g with a maximum $z \in g$, then T is selective.

Proof (a) We first claim that, if a game tree $(N, \supseteq)$ is selective, every chain of the form $\uparrow x \cap \uparrow y$ for $x, y \in N$ has a minimum.[14] Let $x, y \in N$ be such that $x \cap y = \emptyset$ (else the result is obvious) and choose $w \in x$ and $w' \in y$. Because the tree is selective, there is $z \in X$ with $w, w' \in z$ such that $\gamma(z, w) \cap \gamma(z, w') = \emptyset$ by Proposition 4.3(a). Let $z' \in \uparrow x \cap \uparrow y$. Since $\{w, w'\} \subseteq z \cap z'$, either $z' \subset z$ or $z \subseteq z'$ by Trivial Intersection (2.4). If $z' \subset z$, that $w, w' \in z' \in \downarrow z \setminus \{z\}$ would imply $z' \subseteq \gamma(z, w) \cap \gamma(z, w')$ in contradiction

[14] In other words, every selective game tree is well-joined (see Footnote 12).

to $\gamma(z,w) \cap \gamma(z,w') = \emptyset$. Therefore, $z \subseteq z'$ for all $z' \in \uparrow x \cap \uparrow y$ implies together with $z \in \uparrow x \cap \uparrow y$ that $z = \min \uparrow x \cap \uparrow y$.

Suppose $x \in S(N)$ is a strange node. Since $x \neq \inf \uparrow x \setminus \{x\}$, there is a node x' which is a lower bound of $\uparrow x \setminus \{x\}$ but such that $x' \not\subseteq x$. If $x \subset x'$, then $x' = \min \uparrow x \setminus \{x\}$, a contradiction with $x \in S(N)$. By Trivial Intersection this implies $x \cap x' = \emptyset$. Since x' is a lower bound of $\uparrow x \setminus \{x\}$, it follows that $\uparrow x \setminus \{x\} \subseteq \uparrow x \cap \uparrow x'$. By the claim above there exists $z = \min \uparrow x \cap \uparrow x'$. In particular, $z \supseteq x'$ and, therefore, $z \in \uparrow x \setminus \{x\}$. It follows that $\uparrow x \cap \uparrow x' \subseteq \uparrow x \setminus \{x\}$ and, hence, we have equality. But then $z = \min \uparrow x \setminus \{x\}$, in contradiction to $x \in S(N)$.

(b) Assume that T is regular and every history has a continuation with a maximum. Let $w, w' \in W$ be such that $w \neq w'$ and $h = \uparrow \{w\} \cap \uparrow \{w'\}$, so that h is a history. Suppose that h does not have a minimum. By Lemma 5.5 there is $z \in N$ such that $h = \uparrow z \setminus \{z\}$. If $z = \bigcap_{x \in h} x$, then $w, w' \in z$ imply $z \in h$, a contradiction. Therefore $z \subset \bigcap_{x \in h} x$. But then, that h has no minimum and Lemma 4.2(c) imply that $z \in S(N)$, a contradiction.

Therefore, regularity of the tree implies that $h = \uparrow \{w\} \cap \uparrow \{w'\}$ has a minimum $x \in h$. If $\gamma(x,w) \cap \gamma(x,w')$ were nonempty, $\gamma(x,w) = \gamma(x,w')$ would hold by Proposition 4.3(a). But that would imply that there is a node $y \in \downarrow x \setminus \{x\}$ with $w, w' \in y$, so that $y \in h$ by the construction of h. As this contradicts $x = \min h$, $\gamma(x,w) \cap \gamma(x,w') = \emptyset$ follows. ■

The converse of Proposition 5.4(a) is false. This follows from Example 4.11.

Selectiveness is purely a condition on the tree. Proposition 5.4(a) shows that it is a strong one, though. Part (b) gives a sufficient condition for it that is reminiscent of the conditions for playability. In fact, it implies coherence, but not weak up-discreteness.

Selectiveness and coherence are independent properties, however: The third variant of the "hole in the middle" (Example 4.12) is coherent, but not selective. On the other hand, replace the terminal node $\{\infty\}$ of an infinite centipede (Example 4.13) by an inverse infinite centipede (Example 4.15) without root. This yields a selective tree that is not coherent, because all plays that "continue" through the (first) infinite centipede "continue" into the (second) inverse infinite centipede that has no maximum.

Example 5.6 (Differential game and Selectiveness) Reconsider the tree of the differential game (e.g. Examples 4.9 and 4.14). This tree is, in fact, selective. To see this, let $f, g \in W$ with $f \neq g$. If $f(0) \neq g(0)$, then $\gamma(W,f) \neq \gamma(W,g)$. If $f(0) = g(0)$, let $t^* = \sup\{\tau > 0 | f|_{[0,\tau)} = g|_{[0,\tau)}\}$. Then $x_{t^*}(f) = x_{t^*}(g)$ and $\gamma(x_{t^*}(f), f) \neq \gamma(x_{t^*}(f), g)$.

This example shows that selectiveness can at best be a necessary condition for outcome uniqueness, but not a sufficient one. For, even though the tree of the differential game is selective, outcome uniqueness may fail, as shown in Example 5.4. The following is a characterization for when an EDP is an EF in terms of one condition on the tree and another on choices.

Proposition 5.5 *The EDP (T,C) is an EF if and only if T is selective and*

(EDP2′) $x \cap \left[\cap_{i \in J(x)} c_i\right] = \gamma(x,w)$ *for some* $w \in x \cap \left[\cap_{i \in J(x)} c_i\right]$*, for all* $(c_i)_{i \in J(x)} \in \times_{i \in J(x)} A_i(x)$ *and for all* $x \in X$.[15]

Proof "if": Assume selectiveness and (EDP2′). Let $y, y' \in N$ be such that $y \cap y' = \emptyset$ and $w \in y$ and $w' \in y'$. By selectiveness, there is $x \in X$with $w, w' \in x$ such that $\gamma(x,w) \cap \gamma(x,w') = \emptyset$. For each $i \in J(x)$ choose $c_i, c_i' \in A_i(x)$ such that $y \subseteq c_i$ and $y' \subseteq c_i'$ which is possible by (EDP4), since $w, w' \in x$, $w \in y$, $w' \notin y$, $w' \in y'$, and $w \notin y'$ imply $y, y' \in \downarrow x \setminus \{x\}$ by Trivial Intersection (2.4). By the hypothesis and Proposition 4.3(b) $\gamma(x,w) = x \cap \left[\cap_{i \in J(x)} c_i\right]$ and $\gamma(x,w') = x \cap \left[\cap_{i \in J(x)} c_i'\right]$, because $\gamma(x,w) \subseteq x \cap \left[\cap_{i \in J(x)} c_i\right]$ and $\gamma(x,w') \subseteq x \cap \left[\cap_{i \in J(x)} c_i'\right]$ by Lemma 5.2. In particular, for some $i \in J(x)$ it must be the case that $y \subseteq \gamma(x,w) \subseteq c_i$, $y' \subseteq \gamma(x,w') \subseteq c_i'$, and $c_i \cap c_i' = \emptyset$. For, if $c_i \cap c_i' \neq \emptyset$ for all $i \in J(x)$, then (EDP1) implies that $c_i = c_i'$ for all $i \in J(x)$ in contradiction to what has been shown, as $x \in P(c_i) \cap P(c_i')$ by construction.

"only if": Suppose that (EDP3′) holds. Let $w, w' \in W$ be such that $w \neq w'$. By Irreducibilty (2.12) there are $y, y' \in N$ such that $w \in y$, $w' \in y'$, and $y \cap y' = \emptyset$. By (EDP3′) there are $i \in I$ and $c, c' \in C_i$ such that $y \subseteq c$, $y' \subseteq c'$, $c \cap c' = \emptyset$, and $x \in P(c) \cap P(c') \neq \emptyset$, say. By Lemma 5.2 $\gamma(x,w) \subseteq x \cap c$ and $\gamma(x,w') \subseteq x \cap c'$. Since $c \cap c' = \emptyset$, it follows that $\gamma(x,w) \cap \gamma(x,w') = \emptyset$. Because $x \in P(c) \cap P(c')$, it follows that $\uparrow x = \uparrow y \setminus \downarrow c = \uparrow y' \setminus \downarrow c'$ by (4.1), in particular, $w, w' \in x$. Thus, the tree is selective.

Let $x \in X$ and $(c_i)_{i \in J(x)} \in \times_{i \in J(x)} A_i(x)$. By (EDP2) there is $w \in x \cap \left[\cap_{i \in J(x)} c_i\right]$ and by Lemma 5.2, $\gamma(x,w) \subseteq x \cap c_i$ for all $i \in J(x)$. Choose $w' \in x \cap \left[\cap_{i \in J(x)} c_i\right] \setminus \gamma(x,w)$. By Irreducibility (2.12) there are $y, y' \in N$ such that $w \in y$, $w' \in y'$, and $y \cap y' = \emptyset$, so that $y, y' \in \downarrow x \setminus \{x\}$ by Trivial Intersection (2.4). By (EDP3′) there are $i \in I$ and $c, c' \in C_i$ such that $y \subseteq c$, $y' \subseteq c'$, $c \cap c' = \emptyset$, and $x' \in P(c) \cap P(c')$, say. Therefore, $w, w' \in x' \cap x$ implies either $x \subseteq x'$ or $x' \subset x$ by Trivial Intersection. In the latter case $w \in x'$ implies $x' \subseteq \gamma(x,w)$ in contradiction to $w' \in x'$. Therefore, $x \subseteq x'$. If $x \subset x'$ would holds, then $y \cup y' \subseteq x \subseteq \gamma(x',w) \subseteq x' \cap c$ (by Lemma 5.2) would contradict $c \cap c' = \emptyset$. Thus, $x' = x$. But then $c' \cap c_i = \emptyset$ implies $w' \notin c_i$, a contradiction. Hence, $x \cap \left[\cap_{i \in J(x)} c_i\right] \setminus \gamma(x,w) = \emptyset$ yields (EDP2′). ■

This disentangles what (A2) requires on the tree and what on the choices: respectively selectiveness and (EDP2′). For instance, the differential game (Examples 4.9 and 4.14) has been shown to be selective in Example 5.6. Recalling that $\gamma(x_t(f), g) = \left\{h \in x_t(f) \,|\, \exists \tau > t : h|_{[0,\tau)} = g|_{[0,\tau)}\right\}$, it is immediate that this EDP fails (EDP2′), hence, by Proposition 5.5, also (EDP3′). Since outcome uniqueness is satisfied for the one-shot version of the differential game from Example 5.5, this shows that an EF is sufficient, but not always necessary for (A2).

[15] In (EDP2′) the equality may be replaced by weak inclusion $\subseteq$, as Lemma 5.2 and Proposition 4.3(b) imply the reverse inclusion.

Another example where (EDP2$'$) fails, even though the tree is selective, is the "hole in the middle" (Example 4.10), with $[1/4,3/4] \in N$. Make this into an EDP by assigning all nodes except the root plus $[1/4,1/2) \cup (1/2,3/4]$ as choices to a single player. The singleton $\{1/2\}$ is a choice available at $\bar{x} = [1/4,3/4] \in X$, so the decision at $\bar{x}$ is nontrivial. Since $\gamma(\bar{x},1/4) = [1/4,1/2)$ and $\gamma(\bar{x},3/4) = (1/2,3/4]$, the tree is selective, but the construction of choices violates (EDP2$'$).

The next result states that there is indeed a sense in which selectiveness is necessary for (A2).

Proposition 5.6 *Consider a game tree $T = (N, \supseteq)$ with available choices. If T is not selective, then the perfect information EDP $\Pi(T)$ fails outcome uniqueness.*

Proof By Proposition 4.3(a), if the tree is not selective, then there are $w, w' \in W$, $w \neq w'$, such that whenever $w, w' \in x \in X$ then $\gamma(x,w) \cap \gamma(x,w') \neq \emptyset$. Consider $\Pi(T)$ and suppose $w, w' \in x \in X$. By hypothesis $\gamma(x,w) \cap \gamma(x,w') \neq \emptyset$. Then, Proposition 4.3(a) implies $\gamma(x,w) = \gamma(x,w')$. Choose $s \in S$ such that $s(x) = \gamma(x,w) = \gamma(x,w')$ if $w, w' \in x$; $s(x) = \gamma(x,w)$ if $w \in x$ but $w' \notin x$; $s(x) = \gamma(x,w')$ if $w' \in x$ but $w \notin x$; and $s(x)$ arbitrary otherwise. Then, $w \in s(x)$ whenever $w \in x \in X$ and analogously for w'. Hence $w \in R_s(w)$ and $w' \in R_s(w')$. ■

The next corollary follows by combining the "only if"-part of Proposition 5.5 with Proposition 5.4(a).

Corollary 5.3 *The tree of an EF is selective and, hence, regular.*

Going from an EDP to an EF has further implications that concern playability. First, it can be shown that for an EF the set of undiscarded nodes for any strategy combination, $U(s)$, becomes a chain (see Remark 5.1 below). Second, if every strategy combination for an arbitrary EF with tree T induces outcomes after a history h, then so does every strategy for the problem $\Pi(T)$. This says that if an EF is playable resp. everywhere playable, then *every* EF with the same tree is playable resp. everywhere playable. This gives a converse to Proposition 5.2 for an EF.

Proposition 5.7 *Fix a history h for a game tree $T = (N, \supseteq)$. If for an arbitrary EF (T,C) every strategy combination induces outcomes after h, then for the perfect information problem $\Pi(T)$ every strategy combination induces outcomes after h.*

Proof Suppose for some history h there is a strategy s' for $\Pi(T)$ that does not induce an outcome after h. Let h' be a maximal chain in $U^h(s')$ (the undiscarded nodes after h) and $W(h') = \cap_{x \in h'} x$. Fix a play w as follows. If h' has a minimum z (which then cannot be a terminal node by hypothesis), let $w \in s'(z)$. Otherwise, let $w \in W(h')$. If $U^h(s') = \emptyset$ (and hence there is no such chain h'), fix an arbitrary $w \in W(h)$.

Consider now an arbitrary EF (T,C). By Proposition 5.5 T is selective. We construct a strategy combination in (T,C) as follows. For all $y \in N$ such that $w \in y \subseteq W(h)$ and $i \in J(y)$ choose $s_i(y)$ such that $y \cap \left[\cap_{i \in J(y)} s_i(y)\right] = s'(y)$. This is possible by (EDP2$'$) and (EDP4). If $y \in h'$, then of course $s'(y) = \gamma(y,w)$. For any other node specify the strategy combination s arbitrarily. Notice that we determine s

only along a play, which is possible by (EDP4). We claim that $w \notin R_s^h(w)$. For, if it were, by the construction of s, we would obtain $w \in R_{s'}^h(w)$ for $\Pi(T)$, a contradiction.

Let $w' \in W(h)$ with $w \neq w'$. By selectiveness there exists $x \in X$ such that $w, w' \in x$ and $\gamma(x,w) \cap \gamma(x,w') = \emptyset$ (by Proposition 4.3(a)). Notice that necessarily $x \subseteq W(h)$. There are two possibilities. If $x \notin h'$, $x \in D^h(s')$ for $\Pi(T)$, which implies by construction that $x \in D^h(s)$ for (T,C). Hence, $w' \notin R_s^h(w')$. If $x \in h'$, then $s'(x) = \gamma(x,w) \neq \gamma(x,w')$. Since $s'(x) = x \cap \left[\cap_{i \in J(x)} s_i(x)\right]$, it follows that $w' \notin R_s^h(w')$. Since $w' \in W(h)$ was arbitrary, we conclude that s does not induce an outcome in (T,C). ■

Remark 5.1 For an EF, i.e. with (EDP3′) instead of (EDP3), the set of undiscarded nodes for a given strategy combination, $U(s)$, is a chain. For, let $y, y' \in U(s)$ and assume that $y \cap y' = \emptyset$. Choose $w \in y$ and $w' \in y'$ and let $x \in X$ be such that $w, w' \in x$ and $\gamma(x,w) \neq \gamma(x,w')$ which exists, because the tree is selective by Corollary 5.3. By (EDP4), Lemma 5.2, and Proposition 5.5 there are choice combinations $(c_j)_{j \in J(x)}, \left(c_j'\right)_{j \in J(x)} \in \times_{j \in J(x)} A_j(x)$ such that

$$y \subseteq \gamma(x,w) = x \cap \left[\cap_{j \in J(x)} c_j\right] \text{ and } y' \subseteq \gamma(x,w') = x \cap \left[\cap_{j \in J(x)} c_j'\right]$$

with $c_i \neq c_i'$ for at least one $i \in J(x)$ (the latter by the choice of x and selectiveness). By (EDP1), $c_i \cap c_i' = \emptyset$. If $c_i = s_i(x)$ (or $c_i' = s_i(x)$), then $y' \in D(s)$ (or $y \in D(s)$), in contradiction to the assumption. Hence, if $y, y' \in U(s)$, then either $y \subset y'$ or $y' \subseteq y$ by Trivial Intersection (2.4). In other words, $U(s)$ is a chain.

5.5.3 A Uniqueness Result

Up to this point it has been studied which restrictions an EF—replacing (EDP3) by (EDP3′)—impose on the tree and the choices. It will now be shown that an EF is indeed sufficient for (A2).

Theorem 5.4 *Consider an EF as in Definition 5.2 and fix a pure strategy combination $s \in S$. If $w \in R_s(w)$, then (a) $R_s(w) = \{w\}$, and (b) if $w' \in R_s(w')$ then $w' = w$.*

Proof First, we claim that if $w \in R_s(w)$ and $w' \in W \setminus \{w\}$, then there are $x \in X$ and $i \in J(x)$ such that $w \in s_i(x)$ and $w' \notin s_i(x)$. For, by Irreducibility (2.12) there are $y, y' \in N$ such that $w \in y$, $w' \in y'$, and $y \cap y' = \emptyset$. By (EDP3′) there are $i \in I$ and $c, c' \in C_i$ such that $y \subseteq c$, $y' \subseteq c'$, $c \cap c' = \emptyset$, and there is $x \in P(c) \cap P(c')$. Because $w \in R_s(w)$, we have $s_i(x) = c$ and thus $w' \in y' \subseteq c' \neq s_i(x)$ implies that $w' \notin s_i(x)$.

(a) Let $s \in S$, assume that $w \in R_s(w)$, and consider any $w' \in R_s(w)$. If $w' \in W \setminus \{w\}$, then by the claim above there are $x \in X$ and $i \in J(x)$ such that $w \in s_i(x)$ and $w' \notin s_i(x)$. But this contradicts $w' \in R_s(w)$. Hence, $w' = w$.

(b) Let again $s \in S$ and assume $w \in R_s(w)$. Consider any $w' \in W$ such that $w' \in R_s(w')$. If w' were not equal to w, then, again by the claim, there would be $x \in N$ and $i \in J(x)$ such that $w \in s_i(x)$ and $w' \notin s_i(x)$. Since the latter would contradict $w' \in R_s(w')$, it follows that $w' = w$. ■

This result shows that the strengthening of (EDP3) to (EDP3′) is sufficient for (A2). Recall, however, that (EDP3′) is a combination of a condition on the tree, namely selectiveness, and one on choices, namely (EDP2′). The condition on choices, (EDP2′), becomes redundant, though, when the tree is weakly up-discrete.

Proposition 5.8 *An EDP (T,C) with a weakly up-discrete tree $T = (N, \supseteq)$ is an EF if and only if T is selective.*

Proof Let $x \in X$ and $(c_i)_{i\in J(x)} \in \times_{i\in J(x)} A_i(x)$. By (EDP2) there is $w \in x \cap \left[\bigcap_{i\in J(x)} c_i\right]$. By Lemma 5.2, $\gamma(x,w) \subseteq x \cap \left[\bigcap_{i\in J(x)} c_i\right]$ and, as T is weakly up-discrete, $y \equiv \gamma(x,w) \in N$ by Lemma 5.3(c). Suppose there is $w' \in x \cap \left[\bigcap_{i\in J(x)} c_i\right] \setminus \gamma(x,w)$. Then, $y' \equiv \gamma(x,w') \in N$ by Lemma 5.3(c), $y' \subseteq x \cap \left[\bigcap_{i\in J(x)} c_i\right]$ by Lemma 5.2, and $y \cap y' = \emptyset$ by Proposition 4.3(a). By (EDP3) there are $i \in I$ and $c, c' \in C_i$ such that $y \subseteq c$, $y' \subseteq c'$, and $c \cap c' = \emptyset$. Thus, $y' \subseteq c' \cap x$ implies that $x \setminus c \neq \emptyset$ and from $y \subseteq c \cap x$ it follows that $x \setminus c' \neq \emptyset$, i.e. $x \notin \downarrow c$ and $x \notin \downarrow c'$. Thus, $\uparrow x \subseteq \uparrow y \setminus \downarrow c$ and $\uparrow x \subseteq \uparrow y' \setminus \downarrow c'$. By construction of y and y' these inclusions are equalities, though, so that $\uparrow x = \uparrow y \setminus \downarrow c$ and $\uparrow x = \uparrow y' \setminus \downarrow c'$, i.e. $x \in P(c) \cap P(c')$. But then $i \in J(x)$ and $c = c_i \supseteq y' = \gamma(x,w')$ contradicts $c \cap c' = \emptyset$. Thus, $x \cap \left[\bigcap_{i\in J(x)} c_i\right] \setminus \gamma(x,w) = \emptyset$ yields, together with Lemma 5.2, $\gamma(x,w) = x \cap \left[\bigcap_{i\in J(x)} c_i\right]$. As x and $(c_i)_{i\in J(x)}$ were arbitrary, (EDP2′) holds and the statement follows from Proposition 5.5. ■

In the class of weakly up-discrete trees (A2) is, therefore, characterized by selective trees. Note that weak up-discreteness and selectiveness are independent properties. The "hole in the middle," Example 4.12, has a weakly up-discrete tree that is not selective: There is no $x \in X$ such that $1/4, 3/4 \in x$ and $\gamma(x, 1/4) \cap \gamma(x, 3/4) = \emptyset$, because a node $[1/4, 3/4]$ is absent from this tree. For, $1/4, 3/4 \in x$ implies $x \in \{x_t\}_{t=1}^{\infty}$, yet $\gamma(x_t, 1/4) = \gamma(x_t, 3/4) = x_{t+1}$ for all $t = 1, 2, \ldots$ But the first "hole in the middle," Example 4.10, has a selective tree, because $[1/4, 3/4]$ is a node. It is not weakly up-discrete, though, because nodes of the form $[1/4, 1/2)$ and $(1/2, 3/4]$ are still missing, i.e., the chains $\{y_t\}_{t=1}^{\infty}$ and $\{y'_t\}_{t=1}^{\infty}$ have no maxima.

5.6 A Joint Characterization

Combining Theorems 5.2, 5.3, and 5.4 yields a characterization of EDPs that satisfy the desiderata (A1) and (A2) purely in terms of easily verifiable properties of the tree.[16]

[16]Note that Theorem 5.5 implies that the "Lexicographic Centipede" (Example 5.3) fails (A2).

Theorem 5.5 *An EF (T, C) satisfies (A1) and (A2) if and only if the (rooted) game tree $T = (N, \supseteq)$ is regular, weakly up-discrete, and coherent.*

Proof "if": By weak up-discreteness and Corollary 4.1 the tree has available choices. As the tree is weakly up-discrete and coherent by hypothesis, Theorem 5.2 implies that every strategy induces an outcome after every history. In particular, every strategy induces an outcome after the null history that consists only of the root. Hence, (A1) holds true.

To see (A2), observe first that every history without minimum has a continuation with a maximum by coherence. Second, for any history with a minimum all continuations have maxima by Lemma 5.3(b) and weak up-discreteness. Then regularity and Proposition 5.4(b) imply that the tree is selective. Consequently, the EDP (T, C) is an EF by (the "if" part of) Proposition 5.8. It follows from Theorem 5.4 that (A2) holds.

"only if": By Proposition 5.5 the tree T is selective. Hence, it is regular by Proposition 5.4(a). By Proposition 5.7 the game $\Pi(T)$ satisfies (A1). By Corollary 5.1 every EDP defined on T satisfies (A1). Theorem 5.3 then implies that T is up-discrete, hence, weakly up-discrete and coherent by Corollary 5.2. ■

As shown by the examples in Sect. 4.4.1, the three characterizing properties are independent. By Corollary 5.2 a regular tree is up-discrete if and only if it is weakly up-discrete and coherent. Combining this with Theorems 5.3 and 5.5 yields:

Corollary 5.4 *(a) If an EF satisfies (A1) and (A2), then so does every EF with the same tree.*

(b) An EF satisfies (A1) and (A2) if and only if its tree is regular and up-discrete. Furthermore, the EF is then everywhere playable.

5.7 Note: Games in Continuous Time

Examples 5.2 and 5.4 have shown that straightforward specifications of games in continuous time will not be playable. Well-specified profiles of pure strategies might fail to induce outcomes, and, if they do, they might induce a continuum of outcomes rather than a unique one. These problems have been previously pointed out by Anderson (1984), Simon and Stinchcombe (1989), and Stinchcombe (1992), among others. As shown in this chapter (e.g., Example 5.1), they are not exclusive to continuous-time settings. This chapter has now clarified the underlying reasons for these problems, by providing a characterization of the set of extensive forms where every profile of pure strategies generates a unique outcome (and hence a normal-form game can be defined). This characterization can be argued to describe the domain of game theory, because games outside the characterized set cannot be "solved" in any sense of the word. Unsurprisingly, continuous-time games are outside this domain; the key difficulty is that they fail up-discreteness.

There are, however, many economic applications which rely on continuous time. One reason for the interest of economic theory in continuous time is simply the analytical convenience of employing techniques from differential calculus along the time dimension. Other reasons are conceptual. Discrete time sometimes creates artificial phenomena which vanish in continuous time; and it is their vanishing in the latter framework which proves their artificiality in the former. However, the problems pointed out above create serious difficulties with the interpretation of continuous-time applications. For instance, if certain strategy combinations fall out of the framework by virtue of creating outcome existence or uniqueness problems, the meaning of any equilibrium concept becomes questionable, since some deviations might be excluded for merely technical reasons, and not the self-interest of the deviator. Further, if a proper extensive form game cannot be specified for a continuous-time model, notions of "time consistency" cannot rely on subgame perfection or other equilibrium refinements based on backwards induction, since in the absence of a properly formulated extensive form, it is not possible to determine the full collection of subgames capturing the strategic, intertemporal structure of the problem.

The previous literature has dealt with the difficulties related to outcome existence and uniqueness in a variety of ways. The most basic approach is to simply restrict players to suitable classes of "admissible" strategies such that, when profiles thereof are considered, outcome existence and uniqueness is known to obtain. In the case of differential games (Dockner et al. 2000; Friedman 1994), this approach often leads to the specification of a normal-form game, where strategies are required to be e.g. differentiable or integrable functions of some state variable. A different approach is to explicitly impose specific conditions on both strategies and outcomes and then prove that under these conditions every strategy profile induces a unique outcome (Anderson 1984; Bergin and MacLeod 1993; Simon and Stinchcombe 1989; Stinchcombe 1992). This approach was most effectively described by Stinchcombe (1992), who set out to identify a *maximal* set of strategies for a continuous-time game such that every strategy profile induces a unique outcome. The result incorporates elements of a framework introduced by Anderson (1984) and also studied by Bergin and MacLeod (1993), and rests on the basic condition that, given a strategy, it is always possible to identify the player's *next* move.

From a game-theoretic point of view, however, restricting the strategy set is an unsatisfactory approach. On the one hand, since certain strategies are excluded on purely technical grounds, we face the problems with the interpretation of equilibria and time consistency pointed out above. On the other hand, there is a more fundamental, conceptual problem. An extensive form game incorporates a complete description of the possible choices of every player at every decision node. A behavioral strategy is merely a collection of possible "local" decisions at the nodes, and any possible combination thereof is a feasible behavioral strategy. Once the game is specified, there can be no further freedom in the specification of the possible local decisions, since those have already been fixed in the extensive form.

The set of possible behavioral strategies is thus automatically specified once the extensive form is given.

This is not to say that the specification of appropriately restricted strategy sets is an inherently flawed approach. However, one must remember that such restrictions are imposed on technical grounds (e.g. to ensure the existence and uniqueness of outcomes), and hence lack a decision-theoretic justification. In order to understand whether the resulting strategy spaces do reflect an extensive-form structure, it is necessary to establish that they are equivalent to the unrestricted strategy spaces of a well-defined extensive form game. For, if this did not hold, the implied formal object would merely be a (constrained) normal form game with no meaningful time dimension.

Such an equivalence result has recently been rigorously established in Alós-Ferrer and Kern (2015), which we will briefly (and informally) describe here. That article specifies the *Action-Reaction Framework*, which is a class of extensive forms (in the sense of Definition 5.2) that can still be considered well-defined "continuous-time games" but are fully solvable extensive forms, i.e. every strategy profile induces a unique outcome, *without any restriction on the set of strategies.*

Further, Alós-Ferrer and Kern (2015) show that the unrestricted behavioral strategies of the extensive form game are equivalent to a class of strategies which is restricted through specific conditions akin to the ones typically used in the literature. This equivalence hence shows that it is possible to impose strategy restrictions while maintaining an extensive form structure. At the same time, the approach in Alós-Ferrer and Kern (2015) shows that it is possible to use the requirement that strategies are derived from a well-defined extensive form as a criterion to judge the coherence of a restricted class of strategies.

The construction in Alós-Ferrer and Kern (2015) starts by specifying the set of decision paths from a continuous time axis $\mathbb{R}_+$ to a set of action profiles A. Those decision paths capture three properties. First, a player's action revision cannot occur arbitrarily close to a previous action revision. Second, a player's action revision cannot take the form of an instantaneous change which is then abandoned. Third, a player can change action by instant reaction (a shift immediately after t but not at t) only if some other player has changed discontinuously at t (a sudden action revision). Plays are then constructed as decision paths incorporating a record of "inertia times," the amount of time a player sticks to his current action at each t. Finally, the tree is constructed by specifying nodes of three types.

The first type are "action nodes" at which a player might make the decision to initiate a change of action. Such an action node $x_t(w)$ contains the collection of all choices of actions and of inertia times, up to time t, analogously to the nodes of the differential game example (Examples 5.2 and 5.4). The second type are "reaction nodes" which model the possibility of players to react to a change of action initiated by another player. A reaction node contains the collection of all decisions up to time t as well as information about the actions chosen at time t. This allows for players reacting to instant action changes that occurred at time t. The third type are "inertia nodes" where both the actions at t (individual action-change initiations) and the immediate reactions to them have already been decided, but the inertia times still have not.

Alós-Ferrer and Kern (2015) show that this tree is a regular and up-discrete game tree. A natural specification of choices then defines an EF, and hence Theorem 5.5 applies. The key intuition is that the necessary restrictions on players' actions and reactions are built directly into the set of plays, and not imposed *ex post* by restricting the set of strategies. Once this is done in a systematic way, it turns out that the corresponding tree fulfills up-discreteness, and hence the nonexistence and nonuniqueness problems that plague continuous-time formulations disappear.

5.8 Summary

The concept of a non-cooperative game—that is, a game with complete rules—extends well beyond the confines of finite games. Infinitely repeated games, stochastic games (Shapley 1953), and even differential games can be rigorously defined in extensive form. This step, that corresponds to an EDP, verifies that the rules of these games are complete.

Whether the rules of such games are also consistent is a different matter, though. It may well be that at each "when" it is fully specified "who can do what," but at the same time a global specification of such instructions—that is, a strategy combination—may not yield an outcome at all or multiple outcomes. In this chapter we characterized extensive form games that satisfy three global criteria: (A0) Every outcome/play can be induced by some strategy combination; (A1) every strategy combination does induce some outcome/play; (A2) the outcome/play induced by a strategy combination is unique.

An important aspect of the motivation behind these criteria is that (A1) and (A2) are necessary and sufficient for the existence of a function $\phi : S \to W$ that assigns to each strategy combination the play that it induces. This, in turn, is equivalent to the existence of a normal form representation of the game. Once a normal form representation is established, equilibrium concepts (starting with Nash equilibrium) can be defined in a straightforward way.

The first step is to show that (A0) comes for free: It is true for any EDP (Theorem 5.1). Criterion (A1) depends on the tree, in particular if this property is meant to hold after any history. Theorem 5.2 shows that an EDP is everywhere playable if and only if its tree is weakly up-discrete and coherent. Asking for (A2), the uniqueness criterion, takes us to the third milestone in the development: an *extensive form* (EF). The latter imposes an extra restriction on the choice system, (EDP3$'$), instead of (EDP3). In fact, this additional restriction breaks into one condition on the tree, selectiveness, and one purely on the choice system, (EDP2$'$), according to Proposition 5.5. In the class of weakly up-discrete game trees, the second, (EDP2$'$), becomes redundant, though (Proposition 5.8). An EF is finally shown to satisfy (A2) in Theorem 5.4. As a consequence, it is shown that EFs satisfying (A1) and (A2) are fully characterized by three properties of the underlying game tree: regularity, coherence, and weak up-discreteness.

The characterization result allows us to draw a dividing line between those games that can be defined on game trees with the above three properties, and those that cannot. As a rule, almost all games in the literature turn up on the safe side of this line, inclusive of the transfinite cheap-talk game by Aumann and Hart (2003). The only exception concerns differential games. Though the tree of a differential game is regular, it fails to be up-discrete. As a consequence, differential games allow for strategy combinations that do not induce any outcome/play at all—differential games are not "playable." Furthermore, they allow for strategy combinations that induce continua of outcomes/plays—differential games are not "extensive forms."

This, of course, raises the issue of how to interpret existing applications of differential games. One possible reaction is to give up their dynamic interpretation and view them as "one-shot" normal form games. In contrast to extensive form games, in a normal form game strategies are primitives. Therefore, one is free to restrict the strategy sets to those strategies that induce unique outcomes. Another approach is to restrict the strategy sets, but this creates fundamental difficulties with the interpretation of the resulting object. A third one, recently put forward by Alós-Ferrer and Kern (2015), is to incorporate natural restrictions on the space of plays, allowing for the definition of a continuous-time game without imposing any *ex post* constrain on the space of strategies. As shown by Alós-Ferrer and Kern (2015), this can be done systematically in such a way that the resulting tree is up-discrete and the associated extensive decision problem is an EF, hence solving the problems of nonexistence and nonuniqueness of outcomes.

References

Alós-Ferrer C (2002) Individual randomness with a continuum of agents. Mimeo, Vienna

Alós-Ferrer C, Kern J (2015) Repeated games in continuous time as extensive form games. J Math Econ 61:34–57

Alós-Ferrer C, Ritzberger K (2008) Trees and extensive forms. J Econ Theory 43(1):216–250

Alós-Ferrer C, Kern J, Ritzberger K (2011) Comment on 'Trees and extensive forms'. J Econ Theory 146(5):2165–2168

Anderson RM (1984) Quick-response equilibrium. Working papers in economic theory and econometrics, #IP-323, C.R.M, University of California at Berkeley

Aumann RJ (1961) Borel structures for function spaces. Ill J Math 5:614–630

Aumann RJ (1964) Mixed and behavior strategies in infinites extensive games. In: Advances in game theory. Annals of mathematics study, vol 52. Princeton University Press, Princeton, pp 627–650

Aumann RJ (1987) Game theory. In: Eatwell J, Milgate M, Newman P (eds) The new Palgrave, a dictionary of economics, vol 2. Macmillan, London, pp 460–482

Aumann RJ, Hart S (2003) Long cheap talk. Econometrica 71:1619–1660

Baur L, Heindorf L (1997) Initial chain algebras on pseudotrees. Order 14:21–38

Bergin J, MacLeod WB (1993) Continuous time repeated games. Int Econ Rev 34(1):21–37

Dockner E, Jörgensen S, Long NV, Sorger G (2000) Differential games in economics and management science. Cambridge University Press, Cambridge

Feldman M, Gilles C (1985) An expository note on individual risk without aggregate uncertainty. J Econ Theory 35:26–32

Friedman A (1994) Differential games. In: Aumann RJ, Hart S (eds) Handbook of game theory, vol 2. Elsevier, Amsterdam/Princeton, pp 781–799

Hewitt E, Stromberg K (1965) Real and abstract analysis. Graduate texts in mathematics, vol 25. Springer, Berlin/Heidelberg/New York

Koppelberg S (1989) General theory of Boolean algebras. In: Monk J, Bonnet R (eds) Handbook of Boolean algebras, vol 1. Elsevier, Amsterdam/New York, pp 741–773

Koppelberg S, Monk J (1992) Pseudo-trees and Boolean algebras. Order 8:359–577

Piccione M, Rubinstein A (1997) On the interpretation of decision problems with imperfect recall. Games Econ Behav 20:3–24

Rubinstein A (1982) Perfect equilibrium in a bargaining model. Econometrica 50:97–109

Shapley L (1953) Stochastic games. Proc Natl Acad Sci 39:1095–1100

Simon L, Stinchcombe MB (1989) Extensive form games in continuous time, Part I: pure strategies. Econometrica 57:1171–1214

Stinchcombe MB (1992) Maximal strategy sets for continuous-time game theory. J Econ Theory 56:235–265

Chapter 6
Discrete Extensive Forms

Extensive form analysis is basic for non-cooperative game theory, because it provides a tool to verify that the rules of the game are complete. Moreover, popular solution concepts that relate to backwards induction, like subgame perfection, are only defined for extensive form games. Much of the early conceptual work on extensive form games has tacitly adopted finiteness or discreteness assumptions in order to avoid technicalities. Many of these simplifications are not necessary, however, and applied work has soon transcended these limitations. It is, therefore, worth to push the limits of traditional theory, and this chapter is yet another step in this general endeavour.[1]

Previous chapters have focused on the key concept of a *tree* (as introduced by von Neumann and Morgenstern 1944, chp. II.9 and Kuhn 1953). It was shown in Chap. 2 that it involves no loss of generality to view a tree as a collection of subsets of an underlying set of outcomes (with a particular structure), as originally proposed by von Neumann and Morgenstern (1944, pp. 65). Such a formalization entails no finiteness or discreteness assumptions and allows for all known examples of extensive form games, inclusive of infinitely repeated games, stochastic games (Shapley 1953), the transfinite "long cheap-talk" by Aumann and Hart (2003), or "differential games" (decision problems in continuous time).

The previous chapter has investigated when strategies, as derived from an extensive form, satisfy the following three desiderata: (A0) Every outcome/play can be induced by some strategy combination. (A1) Every strategy combination induces an outcome/play. (A2) If a strategy combination induces an outcome, this outcome is unique. The main result of that chapter is that (A1) and (A2) hold true *if and only if* the tree satisfies two conditions: it is "regular" and "up-discrete." (Condition (A0) comes for free.) That is, if games are to be played, restrictions on the tree have to hold. The good news is that this characterization does—almost—not rule out any

[1]The material presented here is an expanded version of Alós-Ferrer and Ritzberger (2013). Section 6.4 is based on Alós-Ferrer and Ritzberger (2016c).

C. Alós-Ferrer, K. Ritzberger, *The Theory of Extensive Form Games*,
Springer Series in Game Theory, DOI 10.1007/978-3-662-49944-3_6

of the classical examples of extensive form games. All aforementioned cases are covered, except for differential games (which are not up-discrete). These results are useful, because they delimit the confines of extensive form analysis in the abstract. In particular, playable extensive forms require an up-discrete tree.

6.1 Preview

This chapter employs the structure of the tree that is *necessary* for playability, but goes beyond that.[2] The *sufficiency* of up-discreteness and regularity reveals that in practise most applications use even more structure. This additional structure, though, often does not seem to simplify the formal treatment. Even the most restrictive definitions of extensive forms in the literature are often involved and notationally overwhelming. The classical definitions by Kuhn (1953) and Selten (1975) employ four objects (exclusive of probabilities for chance moves and the payoff function) and five consistency conditions. The textbook exposition by Mas-Colell et al. (1995, p. 227) employs seven objects (again exclusive of probabilities for chance moves and the payoff function) and three consistency conditions. Both are confined to purely finite cases.

The more general definition of an extensive decision problem (EDP) proposed in Definition 4.1 still consists of two objects (the tree and the set of choices) and four consistency conditions. But it entails no finiteness assumptions, indeed not even a discreteness condition (consequently it covers "differential games"). Yet, even the formalization by Aumann (1964) of infinite extensive form games uses a discreteness assumption, as do infinitely repeated games, infinite bilateral bargaining games à la Rubinstein (1982), or stochastic games.

Therefore, this chapter characterizes the family of discrete extensive forms (Alós-Ferrer and Ritzberger 2013) within the abstract framework outlined above. More precisely, to the properties up-discreteness and regularity a "down-discreteness" property is added. It is shown that the resulting discrete trees are characterized by the existence of an (onto) immediate predecessor function on the set of moves (Theorem 6.1 below). Equivalently, this means that the history of every decision point in the tree can be uniquely reconstructed in a finite number of steps—yet, infinite-horizon games are not excluded.

As a result, an elementary definition of extensive forms (on discrete game trees) is obtained that lends itself easily to applications, but is still more general than the classical ones (Definition 6.2 and Theorem 6.2). In the discrete case the four defining consistency conditions of extensive decision problems collapse to only two properties about the two basic objects. Moreover, the two conditions are very intuitive. One says that the decisions of the relevant players at any given move lead to a unique new node. The other says that players cannot deduce from the

---

[2] Aumann (1964) employs an even more restrictive structure of the tree than what we, later in this chapter, identify as *discrete trees*.

menu of available choices more information than what they have according to their information sets.

A key advantage of the approach presented below, which it inherits from the general approach in this book, is that it is immaterial whether plays or nodes are taken as the primitives. In particular, for discrete trees there is an elementary definition with the nodes as primitives—"simple extensive forms" (Definition 6.4). This is due to the fact that infinite terminal nodes need not be part of the tree (recall Proposition 4.2). In the end, then, one obtains a simple framework for extensive form analysis that still covers almost all relevant cases and preserves the freedom to take plays or nodes as primitives (Theorem 6.4).

The following section introduces discrete trees, characterizes them, and then defines the associated notion of a discrete extensive form. Then we exploit a property of discrete trees to provide a definition and characterization of "simple" trees and associated extensive forms, where nodes (rather than plays) serve as the primitives.

Finally, in the last section, we tackle the condition of perfect recall—a key ingredient for modeling rational behavior which is not implied by the definition of an extensive form. A characterization result (Proposition 6.6) translates the original definition by Kuhn (1953) into conditions on the primitives of an extensive form, and provides a first step towards Kuhn's Theorem on the equivalence of mixed and behavior strategies.

6.2 Discrete Extensive Forms

To begin with recall that a game tree $(N, \supseteq)$ is *regular* if there are no strange nodes, i.e. if $S(N) = \emptyset$. It is *up-discrete* if all (nonempty) chains in N have maxima. By Theorem 5.5 and Corollary 5.4(b) these two properties characterize the class of game trees that satisfy (A1) and (A2) and are, therefore, appropriate for sequential decision theory.

Thus, nodes in a regular game tree are either finite or infinite (recall Definition 4.3). On finite nodes a function p can be defined that assigns to every finite node its *immediate predecessor*. Namely, let $p : F(N) \to X$ be given by

$$p(x) = \min \uparrow x \setminus \{x\} \text{ for all } x \in F(N) \setminus \{W\} \tag{6.1}$$

and $p(W) = W$ by convention. Hence, $x \subset p(x) = \bigcap_{y \in \uparrow x \setminus \{x\}} y$ for $x \in F(N) \setminus \{W\}$.

6.2.1 *Up-Discrete Trees Revisited*

Recall that up-discreteness is equivalent to the chains $\uparrow x$ for $x \in N$ being dually well-ordered (that is, all their subsets have a maximum; see the argument after

Definition 5.1). This condition implies that the set of *immediate successors*, $p^{-1}(y) = \{x \in N | p(x) = y\}$, of a move is nonempty and forms a partition of the move by finite nodes.

Proposition 6.1 *Let $T = (N, \supseteq)$ be an up-discrete game tree. Then, for every move $x \in X$*

(a) the preimage $p^{-1}(x)$ is a partition of x, hence nonempty;
(b) for every $y \in N$ with $y \subset x$ there is $z \in p^{-1}(x)$ such that $y \subseteq z$.

Proof (a) For $w \in x$ the set $M_w(x) = \{y \in N | w \in y \subset x\}$ is nonempty by Lemma 4.1(a) and a chain by Trivial Intersection (2.4). By up-discreteness there is $x_w = \max M_w(x)$ for all $w \in x$. If $w, w' \in x$ are such that $x_w \cap x_{w'} \neq \emptyset$, then with $w'' \in x_w \cap x_{w'}$ the construction implies $x_{w''} = x_w = x_{w'}$. This shows that $M(x) = \{x_w | w \in x\}$ is a partition of x with each element maximal among the nodes in $\downarrow x \setminus \{x\}$. Note that, if y is maximal among the nodes properly contained in x then necessarily $x = \min \uparrow y \setminus \{y\}$, hence, $y \in F(N)$ and $y \in p^{-1}(x) \neq \emptyset$. Reciprocally, if $y \in p^{-1}(x)$ then $y = x_w$ for any $w \in y$. We conclude that $M(x) = p^{-1}(x)$.

(b) Let $z = \max \uparrow y \setminus \uparrow x$, which exists by up-discreteness. ■

Of course, up-discreteness excludes continuous-time examples, like the differential game. On the other hand, up-discreteness imposes a tight connection between choices and nodes, as the following result shows.

Proposition 6.2 *Let $T = (N, \supseteq)$ be an up-discrete game tree and $A(T)$ the set of all $a \in 2^W$ that are unions of nodes. Define*

$$N(a) = \{x \in N | x \subseteq a \text{ and } x \text{ is maximal in } a\} \tag{6.2}$$

for any $a \in A(T)$. Then

(a) $N(a) \subseteq N$ is a partition of a, and
(b) its predecessor set is given by $P(a) = \{p(x) | x \in N(a) \cap F(N)\}$.

Proof (a) For any $w \in a$ let $N_w(a) = \{x \in \downarrow a | w \in x\}$. By Trivial Intersection (2.4) $N_w(a)$ is a chain that is nonempty, because $a \in A(T)$. By up-discreteness there is $x_w = \max N_w(a)$ for all $w \in a$. If $w, w' \in a$ are such that $x_w \cap x_{w'} \neq \emptyset$, then with $w'' \in x_w \cap x_{w'}$ the construction implies $x_{w''} = x_w = x_{w'}$. This shows that $N(a) = \{x_w | w \in a\}$ is a partition of a with each element maximal among the nodes in $\downarrow a$.

(b) Let $x \in N$ be such that $\uparrow x = \uparrow y \setminus \downarrow a$ for some $y \in \downarrow a$, that is, $x \in P(a)$ as in (4.1). It follows that $x \notin \downarrow a$ and $x \in X$, because $y \in \downarrow x \setminus \{x\}$. Let $w \in y$ and consider x_w as in (a). By construction, $p(x_w) = x$, implying that $x_w \in F(N)$. Thus, $P(a) \subseteq \{p(x) | x \in N(a) \cap F(N)\}$. Let $x_w \in N(a) \cap F(N)$ and $x = p(x_w)$. Then $x_w \in \downarrow a$ and $\uparrow x = \uparrow x_w \setminus \downarrow a$, that is, $x \in P(a)$. This completes the verification that $P(a) = \{p(x) | x \in N(a) \cap F(N)\}$. ■

This implies that for any $a \in A(T)$ there exists a coarsest partition by nodes. Applying this result to choices from an EDP yields the following: For any choice $c \in C_i$ for some $i \in I$ there is a set $N(c) \subseteq N$ such that c is a disjoint union over the nodes in $N(c)$ and $P(c) = \{p(x) \,|x \in N(c) \cap F(N)\}$. This property is crucial for going back and forth between trees where plays are taken as the primitive, and trees where nodes are primitive (see Sect. 6.3).

6.2.2 Discrete Trees

To the best of our knowledge all games that have been studied in the literature—with the exception of differential games—satisfy (A1) and (A2), that is, they are regular and up-discrete. As a matter of fact, this is because they usually satisfy even more stringent conditions. In particular, the differential games and the transfinite cheap-talk game by Aumann and Hart (2003) seem to be the only ones which fail the following property that is complementary to up-discreteness.

Definition 6.1 A game tree $(N, \supseteq)$ is **down-discrete** if the chain $\uparrow x \setminus \{x\}$ has an infimum in $E \cup (\uparrow x \setminus \{x\})$ for all $x \in N \setminus \{W\}$. It is **discrete** if it is up-discrete and down-discrete.

That is, a game tree is down-discrete if for all $x \in N \setminus \{W\}$ the chain $\uparrow x \setminus \{x\}$ has either an infimum in the set E of terminal nodes or a minimum. Furthermore:

Proposition 6.3 *A game tree is down-discrete if and only if it is regular and all moves are finite (i.e.* $X \subseteq F(N)$*).*

Proof By definition $\uparrow x \setminus \{x\}$ always has an infimum, so there are no strange nodes. Moreover, if $\uparrow x \setminus \{x\}$ has no minimum, x must be its infimum and, therefore, x is terminal. If, conversely, a game tree is regular and all moves are finite, then for every $x \in X$ there exists $\min \uparrow x \setminus \{x\}$ by the definition of finite nodes; furthermore, any non-finite node $x \in N \setminus F(N)$ must then be terminal and the infimum of $\uparrow x \setminus \{x\}$ by Lemma 4.2(a) and (b), as there are no strange nodes by regularity. ∎

A characterization of game trees that are both up- and down-discrete is given by the following result.

Proposition 6.4 *A game tree* $(N, \supseteq)$ *is discrete if and only if every chain in the set* X *of moves has a maximum and also either an infimum in the set* E *of terminal nodes or a minimum.*

Proof "if:" Let h be a chain and $h' = h \setminus E$. If $h' = \emptyset$, then $h = \{\{w\}\}$ for some $w \in W$ with $\{w\} \in N$ by Lemma 4.1(b). In that case $\{w\} \in N$ provides a maximum for h. Otherwise $h' \neq \emptyset$ and $h' \subseteq X$ implies that h' has a maximum by hypothesis. Since h is a chain, the maximum of h' also provides a maximum for h.

Since by hypothesis every chain in X has either a minimum or an infimum in the set E of terminal nodes, in particular every chain of the form $\uparrow x \setminus \{x\}$ for $x \in N \setminus \{W\}$ has an infimum in $E \cup (\uparrow x \setminus \{x\})$.

"only if:" Up-discreteness means that every chain has a maximum. Let h be an arbitrary chain with $h \subseteq X$. By Boundedness ((2.16) resp. the only-if part of (GT1)) there is $w \in \bigcap_{x \in h} x$. The chain $h' = \uparrow \{w\} \setminus (\bigcup_{y \in h} \uparrow y)$ has a maximum $x = \max h'$

by up-discreteness. By down-discreteness the chain $\uparrow x \setminus \{x\}$ has either a minimum $z \in \uparrow x \setminus \{x\}$ or an infimum $\{w'\} \in E$. If it has a minimum, then $z \in h$ implies that it is also a minimum for h. If $\uparrow x \setminus \{x\}$ has an infimum $\{w'\} \in E$, then by construction $w' = w$, $h' = \{x\}$, and $x = \{w\} \in E$; but then $x = \{w\} = \inf h$ by Lemma 4.2(a). ■

Infinitely repeated games, infinite bilateral bargaining games (Rubinstein 1982), or stochastic games (Shapley 1953) all employ discrete trees, as do Aumann's infinite extensive form games (Aumann 1964). "Long cheap-talk" à la Aumann and Hart (2003) employs a tree that satisfies (A1) and (A2), but is not down-discrete. It has a time axis of order $\omega + 1$, where ω denotes the first infinite (limit) ordinal and $\omega + 1$ its successor according to the usual well-order.

Example 6.1 ($\omega + 1$-centipede) A similar example is obtained by considering the W-centipede (Example 4.8) with $W = \{1, 2, \ldots, \omega, \omega + 1\}$. There, the infimum of the chain $\{x_t \,|\, 1 \leq t < \omega\} = \uparrow x_\omega \setminus \{x_\omega\}$ is the move x_ω. Hence, this tree is not down-discrete, even though it is regular and up-discrete.

An EDP defined on the tree of the previous example will satisfy (A1) and (A2), even though the tree is not discrete—in contrast to differential games that fail (A1) and (A2). So, discreteness is more than what is needed. But it has convenient implications. A particularly striking consequence of discrete trees is captured by the next theorem. It characterizes discrete trees, within the class of regular game trees, by the properties of the *immediate predecessor function* $p : F(N) \to X$.

Theorem 6.1 *For a regular game tree $(N, \supseteq)$ the following statements are equivalent:*

(a) $(N, \supseteq)$ is discrete;
(b) all infinite nodes are terminal ($X \subseteq F(N)$) and the immediate predecessor function $p : F(N) \to X$ is onto and satisfies that for all $x \in F(N)$

$$x \subset p(x) \text{ and if } x \subset y \in N \text{ then } p(x) \subseteq y \subseteq \cup_{t=1}^{\infty} p^t(x) \tag{6.3}$$

where $p^1 = p$ and $p^t = p \circ p^{t-1}$ for all $t = 2, 3, \ldots$;
(c) $X \subseteq F(N)$ and the chain $\uparrow x \cap \downarrow y = \{z \in N \,|\, x \subseteq z \subseteq y\}$ is finite for all $x, y \in F(N)$;
(d) the chain $\uparrow x$ is finite for all $x \in X$.

Proof "(a) implies (b):" Since discreteness entails down-discreteness, that $X \subseteq F(N)$ follows from Proposition 6.3. Let p be defined as in (6.1). Then, $x \subset p(x)$ and if $x \subset y \in N$, then $y \in \uparrow x \setminus \{x\}$ implies $p(x) \subseteq y$, for any $x \in F(N)$.

Moreover, for $x \in F(N)$ consider the chain $\{p^t(x)\}_{t=1}^{\infty}$. Since the tree is up-discrete, it has a maximum $z = p^k(x)$. Therefore, $p(z) = p^{k+1}(x) = p^k(x) = z$ which is a contradiction unless $z = W$. Hence, $\cup_{t=1}^{\infty} p^t(x) = W \supseteq y$ for any $y \in \uparrow x \setminus \{x\}$.

Next it is shown that p is surjective. Because discreteness entails up-discreteness, for any $x \in X$ there exists a maximum y for a maximal chain in $\downarrow x \setminus \{x\}$ (where the latter is nonempty by $x \in X$).[3] Since $y \subset x$, it follows that $p(y) \subseteq x$. If $p(y) \neq x$, then $p(y) \in \downarrow x \setminus \{x\}$ with $y \subset p(y)$ contradicts the construction of y. Therefore, $p(y) = x$ and, because $x \in X$ was arbitrary, p is surjective.

"(b) implies (c):" That $X \subseteq F(N)$ is immediate. Thus, let $x, y \in F(N)$ and $x \subset y$. Then $p(x) \subseteq y \subseteq \cup_{t=1}^{\infty} p^t(x)$ by (6.3). It follows that there must be some $k \geq 1$ such that $p^{k-1}(x) \subset y \subseteq p^k(x)$. But by (6.3) applied to $p^{k-1}(x)$ this implies $y = p^k(x)$. Hence, the chain $\uparrow x \cap \downarrow y$ is contained in the finite chain $\{x, p(x), p^2(x), \ldots, p^k(x) = y\}$ and, therefore, itself finite. If $x, y \in F(N)$ are such that $x = y$, then $\uparrow x \cap \downarrow y = \{x\}$ is finite as well. If $x, y \in F(N)$ are such that $y \subset x$ or $x \cap y = \emptyset$, then $\uparrow x \cap \downarrow y$ is empty and, thus, contains no element.

"(c) implies (d):" This follows by choosing $y = W$.

"(d) implies (a):" First notice that, if the tree is trivial, $N = \{W\}$, there is nothing to prove. If it is nontrivial, then $W \in X$ and, therefore, $W \in F(N)$ by the hypothesis. Let $x \in X, x \neq W$. Since $\uparrow x$ is finite, it follows that the node x is finite, thus $X \subseteq F(N)$. Proposition 6.3 implies that the tree is down-discrete. To prove up-discreteness, let $h \in 2^N$ be any chain in N. Then, for any move $x \in h$ the chain $\uparrow x$ is finite by hypothesis. Enumerate its elements, $x = x_0 \subset x_1 \subset x_2 \subset \ldots \subset x_m = W$, and let t be the largest integer such that $x_t \in h$. We claim that x_t is a maximum for h. Choose any $z \in h$. If $z \subseteq x$ then $z \subseteq x = x_0 \subseteq x_t$. If $x \subset z$ there is $0 \leq \tau < m$ such that $z = x_\tau$ and, therefore, $z = x_\tau \subseteq x_t$. It follows that x_t is a maximum for h. ■

The hypothesis of a regular game tree is necessary for Theorem 6.1. This is illustrated by the examples of "Twins" (Example 2.15) that satisfies both (b) and (c) of Theorem 6.1, but is not regular and, hence, not down-discrete (by the "only if"-part of Proposition 6.3). The following example illustrates that up-discreteness is responsible for the very last part of (6.3), viz. that the root can be reached from any move by iterating the immediate predecessor function.

Example 6.2 (Augmented Inverse Infinite Centipede) As in Example 5.1 consider the negative integers augmented by a "smallest" elements $-\infty$ and call this set W. The corresponding W-centipede is a down-discrete game tree. All nodes are finite (because $\uparrow x_t \setminus \{x_t\} = \{x_\tau\}_{\tau=t-1}^{-\infty}$ and $\uparrow \{t\} \setminus \{t\} = \{x_\tau\}_{\tau=t}^{-\infty}$) and immediate predecessors are given by $p(\{t\}) = x_t$ and $p(x_t) = x_{t-1}$ for all $t = -1, -2, \ldots$ But, if the predecessor function is iterated from any node, the root is never reached, i.e., $\cup_{k=1}^{\infty} p^k(x) \subset W$, even though $p(-\infty) = W$ (that is, even though the immediate predecessor function is surjective). This is so, because no chain in $\downarrow W \setminus \{W\}$ has a maximum.

Most textbooks, of course, simply use trees where all plays have finite length (contain finitely many elements). Say that a game tree $(N, \supseteq)$ has *finite height* if every chain in N has a maximum and a minimum. This definition has no implications

[3] This only requires "weak up-discreteness," see Definition 4.2.

on the number of choices available at any given move (i.e. on the number of successors of a given move). But it does restrict the lengths of plays. For, if there were an infinite play for a game tree $(N, \supseteq)$ with finite height, then this play would have to have a minimum $x \in N$ by definition. Since a game tree with finite height is clearly discrete, Theorem 6.1(c) would then imply that the chain $\uparrow x \cap \downarrow W$ is finite; but, since x is the minimum of the play, $\uparrow x \cap \downarrow W$ equals the play—a contradiction. Hence, for a game tree with finite height all plays have finite length. Conversely, a game tree, where all plays have finite length, is clearly a game tree with finite height.

Moreover, for game trees with finite height all nodes must be finite. For, if all plays have finite length, then all chains of the form $\uparrow x \setminus \{x\}$ (for a node $x \in N$) also do. Hence, all these chains have minima, implying that $N = F(N)$.[4] Consequently, by Theorem 6.1(b), the immediate predecessor function is defined on *all* nodes for game trees with finite height.

6.2.3 Discrete Extensive Forms

In this subsection we show that if one restricts attention to games on discrete trees, a much simpler definition of an EDP can be obtained that is still equivalent to the original one (Definition 4.1) for this class of trees. The key step is to replace the three properties (EDP2), (EDP3), and (EDP4) by a single, but equivalent condition.

Definition 6.2 A **discrete extensive form** (DEF) is a pair (T, C), where $T = (N, \supseteq)$ is a discrete game tree with set of plays W and $C = (C_i)_{i \in I}$ a system consisting of collections C_i (the sets of players' choices) of nonempty unions of nodes (hence, sets of plays) for all $i \in I$, such that

(DEF1) if $P(c) \cap P(c') \neq \emptyset$ and $c \neq c'$, then $P(c) = P(c')$ and $c \cap c' = \emptyset$, for all $c, c' \in C_i$ for all $i \in I$;

(DEF2) $p^{-1}(x) = \left\{ x \cap \left(\cap_{i \in J(x)} c_i \right) \middle| (c_i)_{i \in J(x)} \in A(x) \right\}$, for all $x \in X$;

where $A(x) = \times_{i \in J(x)} A_i(x)$, $A_i(x) = \{c \in C_i \,|\, x \in P(c)\}$ are the choices available to $i \in I$ at $x \in X$, and $J(x) = \{i \in I \,|\, A_i(x) \neq \emptyset\}$ is the set of decision makers at x, which is required to be nonempty, for all $x \in X$.

(DEF1) is identical to (EDP1), that is, the "information set property" that players cannot deduce from the available choices at which move in the information set they are. The second property is also fairly intuitive. (DEF2) say that, at any given move,

[4] The converse is not true, not even in the class of discrete trees. All nodes can be finite, even if the tree does not have finite height. This is, for instance, the case in Example 4.13, when the node $\{\infty\}$, but not the play ∞, is removed from the tree.

the combined decisions of the relevant players lead to a new node, and that any (immediate) successor of the move can be selected by an appropriate combination of the players' decisions. (The intersection with the move $x \in X$ is needed because of potentially large information sets.)

Theorem 6.2 *Let $T = (N, \supseteq)$ be a discrete game tree with set of plays W and $C = (C_i)_{i \in I}$ a system consisting of collections C_i of nonempty unions of nodes from T for all $i \in I$. The pair (T, C) is an EDP if and only if it is a DEF.*

Proof What needs to be shown is that (EDP2), (EDP3), and (EDP4) are equivalent to (DEF2), because (DEF1) is precisely (EDP1).

"if:" Consider some $x \in X$ and $(c_i)_{i \in J(x)} \in A(x)$. By hypothesis there is some $y \in \downarrow x \setminus \{x\}$ such that $P(y) = x$ and $y = x \cap \left(\bigcap_{i \in J(x)} c_i\right)$. As $y \neq \emptyset$, (EDP2) holds.

Next, consider $y, y' \in N$ such that $y \cap y' = \emptyset$. We claim that $\uparrow y \cap \uparrow y'$ has a minimum. If either $y \in X$ or $y' \in X$, this follows from Theorem 6.1(d). Hence, suppose $y, y' \in E$. If there is $z \in \uparrow y \setminus \{y\}$ such that $z \notin \uparrow y' \setminus \{y'\}$, then $z \in X$ and $\uparrow y \cap \uparrow y' \subseteq \uparrow z$ and the desired conclusion follows, because the latter chain is finite by Theorem 6.1(d). Therefore, assume $\uparrow y \setminus \{y\} \subseteq \uparrow y' \setminus \{y'\}$ and analogously for the reverse inclusion, that is, $\uparrow y \setminus \{y\} = \uparrow y' \setminus \{y'\}$. But that implies $y, y' \in S(N)$ in contradiction to Proposition 6.3.

Let $x = \min \uparrow y \cap \uparrow y'$. There are $z, z' \in p^{-1}(x)$ such that $y \subseteq z$ and $y' \subseteq z'$. Let $(c_i)_{i \in J(x)}, (c'_i)_{i \in J(x)} \in A(x)$ be such that $x \cap \left(\bigcap_{i \in J(x)} c_i\right) = z$ and $x \cap \left(\bigcap_{i \in J(x)} c'_i\right) = z'$. Because $z \cap z' = \emptyset$, there is some $i \in J(x)$ such that $c_i \neq c'_i$. By (DEF1) $c_i \cap c'_i = \emptyset$, which completes the verification of(EDP3).

Let $x \supset y \in N$ and $z = \max \uparrow y \setminus \uparrow x$ which exists by up-discreteness. Clearly, $y \subseteq z \in p^{-1}(x)$. Let $(c_i)_{i \in J(x)} \in A(x)$ be such that $z = x \cap \left(\bigcap_{i \in J(x)} c_i\right)$. Then $y \subseteq c_i$ for all $i \in J(x)$ which verifies (EDP4).

"only if:" First, we claim that for any $x \in X$ and $(c_i)_{i \in J(x)} \in A(x)$ the intersection $x \cap \left(\bigcap_{i \in J(x)} c_i\right)$ is an element of $p^{-1}(x)$.

By (EDP2) there is $w \in W$ such that $w \in x \cap \left(\bigcap_{i \in J(x)} c_i\right)$. Let $y = \max \uparrow \{w\} \setminus \uparrow x$. This exists by up-discreteness, because the chain is nonempty by Lemma 4.1(a). By construction $p(y) = x$, so $y \in p^{-1}(x)$. By (EDP4) there are $c''_i \in A_i(x)$ such that $y \subseteq c''_i$ for all $i \in J(x)$. Since $w \in y \cap \left(\bigcap_{i \in J(x)} c_i\right)$, (EDP1) implies that $c''_i = c_i$ for all $i \in J(x)$. Therefore, $y \subseteq x \cap \left(\bigcap_{i \in J(x)} c_i\right)$.

Suppose there is $w' \in \left(x \cap \left(\bigcap_{i \in J(x)} c_i\right)\right) \setminus y$. Then by the same argument as before there is $y' \in p^{-1}(x)$ such that $w' \in y' \subseteq x \cap \left(\bigcap_{i \in J(x)} c'_i\right)$ for some $(c'_i)_{i \in J(x)} \in A(x)$. Because $w' \in y' \setminus y$ and $p(y) = p(y') = x$, the nodes y and y' must be disjoint by Trivial Intersection (2.4). Therefore, by (EDP1) and (EDP3) the choices c_i and c'_i must be disjoint for all $i \in J(x)$. This contradicts the construction of w'. Therefore, $y = x \cap \left(\bigcap_{i \in J(x)} c_i\right)$, completing the proof of the claim.

The claim shows that $\left\{x \cap \left(\bigcap_{i \in J(x)} c_i\right) \middle| (c_i)_{i \in J(x)} \in A(x)\right\} \subseteq p^{-1}(x)$. To prove the reverse inclusion, consider some $y \in p^{-1}(x)$. By (EDP4) there are choices $c_i \in A_i(x)$ such that $y \subseteq c_i$ for all $i \in J(x)$. Hence, $y \subseteq x \cap \left(\bigcap_{i \in J(x)} c_i\right)$. Since

$x \cap \left(\cap_{i \in J(x)} c_i\right) \in p^{-1}(x)$ by the previous claim, it follows from Irreducibility (2.12) that $y = x \cap \left(\cap_{i \in J(x)} c_i\right)$. ■

Theorem 6.2 captures the following quasi-operational interpretation of an extensive form. At every move $x \in X$ each player $i \in J(x)$ is told (by an "umpire") which choices $c \in C_i$ she has available (in the sense that $x \in P(c)$ for $c \in C_i$) and asked to select one of those. No other information is released to players. Given the decisions by all players in $J(x)$, taking the intersection gives a node, which becomes the new "state" of the game.

The simplification associated with the transition from an EDP to a DEF comes at a cost. In particular, transfinite and continuous-time examples are excluded. But other than that nothing is lost. This is the content of the next result.

Theorem 6.3 *If (T, C) is a DEF, then (A1) and (A2) hold.*

Proof By Proposition 6.3, the tree T is regular. Since (T, C) is a DEF, the tree T is up-discrete by definition. Hence, (A1) follows from Theorem 5.3.

Property (A2) will follow from Theorem 5.4 and Proposition 5.5 if (T, C) fulfills two properties: (EDP2') and that T is selective. The first one, (EDP2'), follows from Corollary 5.2 and Proposition 5.8. The second, selectiveness (see Definition 5.3), follows from Proposition 5.4(b) and Lemma 5.4. The first states that a regular tree is selective if every history has a continuation with a maximum. The second implies that the latter property follows from up-discreteness. ■

This theorem shows that DEFs satisfy the basic requirements for a well-founded interpersonal decision theory. Furthermore, the definition of DEF is easier to verify than that of an EDP because the four properties collapse to two.

Corollary 6.1 *Every DEF (T, C) is an EF.*

Proof To show that a DEF (T, C) is an EF it needs to be shown that (EDP3') holds. Let $y, y' \in N$ be such that $y \cap y' = \emptyset$. By down-discreteness the chain $\uparrow y \cap \uparrow y'$ has a minimum $x \in N$, because it must be contained in the set X of moves (see Proposition 6.4). By up-discreteness the chains $\uparrow y \setminus \uparrow x$ and $\uparrow y' \setminus \uparrow x$ have maxima, denoted $z = \max \uparrow y \setminus \uparrow x$ and $z' = \max \uparrow y' \setminus \uparrow x$, which satisfy $z \cap z' = \emptyset$. By (DEF2) and $z, z' \in p^{-1}(x)$ there must be at least one player $i \in J(x)$ such that $y \subseteq z \subseteq x \cap c$, $y' \subseteq z' \subseteq x \cap c'$, and $c \cap c' = \emptyset$ for some $c, c' \in C_i$ with $x \in P(c) \cap P(c')$. ■

Another simplification is already implicit in the definition of an EDP (Definition 4.1). As commented in Sect. 4.2.2, the definition of extensive forms games by Selten (1975) required that the set of decision makers at any move contains exactly one player. In this case (DEF2) reduces to $p^{-1}(x) = \{x \cap c \,|\, c \in A(x)\}$ for any $x \in X$. For simultaneous decisions this leads to cascading information sets. This is allowed but not necessary in the present framework, which encompasses games where several players may decide at the same move.This, for instance, simplifies the extensive form representation of an infinitely repeated game. Furthermore, consider the following examples.

Example 6.3 (Nonatomic one-shot games revisited) Reconsider Example 4.2 Let the player set be a continuum, e.g. $I = [0,1]$. All players choose simultaneously from a set A with at least two elements. This can be formalized by taking as the set of outcomes W the set of functions $f : [0,1] \to A$. Each player $i \in I$ faces the choices $c_i(a) = \{f \in W \,|\, f(i) = a\}$ for all $a \in A$. It is straightforward to check that (DEF1) and (DEF2) are fulfilled. That is, nonatomic games are discrete games. The traditional cascading information sets would give rise to a tree of staggering complexity, when compared to the simplicity of the formalization proposed here. In contrast, in the framework presented here any normal-form game is automatically associated to a natural (trivial) extensive form.

Example 6.4 (Job-Market Signalling) The job-market signalling model by Spence (1973) serves as an example on how the present approach may simplify the analysis of equilibria. In that model chance initially assigns a productivity (type) $\theta \in \{\theta_L, \theta_H\}$, where $0 < \theta_L < \theta_H$, to a potential "worker." The prospective employee then decides on an education level $e \in \mathbb{R}_+$ in order to signal her productivity. This signal, but not the productivity, is observed by "a competitive industry" which then offers a wage $w \in [\theta_L, \theta_H]$. This wage is equal to the expected productivity, given some beliefs conditional on the observed signal. To complete the specification of an extensive form game, assume two firms, who compete à la Bertrand by offering wages to the worker, who in turn chooses among wage offers. In any subgame perfect equilibrium the worker will choose the highest offer and Bertrand competition will imply that both firms offer a wage equal to expected productivity.

This guarantees that the productivities expected by the two firms are equal along the equilibrium path. The natural equilibrium concept for this game, perfect Bayesian equilibrium (PBE), does not force the two firms to hold the same beliefs *off* the equilibrium path, though. The stronger criterion of sequential equilibrium (Kreps and Wilson 1982)—that, strictly speaking, does not apply to this game, since it is only defined for finite games—by contrast, would imply equal beliefs across firms even at unreached information sets—the "common belief property." But PBE does not.[5]

This problem is complicated when simultaneous moves are captured by cascading information sets; see e.g. the graphical representation by Mas-Colell et al. (1995, p. 451). For this example both firms move simultaneously after observing education level e. For the classical representation outcomes are ordered four-tuples

$$(\theta, e, w_1, w_2) \in \{\theta_L, \theta_H\} \times \mathbb{R}_+ \times [\theta_L, \theta_H] \times [\theta_L, \theta_H] = W,$$

assuming that firm 1 "moves first" and ignoring the last (trivial) decision by the worker. When $x_t(\overline{e}) = \{w \in W \,|\, \theta = \theta_t,\, e = \overline{e}\}$ denotes the node where type $t = L, H$

[5] Admittedly, there is no universally accepted definition of PBE for general extensive form games. Mas-Colell et al. (1995, p. 450) avoid this problem by incorporating the equality of beliefs in their definition of PBE for its use in signalling games of the Spence type. From the theoretical point of view this is, of course, unsatisfactory.

has chosen $\overline{e} \in \mathbb{R}_+$, firm 1's beliefs are $\mu_1(\overline{e}) = (\mu_1(x_L(\overline{e})), \mu_1(x_H(\overline{e}))) \in \Delta^1$, where Δ^1 is the one-dimensional simplex. Firm 2 "moves second" at her (infinite) information set

$$g_2(\overline{e}) = \{y_t(\overline{e}, \overline{w}_1) \,|\, t = L, H, \overline{w}_1 \in [\theta_L, \theta_H]\},$$

with $y_t(\overline{e}, \overline{w}_1) = \{w \in W \,|\, \theta = \theta_t, e = \overline{e}, w_1 = \overline{w}_1\}$ denoting the node reached after type $t = L, H$ has chosen $\overline{e} \in \mathbb{R}_+$ and firm 1 has offered $\overline{w}_1 \in [\theta_L, \theta_H]$. Her beliefs are then given by a probability measure $\mu_2(\cdot \,|\overline{e}) : g_2(\overline{e}) \to \mathbb{R}_+$.

This specification is void of economic content, but complicates the "common belief property," because it becomes necessary in a PBE to specify beliefs of firm 2 on the wage offered by firm 1 (and, out of equilibrium, nothing pins down such beliefs). Common beliefs are captured by the statement that the probability mass $\mu_2(\{y_t(\overline{e}, \overline{w}_1) \,|\overline{w}_1 \in [\theta_L, \theta_H]\} \,|\overline{e})$ numerically equals the probability $\mu_1(x_t(\overline{e}))$, for $t = L, H$ for all $\overline{e} \in \mathbb{R}_+$. The beliefs μ_1 and μ_2 are radically different formal objects!

Since the present formulation allows us to replace cascading information sets by players deciding at the same move, the natural representation is much simpler. Outcomes become triplets $(\theta, e, (w_1, w_2)) \in \{\theta_L, \theta_H\} \times \mathbb{R}_+ \times [\theta_L, \theta_H]^2 = W$ and all nodes $x_t(\overline{e})$ are directly followed by terminal nodes (the nodes $y_t(\overline{e}, \overline{w}_1)$ are dispensed with). The information sets $g(\overline{e}) = \{x_t(\overline{e}) \,|\, t = L, H\}$ (common to both firms) consist of two nodes, beliefs for both firms are $\mu_1(\overline{e})$, $\mu_2(\overline{e}) \in \Delta^1$, and the common beliefs property is captured by the statement $\mu_1(\overline{e}) = \mu_2(\overline{e})$ for all $\overline{e} \in \mathbb{R}_+$.

Another implication of Theorem 6.3 is that DEFs can be represented in normal form (without payoffs). This raises issues about equivalences between DEFs. Thompson (1952) introduced four transformations that define equivalence relations on the set of finite extensive form games (see also Dalkey 1953).[6] He showed that these transformations jointly characterize another equivalence relation, given by agreement of the semi-reduced (or pure-strategy reduced) normal form. An important objective for future research is to extend such a characterization to DEFs.

A first difficulty with such an endeavor is that the proofs use finite recursive algorithms. A second, and probably more severe, obstacle is that one of Thompson's transformations, "deletion of irrelevant decision points," relies on payoffs. Then the best that one can hope for is a characterization of the equivalence relation on DEFs given by agreement of triplets of the form (S, W', ϕ), where $S = \times_{i \in I} S_i$ is the space of pure strategy combinations, W' is a quotient space of outcome equivalence classes, and $\phi : S \to W'$ a surjection. (Such a triplet may be called a "normal form," deliberately dropping the word "game.") The conceptual difficulty with such an approach is that the equivalence relation on outcomes would have to imply that all players are indifferent between outcomes in one equivalence class. And this is a statement about the players' preferences, and not about the representation. On the

[6] One of Thompson's transformations, "inflation/deflation," does not preserve perfect recall. This has prompted Elmes and Reny (1994) to drop it and reformulate another one so as to maintain perfect recall.

other hand, no game is complete without a specification of the players' preferences. Therefore, studying Thompson transformations in the context of games represented by DEFs remains an important issue.

6.3 Discrete Games when Nodes are Primitives

Traditional formulations for extensive form games rely on representing the tree by a graph. It turns out that this is encompassed by the present framework. In this section, we aim to show how DEFs translate into a graph representation. However, a few subtleties appear in the transition to a graph representation. Those have to do with the role of terminal nodes.

In a game tree the set of plays can be identified with the set of underlying outcomes, thus making it the natural domain for preferences or payoff functions. Of course, for trees of finite height plays are one-to-one with the set of terminal nodes. On the other hand, for infinite-horizon games this equivalence may break down, since it is not guaranteed that all plays correspond to terminal nodes. For example, in the infinite centipede (Example 2.13) the infinite terminal node $\{\infty\}$ could be removed without affecting the structure of the game tree. This implies that there is some freedom in how one constructs the graph representation. To find a simple representation, one needs to know how far the graph can be "pruned" without affecting the structure of the game.

6.3.1 Simple Trees

Propositions 2.11 and 4.2 show that infinite terminal nodes can be added and removed at will, without affecting the structure of the tree. In particular, by Proposition 4.2, infinite terminal nodes may be suppressed. Doing so allows us to give a definition of discrete trees that does not rely on the concept of a game tree. Such a definition of a "simple tree" as an ordered set of nodes relies on two additional facts. First, it is known from Theorem 2.1 that decision trees as partially ordered sets give rise naturally to trees as collections of subsets of an underlying set (of plays). Second, Theorem 6.1(d) yields that discreteness can be characterized by finiteness of the sets of predecessors of moves. (In the following definition the symbol $\geq$ denotes the predecessor relation.)

Definition 6.3 A **simple tree** $T = (N, \geq)$ is a partially ordered set (the elements of which are called nodes) with a maximum $x_0 \in N$ (called the root) such that

(ST1) for all $x \in N$, the set $\uparrow x = \{y \in N \,|\, y \geq x\}$ is a finite chain;
(ST2) for all $x, y \in N$, if $x > y$ then there exists $z \in N$ such that $x > z$ and neither $y \geq z$ nor $z \geq y$;

where $>$ denotes the strict order derived from $\geq$ (i.e. $x > y \Leftrightarrow x \geq y \neq x$).

For a simple tree, the set of moves is given by $X = \{x \in N \,|\, \exists\, y \in N : x > y\}$. The following result clarifies in which sense the above definition is equivalent to the concept that has been developed in the previous Section.

Proposition 6.5 *(a) Let $T = (N, \geq)$ be a simple tree and W the set of plays (maximal chains) for T. If $W(N) = \{W(x)\}_{x \in N}$ is the collection of all sets $W(x) = \{w \in W \,|\, x \in w\}$ with $x \in N$, then $T' = (W(N), \supseteq)$ is a discrete game tree (with underlying set W) that is order-isomorphic[7] to T with all nodes finite. Further, $(W(N) \cup \{\{w\} \,|\, w \in W\}, \supseteq)$ is a complete game tree (which may have infinite terminal nodes).*

(b) Let $T' = (N, \supseteq)$ be a discrete game tree, and let $F(N)$ be its set of finite nodes. Then, $T = (F(N), \supseteq)$ is a simple tree.

Proof (a) Note that $(W(N), \supseteq)$ is the "image in plays" introduced in Sect. 2.2.4. Conditions (ST1) and (ST2) imply that T is a decision tree (by Corollary 2.1) and that $T' = (W(N), \supseteq)$ is order-isomorphic to T. Next, it is shown that $T'$is a game tree. To see (GT1), let h' be a chain in $W(N)$. By order isomorphism there exists a chain h in T such that $h' = \{W(x) \,|\, x \in h\}$. Since h is contained in some maximal chain $w \in W$ by the Hausdorff Maximality Principle, it follows that $w \in W(x)$ for all $x \in h$, as required by Boundedness (2.16). To see the "if"-part, let h' be a subset of $W(N)$ such that there exists some $w \in W$ with $w \in W(x)$ for all $W(x) \in h'$. It follows that the set h in T such that $h' = \{W(x) \,|\, x \in h\}$ is a chain, thus also h' is a chain.

Now turn to (GT2). Let $w, w' \in W$ with $w \neq w'$. Since w and w' are maximal chains in T, there exist $x, x' \in N$ such that $x \in w$ and $x' \in w'$, but $x \notin w'$ and $x' \notin w$. It follows that $w \in W(x) \setminus W(x')$ and $w' \in W(x') \setminus W(x)$. Thus, (GT2) also holds.

It has been shown that T' is a game tree. By the existence of a maximum T' is rooted. By (ST1) and order-isomorphism $\uparrow W(x)$ is finite and hence $W(x)$ is a finite node for all $W(x) \in W(N)$. Thus T' is regular and rooted, and, by Theorem 6.1, it is discrete.

(b) The existence of a maximum follows from the fact that T' is rooted. (ST1) follows from Trivial Intersection (2.4) and Theorem 6.1. Finally, (ST2) is Separability, which is fulfilled in any game tree (see Lemma 2.8 and Definition 2.4). Hence, for $x, y \in F(N)$ with $x > y$ it follows that there exists $z \in N$ with $x > z$ such that neither $y \geq z$ nor $z \geq y$. If $z \notin F(N)$, then by definition it follows that there must exist $z' \in F(N)$ with $x > z' > z$ such that neither $y \geq z'$ nor $z' \geq y$. ∎

In summary, the transition from simple trees as defined above to (complete) discrete game trees is a mere modelling decision. The only relevant difference is that, in simple trees, infinite terminal nodes are not included.

The order relation $\geq$ in Definition 6.3 cannot directly be replaced by set inclusion, because the resulting object might not be a game tree. The reason is that,

[7]This means that $x \geq y$ if and only if $W(x) \supseteq W(y)$, for all nodes $x, y \in N$.

with such a definition, nodes may contain elements which are irrelevant in the sense that they do not correspond to any play.[8]

Most textbook expositions use trees where, as in simple trees, nodes are taken as the primitives. An exception is Ritzberger (2002), where discrete trees are used with nodes as sets of outcomes as in Definitions 2.4 and 6.1. A prominent example for simple trees in a textbook treatment are the trees used by Osborne and Rubinstein (1994).

Example 6.5 (Osborne-Rubinstein trees as simple trees) Consider the trees introduced in Sect. 2.2.2.1. Let A be an arbitrary set of "actions" and H a set of (finite or infinite) sequences from A such that (a) $\emptyset \in H$, (b) if $(a_\tau)_{\tau=1}^{t} \in H$ (where t may be infinity) and $k < t$, then $(a_\tau)_{\tau=1}^{k} \in H$, and (c) if an infinite sequence $(a_\tau)_{\tau=1}^{\infty}$ satisfies $(a_\tau)_{\tau=1}^{t} \in H$ for every positive integer t, then $(a_\tau)_{\tau=1}^{\infty} \in H$.

The set W of outcomes is given by $W = W_F \cup W_\infty$ where W_F is the set of maximal finite sequences in H and W_∞ is the set of infinite sequences in H.

The ordering on nodes is given by $(a_\tau)_{\tau=1}^{t} \geq (a'_\tau)_{\tau=1}^{k}$ if $t \leq k$ and $a_\tau = a'_\tau$ for all $\tau = 1, \ldots, t$, and $\emptyset \geq h$ for all $h \in H$. Then, $(H, \geq)$ is a (rooted) tree, satisfying (ST1). It may fail (ST2), though, because a given node may only be followed by a single other node.[9] As mentioned in Sect. 2.2.2.1, this can be fixed by adding the following condition: (d) If $(a_\tau)_{\tau=1}^{t} \in H$, then there exists $a'_t \neq a_t$ such that $\left((a_\tau)_{\tau=1}^{t-1}, a'_t\right) \in H$. Let $H' = H \setminus W_\infty$ denoting the set of finite sequences in H. Under condition (d), $(H', \geq)$ is a simple tree.

Let N be formed by the singletons $\{\{w\}\}_{w \in W_\infty}$ together with all sets of the form $W(z) = \{w \in W \,|\, w \geq h\}$ for all finite sequences $h \in H'$ (recall Example 2.6). Then $(N, \supseteq)$ is order-isomorphic to (the completion of) $(H', \geq)$ and a discrete game tree. (Notice that $W(\emptyset) = W$.)

For a simple tree $(N, \geq)$ the immediate predecessor of a node $x \in N \setminus \{x_0\}$ is given by $\hat{p}(x) = \min \uparrow x \setminus \{x\}$, in analogy to the definition of immediate predecessor in game trees. The immediate predecessors of a set $c \subseteq N$ of nodes are given by $\hat{p}(c) = \{\hat{p}(x) \,|\, x \in c\}$. Accordingly, $\hat{p}^{-1}(c) = \cup_{x \in c} \hat{p}^{-1}(x)$ for a set c of nodes. Moves $x \in X$ are defined by the condition $\hat{p}^{-1}(x) \neq \emptyset$.

6.3.2 Simple Extensive Forms

For a set of nodes c in a simple tree $W(c)$ denotes the image of c under the mapping $W(\cdot)$, i.e. $W(c) = \{W(x) \,|\, x \in c\}$ is a set of nodes in the set representation

[8]The resulting tree fails to be its own set representation by plays in the sense of Chap. 2, hence, $(W(N), \supseteq)$ may have a simpler structure than T.

[9]In the terminology of Chap. 2 Osborne-Rubinstein trees may not be *decision trees*; recall Example 2.6.

$(W(N), \supseteq)$ as given by Proposition 6.5(a). In order to distinguish it from the union, $V(c) = \cup_{x \in c} W(x)$ denotes the union.

Definition 6.4 A **simple extensive form** with player set I is a pair (T, C), where $T = (N, \geq)$ is a simple tree and $C = (C_i)_{i \in I}$ consists of collections C_i of sets of nodes such that

(SF1) if $\hat{p}(c) \cap \hat{p}(c') \neq \emptyset$ and $c \neq c'$, then $\hat{p}(c) = \hat{p}(c')$ and $c \cap c' = \emptyset$, for all $c, c' \in C_i$ and all $i \in I$;
(SF2) $\{\{y\} \,|\, y \in \hat{p}^{-1}(x)\} = \left\{\hat{p}^{-1}(x) \cap \left(\cap_{i \in J(x)} c_i\right) \middle| (c_i)_{i \in J(x)} \in A(x)\right\}$, $\forall x \in X$;
(SF3) no choice $c \in C_i$ contains two distinct ordered nodes;
(SF4) for every move $x \in X$ and every choice $c \in C_i$, $\hat{p}^{-1}(x) \setminus c \neq \emptyset$;

where $A(x) = \times_{i \in J(x)} A_i(x)$, $A_i(x) = \{c \in C_i \,|\, x \in \hat{p}(c)\}$ are the choices available to $i \in I$ at $x \in X$, and $J(x) = \{i \in I \,|\, A_i(x) \neq \emptyset\}$ is the set of decision makers at x, which is required to be nonempty, for all $x \in X$.

(SF1) is the direct translation of (DEF1), the "information set property." (SF2) is the version with nodes of (DEF2). Properties (SF3) and (SF4) are required here because now choices are not sets of outcomes anymore. If they were sets of outcomes (as in Definition 4.1) such conditions would be unnecessary.

The combination of properties of (SF1) to (SF3) implies Kuhn's exclusion of absent-mindedness (Kuhn 1953). Absent-mindedness allows the "state" that obtains (the move in the information set at which the player is called upon to choose) to depend on the decision maker's choice.

Lemma 6.1 *Let $T = (N, \geq)$ be a simple tree and let (T, C) be a simple extensive form. If $c \in C_i$ is a choice, then the information set $\hat{p}(c)$ does not contain two distinct ordered nodes.*

Proof Suppose $x, y \in \hat{p}(c)$ with $x > y$ for $c \in C_i$. Note that this implies that $i \in J(x) \cap J(y)$. The chain $\uparrow y \setminus \uparrow x$ is finite by (ST1) and, therefore, has a maximum z, which fulfills $x > z \geq y$. By construction, $z \in \hat{p}^{-1}(x)$. Therefore, (SF2) implies that there is a choice $c' \in C_i$ such that $c' \in A_i(x)$ and $z \in c'$.

If $c = c'$, then $z \in c$. Since $y \in \hat{p}(c)$, there exists a $y' \in \hat{p}^{-1}(y) \cap c$. But then $z \geq y > y'$ and $z, y' \in c$ contradict (SF3). It follows that $c \neq c'$.

Since $x \in \hat{p}(c) \cap \hat{p}(c')$ and $c \neq c'$, (SF1) implies that $\hat{p}(c) = \hat{p}(c')$ and $c \cap c' = \emptyset$. Then, $y \in \hat{p}(c)$ implies $y \subset \hat{p}(c')$. By (SF2) there is a $y'' \in \hat{p}^{-1}(y) \cap c'$. Thus, $z \geq y > y''$ with $z, y'' \in c'$ contradicts (SF3). ■

A property like (SF3) would be redundant if choices were sets of outcomes. For, in that case, if a choice covers the outcomes in a node, then it must cover all outcomes in all successors thereof. This is not the case when choices are sets of nodes.

The proof of Lemma 6.1 does not use property (SF4). The latter is a non-triviality requirement, which forbids that a choice contains all immediate successors of a given move. In such a situation, if choices were sets of outcomes, the choice would have to include that move, and hence the problem cannot arise.

Property (SF4) is necessary for the next Lemma, which allows us to translate the choices from a simple extensive form on a simple tree $T = (N, \geq)$ into choices in the set representation by plays $(W(N), \supseteq)$ given by Proposition 6.5(a). In this representation, nodes are sets of plays (maximal chains) of T, that is, the representation has the set of plays of T as the underlying set. For sets of plays a that are unions of nodes of $W(N)$ Proposition 6.2 gives the decomposition $N(a)$ that consists of nodes in $W(N)$ that are maximal in a (see (6.2)).

Lemma 6.2 *Let $T = (N, \geq)$ be a simple tree, W the associated set of plays (maximal chains), and (T, C) a simple extensive form. If $c \in C_i$ is a choice, then*

(a) $N(V(c)) = \{W(x) \,|\, x \in c\}$, where $V(c) = \cup_{x \in c} W(x)$;
(b) $P(V(c)) = W(\hat{p}(c))$, where $W(Y) = \{W(y) \,|\, y \in Y\}$ for any $Y \subseteq N$.

Proof (a) We first show that, for any $x \in c$, $W(x) \in N(V(c))$. Since $W(x) \subseteq \cup_{x' \in c} W(x') = V(c)$, it remains to show that it is actually maximal in that set. Suppose not. Then, there exists $y \in N$ such that $y > x$ and $W(y) \subsetneq V(c)$. By (SF3) $y \notin c$.

Construct a chain of nodes iteratively as follows. Let $z_0 = y$. For every $k = 1, 2, \ldots$, by (SF4), there exists a $z_k \in \hat{p}^{-1}(z_{k-1})$ such that $z_k \notin c$. The chain $\{z_0, z_1, \ldots\}$ is either infinite or it ends at a terminal node z_K. Let w be a play containing this chain. Then $w \in W(y) \subseteq V(c)$. Therefore there exists $z_w \in c$ such that $w \in W(z_w)$.

We claim that $z_k > z_w$ for all $k = 0, 1, \ldots$ and prove this inductively. Since $y, z_w \in w$ and w is a chain, either $z_w \geq y$ or $y > z_w$. In the first case $z_w \geq y > x$, but $x, z_w \in c$, which contradicts (SF3). Therefore $z_0 = y > z_w$. By induction, suppose $z_{k-1} > z_w$. Since $z_k, z_w \in w$ and w is a chain, either $z_w \geq z_k$ or $z_k > z_w$. In the first case $z_{k-1} > z_w \geq z_k$. Since $\hat{p}(z_k) = z_{k-1}$, by definition of $\hat{p}$, we have that $z_w = z_k$, which contradicts $z_k \notin c$. Hence $z_k > z_w$, completing the argument.

It follows that the chain $\{y, z_1, z_2, \ldots\}$ is contained in $\uparrow z_w$. Since the latter is finite by (ST1), the chain must end in a terminal node z_K, which contradicts $z_K > z_w$.

This shows that the nodes $W(x)$ with $x \in c$ are maximal in $V(c)$. Conversely, let $z \in W(N)$ be maximal in $V(c) = \cup_{x' \in c} W(x')$. Then, there exists some $x \in c$ such that $z \cap W(x) \neq \emptyset$, which by (GT1) in the set representation and maximality of z in $W(x)$ implies $z = W(x)$.

(b) By Proposition 6.2(b) $P(V(c)) = \{P(W(x)) \,|\, x \in c\} = \{W(\hat{p}(x)) \,|\, x \in c\} = W(\hat{p}(c))$. The second equality follows from $P(W(x)) = W(\hat{p}(x))$, which follows from the order isomorphism in Proposition 6.5(a). ■

The following result states that simple extensive forms are the translation of EDPs from discrete game trees to the corresponding simple trees.

Theorem 6.4 *(a) Let (T, C) be a simple extensive form with player set I and $T' = (W(N), \supseteq)$ the set representation given by Proposition 6.5(a). If $C' = (C'_i)_{i \in I}$ with $C'_i = \{V(c) \,|\, c \in C_i\}$ for all $i \in I$, then (T', C') is an EDP on a discrete game tree.*

(b) Let (T', C') be an EDP with player set I on a discrete game tree $T' = (N, \supseteq)$ and $T = (F(N), \supseteq)$ the associated simple tree from Proposition 6.5(b). If

$C = (C_i)_{i \in I}$ *with* $C_i = \{N(c') \cap F(N) \,|\, c' \in C'_i\}$ *for all* $i \in I$*, where* $N(c')$ *is as in Proposition 6.2, then* (T, C) *is a simple extensive form.*

Proof (a) By Theorem 6.2 it only needs to be shown that (DEF1) and (DEF2) hold for (T', C').

To see (DEF1), let $V(c_1), V(c_2) \in C'_i$ for some $i \in I$ be such that $V(c_1) \neq V(c_2)$, hence $c_1 \neq c_2$, and $P(V(c_1)) \cap P(V(c_2)) \neq \emptyset$. We wish to show that $\hat{p}(c_1) \cap \hat{p}(c_2) \neq \emptyset$. To see this, let $W(x) \in P(V(c_1)) \cap P(V(c_2))$. By definition of P, for $k = 1,2$ there exists $W(y_k) \in \downarrow V(c_k)$ such that $\uparrow W(x) = \uparrow W(y_k) \setminus \downarrow V(c_k)$, which in particular implies $W(x) \not\subseteq V(c_k)$. Let $W(z_k)$ be the maximum of the chain $\uparrow W(y_k) \setminus \uparrow W(x)$, which is finite by Theorem 6.1(d). It follows that $p(W(z_k)) = W(x)$ which by order isomorphism (Proposition 6.5(a)) implies $\hat{p}(z_k) = x$.

We claim that $W(z_k)$ is maximal in $V(c_k)$. To see this, suppose there is $W(x') \in W(N)$ such that $W(z_k) \subset W(x') \subseteq V(c_k)$. Since $W(y_k) \subseteq W(z_k)$, it follows that $W(x') \in \uparrow W(y_k)$. If $W(x') \in \uparrow W(x)$, then $W(x) \subseteq W(x') \subseteq V(c_k)$ contradicts that $W(x) \not\subseteq V(c_k)$. Hence $W(x') \in (\uparrow W(y_k)) \setminus (\uparrow W(x))$, contradicting the definition of $W(z_k)$.

By Lemma 6.2(a) $W(z_k) \in N(V(c_k)) = W(c_k)$, hence $z_k \in c_k$. Since $\hat{p}(z_k) = x$ for $k = 1,2$, this shows that $x \in \hat{p}(c_1) \cap \hat{p}(c_2)$. Thus it follows from (SF1) that $\hat{p}(c_1) = \hat{p}(c_2)$ and $c_1 \cap c_2 = \emptyset$.

It remains to show that $V(c_1) \cap V(c_2) = \emptyset$. Suppose there is a play w such that $w \in V(c_1) \cap V(c_2)$. Then there are $x_1 \in c_1$ and $x_2 \in c_2$ such that x_1 and x_2 belong to the play w. This implies that they are ordered, say, $x_1 \geq x_2$. Since $c_1 \cap c_2 = \emptyset$, in fact $x_1 > x_2$. This implies that $x_1 \geq \hat{p}(x_2)$. Since $\hat{p}(c_1) = \hat{p}(c_2)$, there is some $z \in c_1$ such that $\hat{p}(x_2) = \hat{p}(z)$, thus $x_1 \geq z$. If $x_1 = z$ would hold, then $\hat{p}(x_1) = \hat{p}(x_2)$ in contradiction to $x_1 > x_2$. Hence, $x_1 > z$ which contradicts (SF3).

Before turning to (DEF2), note that a choice (set of plays) $V(c) \in C'_i$ is available at a node $W(x)$ in T' if and only if the choice (set of nodes) c is available at the node x of T. This follows from $W(\hat{p}(c)) = P(V(c))$ (Lemma 6.2(b)). Therefore we will not distinguish between availability in (T, C) and in (T', C').

To show (DEF2), we first claim that, given $y \in \hat{p}^{-1}(x)$ such that $\{y\} = \hat{p}^{-1}(x) \cap \left(\cap_{i \in J(x)} c_i\right)$ for some $(c_i)_{i \in J(x)} \in A(x)$, it follows that $W(y) = W(x) \cap \left(\cap_{i \in J(x)} V(c_i)\right)$. The proof is by double inclusion. First let $w \in W(y)$. Since $y \in \hat{p}^{-1}(x)$, it follows that $x \geq y$ and thus $w \in W(x)$. For each $i \in J(x)$, $y \in c_i$ implies that $w \in V(c_i)$, thus $w \in W(x) \cap \left(\cap_{i \in J(x)} V(c_i)\right)$. Conversely, let $w \in W(x) \cap \left(\cap_{i \in J(x)} V(c_i)\right)$. Suppose $w \notin W(y)$. Since $x \in X$ and w is a play, there is a $z \in w \cap \hat{p}^{-1}(x)$. Further, for each $i \in J(x)$ there exists a $z_i \in c_i \cap w$. Then $\hat{p}(z_i) \in w \cap \hat{p}(c_i)$. Since $x \in w \cap \hat{p}(c_i)$, it follows that $x = \hat{p}(z_i)$. Otherwise, $\hat{p}(c_i)$ would contain two distinct ordered nodes, a contradiction to Lemma 6.1. Since $x = \hat{p}(z_i) = \hat{p}(z)$ and $z, z_i \in w$, it follows that $z = z_i$. Therefore $z \in \hat{p}^{-1}(x) \cap \left(\cap_{i \in J(x)} c_i\right) = \{y\}$, a contradiction.

To complete the verification of (DEF2) we now show that

$$p^{-1}(W(x)) = \left\{ W(x) \cap \left(\cap_{i \in J(x)} V(c_i)\right) \,\middle|\, (c_i)_{i \in J(x)} \in A(x) \right\} \tag{6.4}$$

for every $x \in X$. Let $W(y) \in p^{-1}(W(x))$ (proceeding again by double inclusion). This implies $y \in \hat{p}^{-1}(x)$ by the order isomorphism in Proposition 6.5(a). By (SF2), there exists $(c_i)_{i\in J(x)} \in A(x)$ such that $\{y\} = \hat{p}^{-1}(x) \cap \left(\cap_{i\in J(x)} c_i\right)$. By the claim above, it follows that $W(y) = W(x) \cap \left(\cap_{i\in J(x)} V(c_i)\right)$. Conversely, let $a = W(x) \cap \left(\cap_{i\in J(x)} V(c_i)\right)$ for some $(c_i)_{i\in J(x)} \in A(x)$. By (SF2) there exists $y \in \hat{p}^{-1}(x)$ such that $\{y\} = \hat{p}^{-1}(x) \cap \left(\cap_{i\in J(x)} c_i\right)$. By the claim above, $a = W(y)$. This completes the proof of (DEF2).

(b) By Theorem 6.2 (DEF1) and (DEF2) hold for (T', C'). It remains to check (SF1-4) for (T, C). First notice that, if $c' \in C_i'$ for some $i \in I$ and $c = N(c') \cap F(N)$, it follows that $\hat{p}(c) = P(c')$ by Proposition 6.2(b). For,

$$P(c') = \{p(x) \,|\, x \in N(c') \cap F(N)\} = \hat{p}(c),$$

because $c = N(c') \cap F(N)$ and $\hat{p}$ is just the notation for the immediate predecessor when $T = (F(N), \supseteq)$ is considered as a simple tree.

To show (SF1), let $N(c_1') \cap F(N), N(c_2') \cap F(N) \in C_i$ for some $i \in I$ be such that $\hat{p}(N(c_1') \cap F(N)) \cap \hat{p}(N(c_2') \cap F(N)) \neq \emptyset$ and $N(c_1') \cap F(N) \neq N(c_2') \cap F(N)$. This implies $c_1' \neq c_2'$ and $P(c_1') \cap P(c_2') \neq \emptyset$. By (DEF1), it follows that $P(c_1') = P(c_2')$ and $c_1' \cap c_2' = \emptyset$. Hence $\hat{p}(N(c_1') \cap F(N)) = \hat{p}(N(c_2') \cap F(N))$. It remains only to show that $N(c_1') \cap N(c_2') \cap F(N) = \emptyset$. Proceeding indirectly, suppose there exists a node $x \in N(c_1') \cap N(c_2') \cap F(N)$. Then, since x is actually also a node in T', it follows that $\emptyset \neq x \subseteq c_1' \cap c_2'$, a contradiction.

Turn to (SF2). A choice (set of nodes) $N(c') \cap F(N) \in C_i$ is available at a node x in $T = (F(N), \supseteq)$ if and only if the choice (set of plays) c' is available at the node x of $T' = (N, \supseteq)$. This follows from $\hat{p}(N(c) \cap F(N)) = P(c)$. Suppose $y \in p^{-1}(x)$ is such that $y = x \cap \left(\cap_{i\in J(x)} c_i'\right)$ for some $(c_i')_{i\in J(x)} \in A(x)$. For each $i \in J(x)$, $x \in P(c_i') = \hat{p}(N(c_i') \cap F(N))$ and $y \subseteq c_i'$ implies that $y \in N(c_i') \cap F(N)$ by Proposition 6.2. This shows that $y \in p^{-1}(x) \cap \left(\cap_{i\in J(x)} N(c_i')\right) \cap F(N)$. Suppose there is $z \in F(N)$ with $z \in p^{-1}(x) \cap \left(\cap_{i\in J(x)} N(c_i')\right)$. It follows that $z \subseteq x \cap \left(\cap_{i\in J(x)} c_i'\right) = y$, hence $z = y$. This shows that (DEF2) implies (SF2).

Consider (SF3). If for any $c' \in C_i'$ for some $i \in I$ there are $x, y \in N(c') \cap F(N)$ such that $x \supset y$, this contradicts that $N(c')$ is a partition of c' by Proposition 6.2(a).

It remains to show (SF4). Let x be a move in $F(N)$ and $c = N(c') \cap F(N)$ for some $c' \in C_i'$, for some $i \in I$. If $\hat{p}^{-1}(x) \cap c = \emptyset$, (SF4) is void. Therefore, we can assume that $\hat{p}^{-1}(x) \cap c \neq \emptyset$. Then there is $y \in c$ such that $\hat{p}(y) = x$. Suppose $\hat{p}^{-1}(x) \setminus c = \emptyset$. Then, by Proposition 6.2, $z \subseteq c'$ for every $z \in \hat{p}^{-1}(x)$. By Proposition 6.1(a) $x = \cup_{z\in\hat{p}^{-1}(x)} z$, hence, $x \subseteq c'$, which contradicts $y \in N(c')$ (Proposition 6.2). ■

This final theorem allows one to go back and forth between an extensive form representation with a simple tree, where nodes are the primitives, and one with a discrete game tree, where plays are the primitives.

Example 6.6 (Job-Market Signalling revisited) Reconsider Example 6.4, the job market signalling model by Spence (1973). Now the nodes are the primitives, hence identify plays $w \in W$ with terminal nodes $z(\theta, e, (w_1, w_2)) \in Z$, denote the root by

x_0, and keep the notation $x_t(e)$ for the firms' moves. Choices by firms are now sets of nodes, $c_1(\overline{w}_1) = \{z \in Z | w_1 = \overline{w}_1\}$ and $c_2(\overline{w}_2) = \{z \in Z | w_2 = \overline{w}_2\}$. Choices for the worker of type $t = L, H$ are the singletons $c_t(\overline{e}) = \{x_t(\overline{e})\}$. The verification of conditions (SF1), (SF2), (SF3), and (SF4) is straightforward.

6.4 Perfect Recall

There is one property of extensive forms that is crucial for applications, yet is not implied by the definition of an EDP nor a DEF: *perfect recall.* This property was introduced by Kuhn and, according to him, "...is equivalent to the assertion that each player is allowed by the rules of the game to remember everything he knew at previous moves and all of his choices at those moves." (Kuhn 1953, p. 213) As a matter of fact, perfect recall does a little more. It can be shown (see Ritzberger 1999) that it is equivalent to the simultaneous fulfillment of three independent properties: Players never forget what they did; they never forget what they knew; and (for a given player) past, present, and future have an unambiguous meaning.

In a sense an extensive form that fails perfect recall fails to capture rational behavior on the part of the players. Therefore, it appears mandatory to include perfect recall among the assumptions. In this section we prepare the ground for that.

Remark 6.1 There is a subtle, though inconsequential, technical difficulty regarding choices and infinite terminal nodes. Recall that the latter are never actually available as the outcome of a decision, since by definition of infinite nodes they arise only as the limit of infinitely many previous decisions. Suppose that a given choice c is changed by adding to it an infinite terminal node $\{w\}$, where w is a play that does *not* pass through the information set at which the choice is available. This change does not affect the DEF at all. More formally, the set C_i of choices is in principle different from the set $\{c \cap W(P(c)) | c \in C_i\}$, although for all practical purposes they are equivalent.

To see this, suppose $w \notin W(P(c))$, $w \notin c$, where $\{w\} \in E$ is infinite. Then the only element of $\downarrow(c \cup \{w\})$ is $\{w\} \in E$, but $\uparrow\{w\} \setminus \downarrow(c \cup \{w\}) = \uparrow\{w\} \setminus \{\{w\}\}$ has no minimum. Therefore, $P(c \cup \{w\}) = P(c)$.

The converse is also true. If $w \in c \setminus W(P(c))$, then the chain $\uparrow\{w\} \cap \downarrow c$ cannot contain any move. For, if there were $y \in \uparrow\{w\} \cap \downarrow c \cap X$, then the chain $(\uparrow\{w\} \cap \downarrow c) \setminus \{\{w\}\}$ would be nonempty, contained in X, and therefore equipped with a maximum $y \in X$ by Proposition 6.4. Since for a discrete tree every move is a finite node, the chain $\uparrow y \setminus \{y\}$ would have a minimum $x = p(y)$, which would fulfill $\uparrow x = \uparrow y \setminus \downarrow c$ and therefore $x \in P(c)$. But the latter would yield $w \in x \in P(c)$ in contradiction to the hypothesis. Therefore, $\uparrow\{w\} \cap \downarrow c \cap X = \emptyset$, that is, $\uparrow\{w\} \cap \downarrow c = \{\{w\}\}$. If $\{w\} \in E$ were a finite node, the chain $\uparrow\{w\} \setminus \{\{w\}\} = \uparrow\{w\} \setminus \downarrow c$ would have a minimum $x \in X$ that would again satisfy $\uparrow x = \uparrow\{w\} \setminus \downarrow c$, hence, $x \in P(c)$ in contradiction to $w \notin W(P(c))$. Therefore, $\{w\} \in E$ must be infinite and $P(c) = P(c \setminus \{w\})$.

Since an information set $P(c)$ remains unchanged by the addition of a disjoint infinite terminal node or the removal of an infinite terminal node contained in $c \in C_i$ but not in $W(P(c))$, it can be assumed without loss of generality that $c \subseteq W(P(c))$ for all $c \in C_i$ and all $i \in I$. This assumption will henceforth be maintained.

6.4.1 Definition and Characterization

Recall that $\phi : S \to W$ denotes the surjection, identified by Theorem 5.5, which associates to every strategy profile the outcome that it induces. Given a DEF (T,C) with tree $T = (N, \supseteq)$ and a move $x \in X_i$ of player $i \in I$, say that x is *possible* under $s_i \in S_i$ for player i, denoted $x \in \text{Poss}(s_i)$, if there is $s_{-i} \in \times_{j \neq i} S_j \equiv S_{-i}$ such that $\phi(s_i, s_{-i}) \in x$. Similarly, an information set $P(c)$ for $c \in C_i$ for player $i \in I$ is *relevant* under $s_i \in S_i$ for player i, denoted $P(c) \in \text{Rel}(s_i)$, if $P(c) \cap \text{Poss}(s_i) \neq \emptyset$. The following definition is from Kuhn (1953).

Definition 6.5 A DEF (T,C) satisfies **perfect recall** if $P(c) \in \text{Rel}(s_i) \Rightarrow P(c) \subseteq \text{Poss}(s_i)$ for all $c \in C_i$, all $s_i \in S_i$, and all $i \in I$.

Clearly, perfect recall could be defined for each player separately by dropping the last quantifier. Furthermore, the definition may also be rewritten in terms of the function $\phi : S \to W$ as follows: A DEF satisfies *perfect recall* if, for all $x \in P(c)$,

$$x \cap \phi(s_i, S_{-i}) \neq \emptyset \Rightarrow y \cap \phi(s_i, S_{-i}) \neq \emptyset, \forall y \in P(c) \tag{6.5}$$

for all $c \in C_i$, all $s_i \in S_i$, and all $i \in I$, where $\phi(s_i, S_{-i}) = \{\phi(s_i, s_{-i}) \,|\, s_{-i} \in S_{-i}\}$.

A drawback of Definition 6.5 is that it refers to pure strategies, that is, to derived objects, rather than primitives (the tree and choices). The characterization below will give two ways to define perfect recall in terms of the primitives of a DEF, and one way that enables its application to a proof of Kuhn's Theorem. To develop the characterization two auxiliary results are helpful.

Lemma 6.3 *$\phi(s) = w$ if and only if $w \in x \in X_i \Rightarrow w \in s_i(x)$ for all $i \in I$.*

Proof The function $\phi : S \to W$ is defined as the unique fixed point of the map

$$R_s(w) = \cap \{s_i(s) \,|\, w \in x \in X_i, i \in J(x)\} \tag{6.6}$$

that is, $\phi(s) = w$ if and only if $\{w\} = R_s(w)$. Hence, if $w \in x \in X_i \Rightarrow w \in s_i(x)$ for all $i \in I$, then $w = \phi(s)$. To see the converse, assume that for some $i \in I$ there is $x \in X_i \cap \uparrow \{w\}$ such that $w \notin s_i(x)$. Then $w \notin R_s(w)$, hence, $w \neq \phi(s)$. ■

Lemma 6.4 *For a DEF (T,C) and any player $i \in I$: If $x \in P(c)$ for some $c \in C_i$ and $x \subseteq c' \in C_i$, then there is $x' \in P(c')$ with $x \subset x'$.*

Proof Suppose that $x \in P(c)$ for some $c \in C_i$ and $x \subseteq c' \in C_i$. The chain $\uparrow x \backslash \downarrow c'$ (where $\downarrow c' = \{y \in N \,|\, y \subseteq c'\}$) is contained in the set X of moves, the move x does not belong to it, and $x \subset z$ for all $z \in \uparrow x \setminus \downarrow c'$, that is, x provides a lower bound for the chain. Therefore, the chain $\uparrow x \setminus \downarrow c'$ cannot have an infimum in the set E of terminal nodes, as all terminal nodes are singletons and a singleton cannot contain a move. It follows from down-discreteness and Proposition 6.4 that it has a minimum $x' = \min \uparrow x \setminus \downarrow c'$. By (4.1) it follows that $x' \in P(c')$ with $x \subset x'$, as desired. ■

The following result provides three characterizations of perfect recall for a DEF. The first is based on the correspondence $\phi_i : S_i \twoheadrightarrow W$ defined by $\phi_i(s_i) = \phi(s_i, S_{-i})$ for all $s_i \in S_i$. The lower (or weak) inverse ϕ_i^- of the correspondence ϕ_i is given by $\phi_i^-(V) = \{s_i \in S_i | V \cap \phi_i(s_i) \neq \emptyset\}$ for any subset $V \subseteq W$. The other two are alternative definitions of perfect recall proposed in the literature; the first by Selten (1975) and the other by Osborne and Rubinstein (1994).

Proposition 6.6 *For a DEF (T, C) each of the following three statements is equivalent to perfect recall: For all players $i \in I$ and all choices $c, c' \in C_i$ of player i,*[10]

(a) $\phi_i^-(x) = \phi_i^-(y)$ for all $x, y \in P(c)$;
(b) if there is $x \in P(c)$ such that $x \subseteq c'$, then $y \subseteq c'$ for all $y \in P(c)$;
(c) if there are $x \in P(c)$ and $x' \in P(c')$ with $x \subset x'$, then there is a unique $c'' \in A_i(x')$ such that $W(P(c)) \subseteq c''$.

Proof "Perfect recall implies (a):" If $x \in P(c)$ for $c \in C_i$ and $s_i \in \phi_i^-(x)$, then $x \cap \phi(s_i, S_{-i}) \neq \emptyset$ by the definition of ϕ_i^-. Perfect recall then implies that $y \cap \phi(s_i, S_{-i}) \neq \emptyset$ for all $y \in P(c)$, hence, $s_i \in \phi_i^-(y)$. Since $x, y \in P(c)$ enter symmetrically in this argument, the statement follows.

"(a) implies (b):" Suppose for $c \in C_i$ there is $x \in P(c)$ such that $x \subseteq c' \in C_i$. By Theorem 5.1 there is $s \in S$ such that $\phi(s) \in x$. Then, first, $s_i \in \phi_i^-(x)$ implies $s_i \in \phi_i^-(y)$, that is, $y \cap \phi(s_i, S_{-i}) \neq \emptyset$, for all $y \in P(c)$ by (a). Second, because $x \subseteq c' \in C_i$ by hypothesis, there is $x' \in P(c')$ with $x \subset x'$ by Lemma 6.4. Third, $s_i(x') = c'$, because otherwise $s_i(x') \cap c' = \emptyset$ by (DEF1) would contradict $\phi(s) \in x$ by Lemma 6.3.

Consider any $y \in P(c)$. If $y \setminus c' \neq \emptyset$, then by Theorem 5.1 there is $s' \in S$ such that $\phi(s') \in y \setminus c'$. In case $y \setminus c' \neq \emptyset$ there are two possibilities.

First, if there were $y' \in P(c')$ with $\phi(s') \in y'$, then $\phi(s') \in s_i'(y')$ by Lemma 6.3. On the other hand, by the definition of a pure strategy $s_i'(y') = s_i'(x') \neq c'$, because $\phi(s') \notin c'$. This yields $s_i'(x') \cap c' = \emptyset$ by (DEF1), which implies from $x \subseteq c'$ that $x \cap \phi(s_i', S_{-i}) = \emptyset$, that is, $s_i' \notin \phi_i^-(x)$, even though $\phi(s') \in y \in P(c)$, in contradiction to (a).

This leaves the second possibility, that $P(c') \cap \uparrow \{\phi(s')\} = \emptyset$. In this case let $s_i'' \in S_i$ be such that $s_i''(z) = s_i'(z)$ for all $z \in X_i \setminus P(c')$ but $s_i''(z) \neq c'$ for all $z \in P(c')$. From $P(c') \cap \uparrow \{\phi(s')\} = \emptyset$ it follows that $\phi(s_i'', s_{-i}') = \phi(s') \in y \in P(c)$, hence,

[10]If perfect recall were defined as a property of player i's choice set alone, as it is possible, the first quantifier could be dropped.

$s_i'' \in \phi_i^-(y)$. But $x \cap \phi(s_i'', S_{-i}) = \emptyset$, because $s_i''(x') \cap c' = \emptyset$ by (DEF1) and $x \subseteq c'$, which once again contradicts (a).

Therefore, $y \setminus c' = \emptyset$ or, equivalently, $y \subseteq c'$. Since $y \in P(c)$ was arbitrary, statement (b) follows.

"(b) implies (c):" Suppose there are $x \in P(c)$ and $x' \in P(c')$ with $x \subset x'$. By up-discreteness the chain $\uparrow x \setminus \uparrow x'$ has a maximum $z = \max \uparrow x \setminus \uparrow x' \in X$. Since $z \in p^{-1}(x')$, by (DEF2) there is $c'' \in A_i(x')$ such that $z = x' \cap c'' \cap \left[\bigcap_{j \in J(x') \setminus \{i\}} c_j\right]$ for some choice combination $(c_j)_{j \in J(x') \setminus \{i\}} \in \times_{j \in J(x') \setminus \{i\}} A_j(x')$. It follows from $x \subseteq z \subset x'$ and $z \subseteq c''$ that $x \subseteq c''$. By (b) this implies that $y \subseteq c''$ for all $y \in P(c)$, hence, $W(P(c)) \subseteq c''$. That $c'' \in A_i(x')$ is unique follows from (DEF1), because any $\hat{c} \in A_i(x') \setminus \{c''\}$ is disjoint from c'' and, therefore, cannot cover $x \in P(c)$.

"(c) implies perfect recall:" Suppose for $s_i \in S_i$ there is $s_{-i} \in S_{-i}$ such that $\phi(s_i, s_{-i}) \in x \in P(c)$ for some $c \in C_i$, that is, $P(c) \in \text{Rel}(s_i)$. There are two cases to consider. Either player i has a decision point that properly contains x or not.

Suppose first that there is $x' \in \uparrow x \setminus \{x\}$ with $x' \in P(c')$ for some $c' \in C_i$. Then by (c) there is a unique $c'' \in A_i(x')$ such that $W(P(c)) \subseteq c''$. It follows from Lemma 6.3 and the definition of a pure strategy that $s_i(y') = c''$ for all $y' \in P(c')$ and, therefore, $y \cap \phi(s_i, S_{-i}) \neq \emptyset$, that is, $y \in \text{Poss}(s_i)$, for all $y \in P(c)$, as required by perfect recall.

Otherwise, if $X_i \cap (\uparrow x \setminus \{x\}) = \emptyset$, then indeed $X_i \cap (\uparrow y \setminus \{y\}) = \emptyset$ for all $y \in P(c)$. For, if there were $y \in P(c)$ and $y' \in X_i$ with $y \subset y'$, then by (c) there would by a unique $c'' \in A_i(y')$ such that $W(P(c)) \subseteq c''$, in particular, $x \subseteq c''$. By Lemma 6.4 there would be $x' \in P(c'')$ with $x \subset x'$, in contradiction to the hypothesis. Hence, indeed $X_i \cap (\uparrow y \setminus \{y\}) = \emptyset$ for all $y \in P(c)$. But then $y \cap \phi(s_i, S_{-i}) \neq \emptyset$ for all $y \in P(c)$, that is, $P(c) \subseteq \text{Poss}(s_i)$, as required by perfect recall. ■

Since to a certain extent statement (a) is a rewriting of Kuhn's definition, Proposition 6.6 essentially summarizes known definitions of perfect recall. It follows that any one of the definitions from the literature can be used in applications. This also holds if the game under scrutiny has an infinite horizon and/or large action sets.

Perea (2001, Definition 2.1.2) gives a very intuitive definition of perfect recall for the case of finite games which also extends to the infinite case in a straightforward way.

Corollary 6.2 *A DEF (T, C) satisfies perfect recall if and only if for each $i \in I$, each $c \in C_i$, and each $x, y \in P(c)$, the path from the root to x implies the same collection of player i's choices as the path from the root to y.*

Proof The collection of choices implied by a path of play from the root to x is

$$Ch(x) = \left\{c' \in C_i \;\middle|\; c' \supseteq x \text{ and } \exists y \in X_i \text{ with } x \subset y, y \in P(c')\right\}.$$

It is then straightforward that $Ch(x) = Ch(y)$ if and only if condition (a) in Proposition 6.6 holds. ■

6.4.2 A Choice-Based Definition

In addition to the ones discussed above, there is one yet unknown definition, that refers to a basic tree structure of the set of choices.

Corollary 6.3 *A DEF (T,C) has perfect recall if and only if the set of choices C_i satisfies Trivial Intersection, i.e.,*

$$\textit{if } c\cap c' \neq \emptyset \textit{ then either } c \subset c' \textit{ or } c' \subseteq c \tag{6.7}$$

for all $c,c' \in C_i$ and all $i \in I$.

Proof "if:" It suffices to demonstrate that statement (b) from Proposition 6.6 holds under the hypothesis. Hence, suppose that, for $c,c' \in C_i$ and $i \in I$, there is $x \in P(c)$ such that $x \subseteq c'$. Since $x \subseteq c'$ and $x \setminus c \neq \emptyset$ by (4.1), there is $w' \in c' \setminus c$. Therefore, $c \subset c'$ by $\emptyset \neq x \cap c \subseteq c' \cap c$ and the hypothesis of Trivial Intersection for C_i. Hence, $c' \in C_i$ cannot be available at moves in $P(c)$ by (DEF1). Since $x' \cap c \neq \emptyset$ for all $x' \in P(c)$ by (4.1), it follows that $x' \cap c' \neq \emptyset$ for all $x' \in P(c)$. Suppose there were $x' \in P(c)$ such that $x' \setminus c' \neq \emptyset$. Because there is a node $y' \in p^{-1}(x')$ with $y' \subseteq c$ by (DEF2) and by $c \subset c'$ this node would satisfy $y' \in \downarrow c'$ and it would follow from (4.1) that $x' \in P(c')$, a contradiction. Therefore, $x' \subseteq c'$ for all $x' \in P(c)$, as required.

"only if:" Assume perfect recall. If $c \cap c' \neq \emptyset$ for $c,c' \in C_i$, then the two choices must be available at different information sets by (DEF1). Furthermore there must be moves $x \in P(c)$ and $x' \in P(c')$ such that $x \subset x'$, say. For, if $w \in c \cap c'$, then $w \in W(P(c)) \cap W(P(c'))$, because choices partition the (set of plays passing through the) information sets at which they are available, and $x,x' \in \uparrow \{w\}$ by Lemma 6.4. It follows from Proposition 6.6(c) that $W(P(c)) \subseteq c'$ and from $c \subset W(P(c))$ (see Remark 6.1) that $c \subset c'$. ■

The proof of the only-if part of this Corollary in fact demonstrates a stronger property than Trivial Intersection for choices: Perfect recall implies that choices and information sets *jointly* satisfy Trivial Intersection, that is, for all $b,b' \in B_i \equiv C_i \cup \{W(P(c)) \,|c \in C_i\}$,

$$\text{if } b \cap b' \neq \emptyset \text{ then either } b \subset b' \text{ or } b' \subseteq b, \tag{6.8}$$

for all $i \in I$. Since the latter implies that choices alone satisfy Trivial Intersection, (6.7), it is again equivalent to perfect recall by the if part of Corollary 6.3. Furthermore, under perfect recall information sets alone, $\{W(P(c)) \,|c \in C_i\}$, clearly also satisfy Trivial Intersection. Yet, the proof of the if part of Corollary 6.3 reveals that the equivalence of both (6.7) and (6.8) to perfect recall is due to the tree structure. In particular, in the absence of a tree structure the appropriate definition of perfect recall may be (6.8).

Corollary 6.4 *If* (T,C) *is a DEF with perfect recall, then for each choice* $c \in C_i$ *the set* $\{c' \in C_i \,|\, c \subseteq c'\}$ *of i's choices that come before c is a finite chain, for all players* $i \in I$.

Proof By the only-if part of Corollary 6.3 the set $\{c' \in C_i \,|\, c \subseteq c'\}$ is a chain. Since each $x \in P(c)$ is a finite node by Theorem 6.1(c), x has only finitely many predecessors $x' \in P(c') \cap \uparrow x$ with $c \subseteq c' \in C_i$. As these predecessors along a play $w \in c$ correspond one-to-one to choices $c' \in C_i$, the latter also form a finite chain. ■

6.4.3 Some Implications

In applications of game theory perfect recall is prominent because of Kuhn's Theorem (Kuhn 1953). Kuhn's Theorem characterizes perfect recall by the equivalence of mixed and behavioral strategies (Ritzberger 2002, Section 3.4) for finite games. An important step in the proof of that equivalence is to show that the conditional probability of a choice given a decision point (where this choice is available) depends only on the information set and not on the particular move therein. Provided conditional probabilities exist, this step follows from Proposition 6.6(a).

More precisely, suppose there is a probability measure σ_i over pure strategies S_i that captures the notion of a mixed strategy. Suppose also that at a given move x the set $\phi_i^-(x)$ of strategies potentially leading to x is measurable in the appropriate σ-algebra and $\sigma_i(\phi_i^-(x)) > 0$. Let x' be another move such that $x, x' \in P(c)$ for some choice $c \in C_i$. By Proposition 6.6(a) it follows that $\phi_i^-(x) = \phi_i^-(x')$. The set of strategies potentially leading to x (and x') that then select the choice c is given by $\phi_i^-(x \cap c) = \phi_i^-(x) \cap S_i(c)$ where

$$S_i(c) = \{s_i \in S_i \,|\, \exists x \in X_i : s_i(x) = c\} \tag{6.9}$$

and hence $\phi_i^-(x \cap c) = \phi_i^-(x' \cap c)$. Provided this latter set is also measurable, we immediately obtain that

$$\frac{\sigma_i(\phi_i^-(x \cap c))}{\sigma_i(\phi_i^-(x))} = \frac{\sigma_i(\phi_i^-(x' \cap c))}{\sigma_i(\phi_i^-(x'))}. \tag{6.10}$$

This equation says that the conditional probability of choice c given x equals the conditional probability of choice c given x', as desired.

This, of course, takes very strong hypotheses. As observed by Aumann (1964), if the set of pure strategies is "too large" it is by no means true that a randomization among an arbitrary subset of pure strategies does induce a well-defined probability distribution over outcomes (plays in W). Indeed, Aumann (1961) provides a characterization showing that, even for two-stage games, this requires the set of pure strategies over which the mixed strategy randomizes to be "relatively small" (see also Rao 1971). Using an extraneous probability space $(B, \mathscr{B}, \lambda)$ on which further assumptions can be imposed, bypasses this unsolvable problem (see Aumann 1963). The further assumptions on $(B, \mathscr{B}, \lambda)$ can be used to guarantee existence of probability distributions over outcomes for those randomizations which are still allowed under measurability. But one should be aware of the fact that, for arbitrarily large games, this is indeed a restriction. For, if $\mathscr{B}$ is "small" while the set of pure strategies is "large", then the measurability requirement might rule out otherwise interesting randomizations.[11]

Remark 6.2 To be more precise, let (Ω, Σ, μ) be a probability space such that Ω is either finite or countable with the discrete σ-algebra or isomorphic to the unit interval. Let $(W, \mathscr{W})$ be a measurable space. The probability space (Ω, Σ, μ) captures chance or other players and $(W, \mathscr{W})$ outcomes. Your own strategy space S is the set of all $(\Sigma, \mathscr{W})$-measurable functions $s : \Omega \to W$. If you wish to randomize your strategy choice, you will be interested in the subset $F \subseteq S$ such that there is a σ-algebra $\mathscr{S}_F$ on F for which the "outcome function" $\rho_F : F \times \Omega \to W$, defined by $\rho(s, \omega) = s(\omega)$ for all $(s, \omega) \in F \times \Omega$, is jointly measurable. For, if you randomize over F according to the probability measure ν (on $\mathscr{S}_F$) and $A \in \mathscr{W}$ can be assigned a probability, then $\Pr(s(\omega) \in A) = (\mu \times \nu)\left(\rho_F^{-1}(A)\right)$, hence $\rho_F^{-1}(A)$ must belong to the product σ-algebra on $F \times \Omega$.

Aumann (1961) proves a characterization theorem for such a subset $F \subseteq S$ (under the hypothesis that both (Ω, Σ) and $(W, \mathscr{W})$ have countable generating families). This theorem relates the "sizes" of (Ω, Σ) and $(W, \mathscr{W})$ to how large F can be. If both Ω and W are finite, then randomization can apply to all strategies, that is, $F = S$ actually holds. On the other hand, if both Ω and W are continua, then the best that can be achieved is a proper subset, $F \subset S$. In particular, if both Ω and W are like unit intervals (with the Borel algebra), it is impossible to randomize over *all* functions from the unit interval to itself. Aumann (1963) shows that the maximal subset F over which randomization works is as if one uses an extraneous (standard) randomization device: Instead of mixing over functions $s : \Omega \to W$ you use a "roulette wheel" $(B, \mathscr{B}, \lambda)$ and select a random variable $\sigma : B \to S$ that picks a strategy for you.

Even with a "reasonable" set of pure strategies for a given player, it is not true that a mixed strategy (with measurability assumption included) always induces a well-defined probability distribution over outcomes for any given pure strategy combination of the other players. To see this point, consider the following game, which is also adapted from Aumann (1964).

Example 6.7 There are two players, 1 and 2, engaged in ultimatum bargaining. Player 1 can propose a split of a surplus, which is any number from the interval $[0,1]$. Player 2 observes the proposal and responds by either accepting (1) or rejecting (0) the split. The set of pure strategies of player 1 is $S_1 = [0,1]$ and the set of pure strategies of player 2 is the set S_2 of all functions of the form $s_2 : [0,1] \to \{0,1\}$.

[11]For instance, if B is finite, no matter how large the set of pure strategies is, the mixed strategy will only be allowed to randomize among finitely many pure strategies.

The set of plays can then be taken to be $W = [0,1] \times \{0,1\}$. The requirement for the σ-algebra on W boils down to the requirement that singletons and sets of the form $\{r\} \times \{0,1\}$, for $r \in [0,1]$, are measurable. This is fulfilled if one e.g. takes the product of the Borel σ-algebra on $[0,1]$ and the discrete σ-algebra on $\{0,1\}$. The outcome function ϕ is explicitly given by $\phi(s_1, s_2) = (s_1, s_2(s_1))$, which induces the Borel σ-algebra on S_1. The induced σ-algebra on S_2 is, of course, more complex. Arbitrary randomizations over S_2 pose a problem, as observed above. But surely there will be no problem for player 1 to, say, randomize uniformly if player 2 adopts a given pure strategy?

This is not the case. In particular, take a non-Borel set A of $[0,1]$ and consider the indicator function $\mathbf{1}_A \in S_2$. This is a pure strategy, thus, for a pure decision theorist, no restriction can rule it out. Now suppose player 1 randomizes uniformly over $S_1 = [0,1]$. What is the induced distribution over outcomes? Clearly, the set $[0,1] \times \{1\}$ should be measurable for any reasonable model of the game. But $\phi_1^{-1}([0,1] \times \{1\}) = A$, which is by construction not measurable. Thus the uniform randomization of player 1 (which is a well-defined random variable) does not induce a distribution over the set of outcomes.[12]

A generalization of Kuhn's theorem (Kuhn 1953), therefore, poses serious technical difficulties. On the other hand, given that DEFs satisfy no-absent-mindedness (see Proposition 4.1), it is not difficult to show that mixed strategies are at least as powerful as behavior strategies. That is, for each probability distribution on plays induced by a behavior strategy combination there is a mixed strategy combination that induces the same distribution—irrespective of whether or not perfect recall obtains.

To make this precise we follow Aumann (1964). Fix a "standard" probability space $(\Omega, \Sigma, \lambda)$ (a "roulette wheel"), i.e., such that Ω is either finite or countable with the discrete σ-algebra or isomorphic to the unit interval. Endow the product space S of pure strategy profiles with a σ-algebra $\mathscr{S}$ such that the outcome function $\phi : S \to W$ is $(\mathscr{S}, \mathscr{W})$-measurable. Finally, for each player $i \in I$ endow the space S_i of pure strategies with the σ-algebra $\mathscr{S}_i$ given by the projection σ-algebra onto S_i.

A *mixed strategy* for player $i \in I$ then is a $(\Sigma, \mathscr{S}_i)$-measurable function $\sigma_i : \Omega \to S_i$. Denote by M_i the set of all mixed strategies of player $i \in I$ and by $M = \times_{i \in I} M_i$ the space of all mixed strategy profiles. The interpretation of a mixed strategy of player $i \in I$ is that i picks the set $\vartheta \in \Sigma_i$ of pure strategies with probability $\lambda\left(\sigma_i^{-1}(\vartheta)\right)$. By varying the function σ_i the player chooses this probability. Of course, the same caveat as before applies. If S_i is too large as compared to Ω, only "few" pure strategies can be chosen with positive probability.

Another, piecemeal approach to randomized strategies is to allow players to randomize among available choices, independently at each information set. Formally, for each player $i \in I$ endow the set C_i of choices with a σ-algebra $\mathscr{C}_i$ and let B_i denote

[12] It may be tempting to replace the Borel σ-algebra by the discrete σ-algebra. But, of course, it is known that on the class of all subsets of the unit interval there exists no probability measure which assigns the value zero to all singletons; see Birkhoff (1973, p. 187) or Billingsley (1986, p. 41).

the set of all $(\Sigma, \mathscr{C}_i)$-measurable functions $b : \Omega \to C_i$. A *behavior strategy* for player $i \in I$ is a function $\beta_i : X_i \to B_i$, whose values are denoted $b_{ix} = \beta_i(x) : \Omega \to C_i$ for all $x \in X_i$, such that, for all $x, y \in X_i$, (a) $b_{ix}(\Omega) \subseteq A_i(x)$, (b) if $y \in P(b_{ix}(\omega))$ for some $\omega \in \Omega$, then $b_{iy} = b_{ix}$, and (c) if there is no $c \in C_i$ with $x, y \in P(c)$, then $\lambda\left(b_{ix}^{-1}(\vartheta) \cap b_{iy}^{-1}(\vartheta')\right) = \lambda\left(b_{ix}^{-1}(\vartheta)\right)\lambda\left(b_{iy}^{-1}(\vartheta')\right)$ for all $\vartheta, \vartheta' \in \mathscr{C}_i$.

Condition (a) states that if $c \in b_{ix}(\Omega) = \{b_{ix}(\omega) \in C_i \,|\, \omega \in \Omega\}$, then $c \in A_i(x) = \{c \in C_i \,|\, x \in P(c)\}$, i.e. $x \in P(c)$, for all $x \in X_i$; that is, it ensures that the random variable b_{ix} is supported on choices that are available at $x \in X_i$. Condition (b) demands that the same random variable b_{ix} is assigned to all moves y in the information set that contains x; hence, the behavior strategy β_i does not use more information than what the player has. Finally, condition (c) imposes independence on decisions at distinct information sets. Denote by $\mathscr{B}_i$ the set of all behavior strategies of player $i \in I$, and by $\mathscr{B} = \times_{i \in I} \mathscr{B}_i$ the space of all behavior strategy profiles.

The interpretation of the probability $\lambda\left(b_{ix}^{-1}(\vartheta)\right)$ is as the *conditional* probability that player i takes a choice in the set $\vartheta \in \mathscr{C}_i$ *given* that move $x \in X_i$ has materialized. Player i decides on this conditional probability by choosing the function (random variable) $b_{ix} \in B_i$. By condition (b) these decisions are perfectly correlated across all moves in the information set that contains x, but independent across different information sets by condition (c). Hence, while mixed strategies pick functions from decision points to choices potentially at random, behavior strategies pick choices at each decision point (again potentially at random), independently across different information sets. Because of the independence inherent in behavior strategies, they are, in general, less powerful than mixed strategies are.

Proposition 6.7 *Let (T, C) be a DEF. If the behavior strategy profile $\beta \in \mathscr{B}$ induces the probability measure $\mu : \mathscr{W} \to [0, 1]$ on the measurable space $(W, \mathscr{W})$, then there exists a mixed strategy combination $\sigma \in M$ that also induces μ.*

Proof To begin with, define for the behavior strategy profile $\beta \in \mathscr{B}$ and each player $i \in I$ the function $f_i : X_i \times \Omega \to C_i$ by $f_i(x, \omega) = b_{ix}(\omega) = \beta_i(x)(\omega)$ for all $(x, \omega) \in X_i \times \Omega$, and let $f = (f_i)_{i \in I}$ denote the associated profile. Let $\phi : S \to W$ be the surjection that assigns to each pure strategy combination $s \in S$ the play that it induces, as in (6.6). This function exists by Theorems 5.2 and 5.5. Observe that for each fixed $\omega \in \Omega$ the function $f(\cdot, \omega) : X_i \to C_i$ is a pure strategy combination, i.e. $f(\cdot, \omega) \in S$. For, (a) guarantees that $f_i(\cdot, \omega)^{-1}(c) \subseteq P(c)$ and (b) ensures that $f_i(\cdot, \omega)^{-1}(c) \supseteq P(c)$ for all $c \in C_i$ and all $i \in I$. Hence, β induces μ if $\mu(V) = \lambda(\{\omega \in \Omega \,|\, \phi(f(\cdot, \omega)) \in V\})$ for all $V \in \mathscr{W}$. For fixed $\omega \in \Omega$ a mixed strategy profile $\sigma \in M$ is a pure strategy combination by definition, $\sigma(\omega) \in S$. Hence, σ induces μ if $\mu(V) = \lambda(\{\omega \in \Omega \,|\, \phi(\sigma(\omega)) \in V\})$ for all $V \in \mathscr{W}$.

Given the behavior strategy profile $\beta \in \mathscr{B}$, construct a mixed strategy profile $\sigma \in M$ by setting $\sigma(\omega) = f(\cdot, \omega) \in S$ for each $\omega \in \Omega$. Then by construction, if β induces μ, then $\sigma \in M$ also does, and the statement is verified. ■

This result is made possible by the fact that no-absent-mindedness is an integral part of the approach at hand. Specifically, the proof of Proposition 6.7 implicitly relies on the result (Proposition 4.1) that in a DEF every play can pass through an information set at most once, viz. no-absent-mindedness.

Proposition 6.7 is the generalization to large games of one of the implications contained in Kuhn's theorem (Kuhn 1953), which in its most general form states that behavior strategies and mixed strategies are equivalent if and only if the game satisfies perfect recall (for an alternative approach to the proof of Kuhn's theorem, see von Stengel 2002, Corollary 4.2). Without perfect recall, the converse of Proposition 6.7 (i.e., that behavior strategies are as powerful as mixed strategies), does not hold, because of condition (c) in the definition of behavior strategies (independence). This can lead to problems in applications, where one often determines optimal choices locally, at each information set separately, and pastes together a solution to the overall game from these local solutions. For instance, an "equilibrium in behavior strategies" may not be an equilibrium at all due to a profitable deviation in mixed strategies, or there may be equilibria in mixed strategies which cannot be reproduced in behavior strategies (see, e.g., the remarkable example by Wichardt 2008).

This explains the importance of the characterization embodied in Kuhn's theorem. Kuhn (1953) established this result for the finite case only. This was because the general case poses the technical difficulties sketched above. It took over 10 years until Aumann (1964) proved that under perfect recall behavior strategies are as powerful as mixed strategies, in a set-up with an infinite horizon and action sets that are homeomorphic to the unit interval. This version states only the sufficiency part of Kuhn's theorem, and it does so only for a fixed strategy profile among the opponents. The latter appears acceptable for practical purposes. The former, the necessity of perfect recall, was proved another 10 years later by Schwarz (1974) under some additional (mild) measurability assumptions.

6.5 Summary

This chapter identifies what is really needed for applications. For, when an abstract framework has been sufficiently explored, it is often helpful to turn around and look for extra structure that does not limit its practical applicability, but simplifies the model and adds transparency. The necessary and sufficient conditions for an extensive form to be playable fall short of the structure that has been employed in the models commonly used in the literature. That is, practical applicability is preserved when a discreteness property is added.

The result is a greatly simplified framework that allows almost everything that is relevant in practise and still is more general than the classical definitions. The characterization of discrete trees yields the existence of an (onto) immediate predecessor function (Theorem 6.1). This, in turn, allows us to collapse the abstract definition of an extensive form to two simple and intuitive properties: that the decisions of the relevant players at any given move lead to a new node (the new "state" of the game), and that players cannot deduce from the available choices where they are in an information set (Definition 6.2 and Theorem 6.2).

The freedom to take nodes or plays as the primitives of the formalization is preserved in these steps. In the end it can be shown that under discreteness one can take nodes as primitives—like in many textbook expositions—and still go back and forth between this and a model, where plays are the primitives (Theorem 6.4). The only cases that are precluded by discreteness are continuous time problems and transfinite games.

In the last section of this chapter the framework of a discrete extensive form is used to characterize the key concept of perfect recall. It is shown (Proposition 6.6) that for the discrete case Kuhn's definition of perfect recall indeed agrees with other definitions that have been proposed in the literature. Moreover, we discuss the potential for a proof of Kuhn's Theorem, which states that perfect recall is equivalent to mixed and behavioral strategies being equally powerful. This raises a number of measure-theoretic issues, though, that are beyond the scope of the present treatment.

References

Alós-Ferrer C, Ritzberger K (2013) Large extensive form games. Econ Theory 52(1):75–102
Alós-Ferrer C, Ritzberger K (2016c) Characterizations of perfect recall. Int J Game Theory. doi:10.1007/s00182-016-0534-x
Aumann RJ (1961) Borel structures for function spaces. Ill J Math 5:614–630
Aumann RJ (1963) On choosing a function at random. In: Wright F B (ed) Ergodic theory. Academic Press, New Orleans, pp 1–20
Aumann RJ (1964) Mixed and behavior strategies in infinites extensive games. In: Advances in game theory. Annals of mathematics study, vol 52. Princeton University Press, Princeton, pp 627–650
Aumann RJ, Hart S (2003) Long cheap talk. Econometrica 71:1619–1660
Billingsley P (1986) Probability and measure. Wiley, New York
Birkhoff G (1973) Lattice theory. American mathematical society, vol XXV, 3rd edn. Colloquium Publications, Providence
Dalkey N (1953) Equivalence of information patterns and essentially determinate games. In: Kuhn H, Tucker A (eds) Contributions to the theory of games. Annals of mathematics study, vol II, vol 28, Princeton University Press, Princeton, pp 217–243
Elmes S, Reny P (1994) On the strategic equivalence of extensive form games. J Econ Theory 62:1–23
Kreps D, Wilson R (1982) Sequential equilibria. Econometrica 50:863–894
Kuhn H (1953) Extensive games and the problem of information. In: Kuhn H, Tucker A (eds) Contributions to the theory of games, vol II. Annals of mathematics study, vol 28. Princeton University Press, Princeton, pp 217–243
Mas-Colell A, Whinston M, Green J (1995) Microeconomic theory. Oxford University Press, Oxford
Osborne MJ, Rubinstein A (1994) A course in game theory. The MIT Press, Cambridge
Perea A (2001) Rationality in extensive form games, theory and decision library, series C, vol 29. Kluwer Academic, Boston/Dordrecht/London
Rao BV (1971) Borel structures for function spaces. Colloq Math 23:33–38
Ritzberger K (1999) Recall in extensive form games. Int J Game Theory 20:69–87

Ritzberger K (2002) Foundations of non-cooperative game theory. Oxford University Press, Oxford
Rubinstein A (1982) Perfect equilibrium in a bargaining model. Econometrica 50:97–109
Schwarz G (1974) Ways of randomizing and the problem of their equivalence. Isr J Math 17:1–10
Selten R (1975) Reexamination of the perfectness concept for equilibrium points in extensive games. Int J Game Theory 4:25–55
Shapley L (1953) Stochastic games. Proc Natl Acad Sci 39:1095–1100
Spence AM (1973) Job market signaling. Q J Econ 83:355–377
Thompson FB (1952) Equivalence of games in extensive form. RAND research memorandum, vol 759. Rand Corp., Santa Monica
von Neumann J, Morgenstern O (1944) Theory of games and economic behavior. Princeton University Press, Princeton
von Stengel B (2002) Computing equilibria for two-person games. In: Aumann RJ, Hart S (eds) Handbook of game theory, vol 3. Elsevier, Amsterdam, pp 1723–1759
Wichardt PC (2008) Existence of Nash equilibria in finite extensive form games with imperfect recall: a counterexample. Games Econ Behav 63(1):366–369

Chapter 7
Equilibrium

A prime motivation for the present work is that extensive form games with large action spaces and/or an infinite horizon are pervasive in economics. Examples include oligopoly models à la Cournot (1838), Bertrand (1883), or von Stackelberg (1934), infinite bilateral bargaining (Rubinstein 1982), or stochastic games (Shapley 1953). In contrast, the theory of games is best understood when the representation of the game is finite. The previous chapters have extended this understanding to large games, with infinitely long plays and/or continuum action spaces. Yet, up to this point the extensions have concerned only the representation of the game. Now we turn to solutions for the first time. This raises new issues, as solutions involve optimization and equilibrium. On the other hand, discrete game trees and extensive forms provide a framework that is simultaneously rich enough and sufficiently simple to tackle these issues.

The first question, of course, is how to begin. The necessity to consider optimal behavior and equilibrium calls for additional structure. But what is a reasonable criterion to identify which structure needs to be added? In this chapter we argue for a minimal criterion.[1] More precisely, we would like to identify, for a given discrete game tree, what is the minimal added structure needed for existence of subgame perfect equilibria in perfect information games defined on this tree. In other words, rather than formulating a set of sufficient conditions for an existence theorem, our aim is to also identify which are the *necessary* conditions that need to be imposed if one takes the existence of subgame perfect equilibria as a postulate.

The spirit of the exercise is similar to the imposition of (A0), (A1), and (A2) in Chap. 5. On the other hand, the present endeavour is less ambitious. In particular, we do not ask for the existence of equilibria in arbitrary imperfect information games defined on a game tree. We are content with understanding which

[1]This chapter is based on Alós-Ferrer and Ritzberger (2015), Alós-Ferrer and Ritzberger (2016a), and Alós-Ferrer and Ritzberger (2016b).

C. Alós-Ferrer, K. Ritzberger, *The Theory of Extensive Form Games*,
Springer Series in Game Theory, DOI 10.1007/978-3-662-49944-3_7

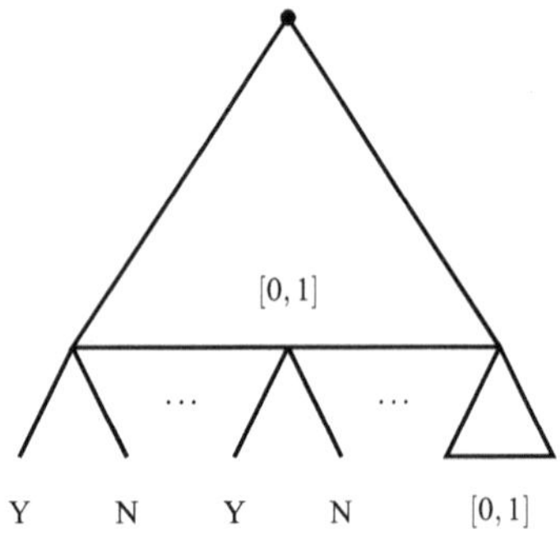

Fig. 7.1 Graphical representation of Example 7.1. Player 1 chooses a number a in $[0,1]$. For any $a<1$, player 2 either agrees ($b=1$) or disagrees ($b=0$). For $a=1$, player 2 may choose any number $b\in[0,1]$

additional structure ensures the existence of (subgame perfect) equilibria for *perfect information* games.

Sufficient conditions for existence of subgame perfect equilibria in perfect information games (Kuhn 1953) have been provided for different classes of large extensive form games by Fudenberg and Levine (1983), Harris (1985a), and Hellwig and Leininger (1987). *Necessary* conditions have not yet received attention. The contribution of the present chapter is to identify conditions which are both necessary *and* sufficient. That is, we aim at *characterizing* with which structure a potentially large game tree has to be endowed for subgame perfect equilibria to exist for all perfect information games on that tree (for continuous preferences and suitable assignments of decision points).

7.1 Preview

The conditions, of course, concern the topology on the domain of preferences viz. the set of plays of the game tree. It will be shown that the existence of subgame perfect equilibria in perfect information games necessarily requires two natural conditions on the topology that relate nodes and plays in the tree.

The criterion that all perfect information games on the tree ought to have subgame perfect equilibria is a modest one. For, it shuts down all complications that may arise from imperfect information. Still, it already implies more than one may expect. In particular, this criterion excludes the following example.

Example 7.1 (Yes, No, or More) There are two players, $i=1,2$. Initially player 1 chooses a number $a\in[0,1]$ from the unit interval. Then player 2 sees a. If $a<1$, player 2 may either agree, $b=1$, or disagree, $b=0$. If $a=1$, player 2 also chooses a number $b\in[0,1]$ from the unit interval. Once a and b have been chosen, the game terminates. See Fig. 7.1 for a graphical representation of the game tree. Payoffs are $U_1(a,b)=ab$ and $U_2(a,b)=\min\{b,1/2+a-b\}$. The set of plays can be endowed with the relative Euclidean topology of the plane, so that it is compact and payoff functions are continuous. If player 1 chooses $a<1/2$, then player 2's best reply is

$b = 0$, if $a = 1/2$ player 2 is indifferent, if $1/2 < a < 1$ player 2's best reply is $b = 1$, and if $a = 1$ player 2's best reply is $b = 3/4$. Therefore, player 1 obtains zero for $a < 1/2$, not more than $1/2$ at $a = 1/2$, payoff $U_1(a,1) = a > 1/2$ when choosing $a \in (1/2, 1)$, and $3/4$ at $a = 1$. This "value function" has no maximum and the game no subgame perfect equilibrium.

In the last example, player 2's "action set" explodes from binary choice (if player 1 chooses $a < 1$) to a continuum of options (if player 1 chooses $a = 1$). The next example illustrates that the conceptual problems that need to be tackled do not reduce to whether the action sets of players vary in a regular way. Rather, one needs to address properties of the tree. If the structure of the tree is irregular enough, existence may fail even if compactness holds and all preferences are continuous, even if each player only acts at exactly one move.

Example 7.2 (Square example revisited) Consider a 2-player perfect information game where player 1 picks at the root of the tree either a pair $(a,b) \in [0,1]^2$ with $a < 1$ or sets $a = 1$ and gives the move to player 2 (this is Example 1.2 from the Introduction). In the former case ($a < 1$) the game ends with payoffs $U_1(a,b) = ab$ and $U_2(a,b) = 1 - b$. In the latter case ($a = 1$) player 2 receives the opportunity to choose $b \in [0,1]$ and, once b has been selected, the game ends with payoffs $U_1(1,b) = b$ and $U_2(1,b) = 1 - b$. The set W of possible outcomes (plays) can be identified with the unit square $W = [0,1]^2$ and endowed with the relative Euclidean topology of the plane. Then W is compact and payoff functions are continuous. Since player 2, if called upon to move, will always choose $b = 0$, player 1's "value" function has no maximum and the game no subgame perfect equilibrium.

The examples above show that asking for existence of subgame perfect equilibrium in perfect information games defined on a given tree *does* constrain the topological structure imposed on the plays of the tree—compactness and continuity of preferences are not enough. (Example 7.2 also illustrates a few other points to which we will return below.) In fact, we identify two conditions that are *necessary* for existence of equilibrium in perfect information games. The first is simply that nodes in the tree are closed as sets of plays. The second condition demands that the function assigning to each (finite) node of the tree its immediate predecessor is an open map in the appropriate topology on nodes. Henceforth topologies satisfying these two conditions are called "tree topologies."

In fact, these conditions are also *sufficient* for existence of equilibrium in perfect information games. This is shown by exhibiting an algorithm that differs from Kuhn's (1953) finite one and from its generalizations by Fudenberg and Levine (1983), Harris (1985a), or Hellwig and Leininger (1987). In particular, this algorithm iterates the players' expectations of future players' behavior. They start naively, as if they had full control over the entire future. At the next iteration step they take into account the behavior of future players from the previous iteration step, and so on. This procedure has a limit at which all players correctly anticipate how later players will behave—a "rational expectations" limit. And this limit contains a subgame perfect equilibrium.

Putting together these two results yields a characterization. Any (compact and perfectly normal) topology on the set of plays of a (discrete) game tree admits a subgame perfect equilibrium for every (well-behaved) perfect information game defined on this tree *if and only if* it is a tree topology. Thus, tree topologies are precisely the structure that needs to be added in order to do equilibrium analysis in large (extensive form) games.

The virtue of a characterization is that it pins down precisely which conditions are needed and which are not. For example, the present result shows that no condition on upper hemi-continuity of the successor correspondence is needed to establish existence of subgame perfect equilibria. Furthermore, the characterization shows that without a tree topology there is a well-behaved perfect information game that has no subgame perfect equilibrium, where the game is well-behaved if preferences are continuous and decision points are assigned such that players locally face well posed optimization problems.

The characterization is based on two stronger results. The two defining conditions are necessary if the topology is *perfectly normal* (which is always the case if it is metrizable), but not necessarily compact. So, necessity only takes a particular separation axiom. The two conditions are sufficient if the topology is *compact*, but the separation axiom can be weakened to a separated topology (Hausdorff). Therefore, the two results underlying the characterization have merits in their own right.

One may guess that the hypothesis of compactness in the sufficiency result is in fact also necessary, at least for the relevant case of a perfectly normal space. Strikingly, the underlying issue is undecidable under the standard axioms of set theory (see Sect. 7.7.3 for details).

In the build-up towards the main result a few surprises arise. In particular, in infinite games backwards induction may not be equivalent to subgame perfection (see Example 7.6). Theorem 7.1 below gives a sufficient condition for the equivalence of backwards induction and subgame perfection: preferences are lower semi-continuous on plays. In that case a one-shot deviation principle applies, which thus turns out, somewhat unexpectedly, to require a topological assumption. A second point that may come as a surprise is that multiple subgame perfect equilibria may arise in large but generic perfect information games (see Example 7.16).

To gain intuition, Sect. 7.2 discusses several examples and illustrates the advantages of the set-tree approach over other formalizations, with an eye on equilibrium analysis. Section 7.3 introduces the basic objects of analysis, (well-behaved) perfect information games and topologies that admit equilibrium analysis. It also contains the one-shot deviation principle which holds that subgame perfection and backwards induction are equivalent. Section 7.4 presents the characterization result that identifies the necessary and sufficient conditions for a topology to admit equilibrium analysis ("tree topology"). This is a consequence of two stronger results. The first one (Sect. 7.5) is the necessity result for perfectly normal but not necessarily compact topologies. The second one (Sect. 7.6) is the existence result that compact separated tree topologies are sufficient to admit equilibrium analysis. Section 7.7 discusses a few fine points, and Sect. 7.8 summarizes.

7.2 Motivating Examples: Nodes as Sets Versus Plays as Sequences

Many alternative definitions of extensive form games have been introduced in the literature. In this chapter our goal is to analyze the existence of subgame perfect equilibria on the basis of topological properties of preferences. Preferences are defined on ultimate outcomes of the game, and hence our approach to modeling trees and games (which, we remind the reader, follows the original approach of von Neumann and Morgenstern 1944, Section 8) becomes particularly convenient, because the specification starts with ultimate outcomes (plays). This is in contrast, for instance, to the "graph approach" popularized by Kuhn (1953). It also has to be mentioned that other models have been used in the literature, in particular when large games are at stake. For example, Harris (1985a) proposed a sequence-approach that was later popularized by Osborne and Rubinstein (1994), and which has been frequently used to analyze equilibrium existence in large games. The present subsection illustrates why our set-tree approach is most convenient for the present purposes.

It is useful to informally summarize the sequence approach at this point. Imagine an infinite-horizon game with a clear stage structure as follows. Each period, one and only one player makes a decision, choosing an action from a fixed action set A_t, which only depends on the period, t. It is then natural to consider a product structure, where plays are the elements of the product set $A = \times_{t \geq 0} A_t$. If one assumes that every one of the action sets A_t is compact, Tychonoff's Theorem then guarantees that the product space A is also compact. The sequence approach is merely the extrapolation of this construction to more general games. Depending on the structure of the tree, several players might be able to make choices at the same stage, giving rise to different action sets A_{it}. Under the sequence approach, one still assumes that each of those action sets is a compact set and builds the product set, applying Tychonoff's theorem to obtain compactness of the latter.

This approach creates three major difficulties. First, as we will illustrate below, it creates an artificial space which might be far larger than the actual space of outcomes. That is, unless the game has a simple stage structure with just one active player per period and no possibility of ending the game before the "last period," the product space will generally contain many sequences which do not correspond to any play. Second, as a consequence of this, Tychonoff's theorem is not enough to guarantee compactness of the space of plays, because the latter is a subset of the former. To remedy this problem, one needs an additional technical assumption, namely that the set of plays $W \subseteq A$ is a closed subset of the product space (as assumed by Harris 1985a). Third, the product topology as a construction is often convenient, but for infinite-horizon games it rarely relates to the structure of the problem described in the game. As noticed already by Sobel (1975, p. 974), there is "no apparent connection" between the product topology and the continuity properties that one needs to impose on the players' preferences.

Harris (1985a) provided a remarkable equilibrium existence result following the sequence approach. That result assumes that the action sets are Hausdorff and compact, and that the set of outcomes is a closed subset of the product of the action sets. Further, since the game does not need have a stage structure, additional assumptions are needed on the "action correspondence" which assigns action sets to players' decision points. The assumptions on the action sets imply that this correspondence is upper hemi-continuous. Harris (1985a) further assumes lower hemi-continuity, that is, the action correspondence needs to be fully continuous. Under those assumptions Harris (1985a) shows existence of subgame perfect equilibria whenever the players' payoff functions are continuous.

In contrast, our Theorem 7.4 below follows the set-tree approach. In doing so, it is uncovered that several of the assumptions just mentioned are imposed by the sequence approach, but are actually unnecessary. Our theorem does not require the action sets to be Hausdorff, or the action correspondence to be upper hemi-continuous. We will return to those differences in Sect. 7.7. The purpose of the present section is to illustrate why the set-tree approach is more natural, and why the sequence approach creates artificial difficulties.

The next example shows how distinct formal approaches can lead to substantial differences, not only when modeling games, but also when analyzing them. The example is a game that is easily captured by our approach and where subgame perfect equilibria always exist (for continuous preferences). However, if the game is modeled by other approaches, the result is a less natural construction that creates formal difficulties. Even worse, as a consequence of those, existence results from the literature fail to apply.

Example 7.3 (Four Intervals) Two players jointly decide a real number $a \in [0,1]$ and a natural number $n \in \{1,2,3,4\}$. The rules are as follows. Player 1 can either choose $a < 1$ and let player 2 choose $b \in \{1,2,3,4\}$, or fix $a = 1$ and simultaneously choose "Low" or "High." In the latter case, if player 1 chooses Low player 2 will be able to choose $b \in \{1,2\}$, while if player 1 chooses High player 2 will be able to choose $b \in \{3,4\}$. The natural set of outcomes is $W = [0,1] \times \{1,2,3,4\}$, which can be endowed with the relative Euclidean topology (which coincides with the product of the Euclidean topology on $[0,1]$ and the discrete topology on $\{1,2,3,4\}$). The set of nodes is also easy to construct. The root is simply W itself. The "terminal" nodes are those of the form $\{(a,n)\}$ with $(a,n) \in W$. There is a continuum of intermediate nodes (where player 2 decides). Those are the nodes of the form $\{\{a\} \times \{1,2,3,4\} \mid a < 1\}$, plus the two distinguished nodes $x_L = \{(1,1),(1,2)\}$ and $x_H = \{(1,3),(1,4)\}$. The set of nodes is

$$N = \left\{W, x_L, x_H, (\{a\} \times \{1,2,3,4\})_{a \in [0,1)}, (\{w\})_{w \in W}\right\}$$

once again ordered by set inclusion.

The sequence approach is cumbersome in this example. Following the pioneering work in this approach (Harris 1985a), one examines each stage of the game and endows each player with a fixed universal set of actions for that stage. In Harris's (1985a) approach this is without loss of generality, because a later restriction reduces the resulting product set to the subset of plays, incorporating all constraints given by the game. In the present example the natural choice of the (compact) action set for player 1 is $A_1 = [0,1] \times \{Low, High\}$, since this player may (but need not) choose both a and either Low or High. For player 2 the natural choice is $A_2 = \{1,2,3,4\}$. This results in the product space $A = A_1 \times A_2 = [0,1] \times \{Low, High\} \times \{1,2,3,4\}$, which is strictly larger than the actual set of outcomes W. Again following Harris (1985a), the structure of the game is recovered by restricting attention to the set H of plays, a subset of A. First, all plays must be included where player 1 sets $a < 1$, which yields the subset $H_1 = \{((a, Low), n) \mid a < 1\}$. In this set the coordinate Low is merely a marker which does not constrain the future choices of player 2. (We could instead use High as a marker, or a third marker which would complicate things even further, but some marker is needed because the sequence approach is based on a product construction.) Second, all plays must be included where player 2 sets $a = 1$, which must then incorporate the appropriate constraint on player 2's choice. This yields the set $H_2 = \{((1, Low), 1), ((1, Low), 2), ((1, High), 3), ((1, High), 4)\}$. The set of plays is the subset of $[0,1] \times \{Low, High\} \times \{1,2,3,4\}$ given by $H = H_1 \cup H_2$.

The fact that the sequence approach results in artificially enlarged product sets creates a conceptual difficulty in games which do not have a straightforward stage structure, as Example 7.3. In his existence theorem Harris (1985a) assumes that the action sets A_i are compact and separated, which is unproblematic with the construction given above and the (natural) Euclidean topologies on the sets A_i. Yet, a further, crucial assumption requires the set H to be a closed subset of A. This property is not guaranteed by the construction. Consider the sequence $w_n = ((a_n, Low), 4)$ with $0 \leq a_n < 1$ for all $n = 1, 2, \ldots$ and $a_n \to_{n\to\infty} 1$. Then $w_n \in H$ for all $n = 1, 2, \ldots$, but $w_n \to_{n\to\infty} ((1, Low), 4) \notin H$. Hence the existence theorem in Harris (1985a) does not apply because its assumptions are not fulfilled: compactness has been lost purely because of the construction. However, this game has a subgame perfect equilibrium for any assignment of continuous payoffs to players. This can be shown by applying the existence theorem in Sect. 7.6 below, whose assumptions do cover this example. We show it here directly.

Let U_1 and U_2 be continuous payoff functions for players 1 and 2. For each $a \in [0,1]$ (including $a = 1$), let $F(a)$ be the number n that maximizes U_2 in the set $\{(a,1), (a,2), (a,3), (a,4)\}$. The correspondence $F : [0,1] \mapsto \{1,2,3,4\}$ is u.h.c. by the maximum theorem. Further, since both the domain and the codomain are metrizable and compact, the graph of F is a closed subset of W, hence compact. Let (a^*, n^*) maximize the continuous function U_1 on the graph of F. Let f be a selection from F such that $f(a^*) = n^*$. Further, let $n^{Low} \in \{1,2\}$ and $n^{High} \in \{3,4\}$ maximize $U_2(1,n)$ on the sets $\{1,2\}$ and $\{3,4,\}$, respectively.

Note that $f(1) \in \{n^{Low}, n^{High}\}$. Suppose $f(1) = n^{Low}$ (the argument if $f(1) = n^{High}$ is symmetric). Hence, we have that $U_1(a^*, n^*) \geq U_1(1, n^{Low})$. Suppose that, additionally, $U_1(a^*, n^*) \geq U_1(1, n^{High})$. Then there is a subgame perfect equilibrium where player 1 chooses a^*, player 2 chooses $f(a)$ for each $a < 1$, n^{Low} after $(1, Low)$, and n^{High} after $(1, High)$. Alternatively, suppose $U_1(a^*, n^*) < U_1(1, n^{High})$. Then there is a subgame perfect equilibrium where player 1 chooses $(1, High)$ and player 2 chooses as above.

Example 7.3 illustrates that the sequence approach can artificially enlarge the relevant set in such a way that seemingly technical conditions are violated and the analysis of a game becomes unnecessarily complex. This may lead to a failure of existence theorems based on the sequence approach, even though existence holds and is readily obtained under our approach (this example is covered by the existence theorem in Sect. 7.6). The next example revisits Example 7.2 and shows that the problems of the sequence approach can actually lead to the consideration of artificial product sets that are far larger than the relevant outcome sets.

Example 7.4 (Square example, continued) Recall Example 7.2. Player 1 can either decide on the whole pair (a,b) with $a<1$, or fix $a=1$ and let player choose b. Constructing the set of nodes is again simple. Let again $W=[0,1]^2$ be the set of plays. The root and the "terminal" nodes are as in the previous example, but this time there is only one intermediate node (player 2's only decision point), namely the set $\{1\}\times[0,1]$ where $a=1$ has been fixed but b remains to be decided. The set of nodes is hence

$$N=\left\{W,\{1\}\times[0,1],(\{w\})_{w\in W}\right\}$$

again ordered by set inclusion.

The sequence approach is again cumbersome in this example. Following Harris (1985a), the natural choice of the action set for player 1 is $A_1=[0,1]^2$, since this player may (but need not) choose both a and b. For player 2 the action set is clearly $A_2=[0,1]$. This results in the product space $A=A_1\times A_2=[0,1]^3$. That is, the sequence approach turns a two-dimensional object into a three-dimensional one. To recover the structure of the game we must consider the set H of plays (viewed as sequences) as a subset of A. In the present case this must include all plays where player 1 has fixed both a and b, and player 2 does not get to choose. This requires the subset $H_1=\{((a,b_1),b_2)\mid b_1=b_2,\ a<1\}$ to be one part of the set of plays. The fact that player 2 does not decide on b is incorporated in H_1 by the restriction that $b_2=b_1$. The set $H\subseteq A$ must also include all plays where player 1 has chosen $a=1$ and player 2 decides on $b\in[0,1]$. This gives the set $H_2=\{((a,0),b)\mid a=1, b\in[0,1]\}$. The coordinate 0 in $(a,0)$ is an arbitrary marker indicating that player 1 does not actually decide on b. Any other marker would also do, but it would be incorrect to write (a,b), because if $a=1$ player 2's choice of b is unconstrained by player 1's decision. The set of plays finally is the subset of $[0,1]^3$ given by $H=H_1\cup H_2$.

The point of this example is that the phenomenon illustrated in Example 7.3 is not peculiar to that example. The sequence approach unnecessarily blows up the relevant outcome space into a potentially large product set (even requiring additional dimensions). Ultimately, the reason for the difficulties above is that the sequence approach naturally leads to a Tychonoff construction, which first identifies the action sets at each stage, then imposes topological constraints on those, and finally takes the product set. The construction then needs to be "patched up" by adding additional assumptions that need not be guaranteed in well-behaved examples. In Example 7.3 existence of equilibria holds, but previous existence theorems do not apply, because the construction seems to destroy compactness (even though the set of plays is compact in the natural topology). In Example 7.4 an analogous argument shows that the set H is not a closed subset of A. Indeed, consider the sequence $w_n = ((a_n, 1), 1/2)$ with $0 \leq a_n < 1$ for all $n = 1, 2, \ldots$ and $a_n \to_{n\to\infty} 1$. Then $w_n \in H$ for all $n = 1, 2, \ldots$, but $w_n \to_{n\to\infty} ((1,1), 1/2) \notin H$. Hence the approach of Harris (1985a) does not work. However, this obscures the fact that failure of compactness is *not* the reason for the failure of existence in Example 7.4. Compactness assumptions are immediately fulfilled under our approach, for the outcome space $W = [0,1]^2$ is compact under the natural topology, and this is the set of interest. As will be discussed below, the reason for the failure of existence in this example is the failure of a completely different condition that concerns the structure of the tree and not of the outcome space. In summary, what Example 7.4 shows is that the search for *necessary* topological conditions is better undertaken under our approach, for otherwise one needs to consider additional, extraneous conditions relating the product set of actions and the set of plays viewed as a subset thereof. An analysis of which conditions are *necessary* cannot be carried out under the sequence approach, because artificial conditions have been introduced there that obscure the analysis.

7.3 Perfect Information and Backwards Induction

The analysis will assume a (rooted) complete discrete game tree. The first assumption, completeness, is without loss of generality by Propositions 2.11 and 4.2. The second assumption, discreteness, allows us to use the machinery developed in the last chapter. The following is implied by Definitions 2.4 and 6.1 together with Proposition 6.4 and the fact that for complete game trees Irreducibility (GT2) is automatically true:

Proposition 7.1 *A W-poset $(N, \supseteq)$ is a complete discrete game tree if and only if $\{w\} \in N$ for all $w \in W$, and*

(DGT1) *$h \subseteq N$ is a chain if and only if there is $w \in W$ such that $w \in \bigcap_{x \in h} x$,*
(DGT2) *every chain in the set $X = N \setminus \{\{w\}\}_{w \in W}$ of moves has a maximum and either an infimum in the set $E = \{\{w\}\}_{w \in W}$ of terminal nodes or a minimum.*

As before, a complete discrete game tree is *rooted* if $W \in N$. Furthermore, recall that for a discrete game tree all nodes are either finite or infinite by Proposition 6.3. The set of infinite nodes is denoted by $I(N)$, and the set of finite nodes together with the root $W \in N$ by $F(N)$. On the latter set the function $p : F(N) \to X$ can be defined that assigns to every finite node its *immediate predecessor*. Namely, for each $x \in F(N) \setminus \{W\}$ let

$$p(x) = \min \uparrow x \setminus \{x\} \tag{7.1}$$

and $p(W) = W$ by convention. Hence, $x \subset p(x) = \cap \{y | y \in \uparrow x \setminus \{x\}\}$ for all $x \in F(N) \setminus \{W\}$. For all $t = 1, 2, \ldots$ define $p^t = p \circ p^{t-1}$ where p^0 is the identity.

For a discrete game tree every move is a finite node, $X \subseteq F(N)$, by Theorem 6.1(c). As a consequence, the nodes of a discrete game tree $(N, \supseteq)$ can be partitioned into *slices*, defined as the sets of nodes reached after a fixed number of iterations of p^{-1} from the root. The reason to replace the more popular but also more restrictive term "stage" with the term "slice" is that the latter may contain terminal nodes. Formally, the slices $Y_0, Y_1, \ldots$ are constructed follows. Let $Y_0 = \{W\}$ and $Y_t = \{x \in N \mid p(x) \in Y_{t-1}\} = \left\{x \in N \mid p^{t-1}(x) \subset p^t(x) = W\right\}$ for all $t = 1, 2, \ldots$ Nodes in the slice Y_t are pairwise disjoint: For, if $x \cap y \neq \emptyset$, then by Trivial Intersection (2.4) either $y \subset x$ or $x \subseteq y$, hence either $y \in Y_t$ implies $x \in Y_{t-k}$ for some $k > 0$ or $x \in Y_t$ implies either $y = x$ or $y \in Y_{t-k}$ for some $k > 0$. Infinite (terminal) nodes $\{w\} \in I(N)$ do not belong to any slice.

For instance, in Example 7.2 the slice Y_1 consists of terminal nodes $\{(a,b)\} \in N$ with $a < 1$ plus the move $\{1\} \times [0,1]$. The slice Y_2 consists of the terminal nodes $\{(1,b)\} \in E$ only. In Example 7.3 the slice Y_1 consists of moves of the form $\{a\} \times \{1,2,3,4\}$ with $a < 1$ plus the two distinguished nodes $x_L = \{1\} \times \{1,2\}$ and $x_H = \{1\} \times \{3,4\}$.

In this chapter, $W : N \twoheadrightarrow W$ denotes the correspondence that assigns to every node, viewed as an element of the discrete game tree $(N, \supseteq)$, the set of its constituent plays, that is, the node itself viewed as a set of plays, i.e. $W(x) = x$ for all $x \in N$. On a fixed slice Y_t the correspondence $W(\cdot)$ is the inverse of the function $\text{proj}_t : W(Y_t) \to Y_t$ uniquely defined by $w \in \text{proj}_t(w)$ for all $w \in W(Y_t)$. For a set $Y \subseteq N$ of nodes write $W(Y) = \cup_{x \in Y} x \subseteq W$ for the union, and refer to $W(Y)$ as the *plays passing through* Y.

7.3.1 Perfect Information Games

Whether or not strategy combinations induce outcomes and do so uniquely is purely a matter of the tree. This has been shown in Chap. 5. Solution concepts are a different matter, though. Studying equilibrium entails maximization problems. And optimization naturally leads to topological considerations. Until Alós-Ferrer and Ritzberger (2015) no study had asked which topological assumptions are necessary for equilibrium analysis. Therefore, this chapter considers topologies τ imposed on the set W of plays and asks which properties of τ are necessary and sufficient in order to confront the players with well defined optimization problems. To focus on this issue, the class of games is restricted to those with *perfect information*. Without this assumption further issues about the information structure arise which are not topological in nature (but instead measure-theoretic). In fact, even with perfect information it is known that a subgame perfect equilibrium may fail to exist if chance moves destroy continuity of payoff functions. The following example (from Luttmer and Mariotti 2003) illustrates this.

Example 7.5 (Luttmer and Mariotti's example) Let $M \subset \mathbb{R}^3$ denote the graph of the correspondence defined by $(a_1,a_2) \in [0,1]^2 \mapsto [a_1+a_2-2, 2-a_1-a_2]$ and let the set W of plays be given by $W = M \times \{1,2,3\}$. The tree $(N, \supseteq)$ is given by the set N of nodes consisting of the root $W \in N$ and the terminal nodes $(\{w\})_{w \in W}$ together with the following four sets of nodes:

$$X^1 = \{\{w \in W \,|\, w_1 = a_1\} \,|\, a_1 \in [0,1]\},$$

$$X^2 = \left\{\{w \in W \,|\, (w_1,w_2) = (a_1,a_2)\} \,\middle|\, (a_1,a_2) \in [0,1]^2\right\},$$

$$X^3 = \{\{w \in W \,|\, (w_1,w_2,w_3) = (a_1,a_2,a_3)\} \,|\, (a_1,a_2,a_3) \in M\},$$

$$X^4 = \{x \cap \{w \in W \,|\, w_4 \in \{2,3\}\} \,|\, x \in X^3\}.$$

For the five "stages" of the game there are four personal players, $I = \{1,\ldots,4\}$, with choices $C_1 = X^1$, $C_2 = X^2$, $C_3 = X^4 \cup \{\{w\} \in E \,|\, w_4 = 1\}$, and, finally, $C_4 = \{\{w\} \in E \,|\, w_4 \in \{2,3\}\}$. The choices in X^3 (except the one with $a_1 = a_2 = 1$, hence $a_3 = 0$) belong to chance and involve mixing uniformly over the interval $[w_1+w_2-2, 2-w_1-w_2]$.

That is, player 1 begins by choosing a number from the unit interval. Then player 2 sees that and also chooses a number from the unit interval. Conditional on the first two players' choices chance then mixes uniformly. Then player 3 sees the first two players' choices and the realization of chance's mixing and chooses between ending the game ($w_4 = 1$) or giving the move to player 4. If called upon to choose, player 4 has a binary choice, between ending the game with $w_4 = 2$ or with $w_4 = 3$. See Fig. 7.2 for a representation of a subgame starting at $x \in X^3$.

Payoffs are

$$u((w_1,w_2,w_3,1)) = (w_1, 2w_2, 1, 1),\; u((w_1,w_2,w_3,2)) = (0,0,0,w_3)$$
$$\text{and } u((w_1,w_2,w_3,3)) = (2w_1, w_2, 2, 0).$$

If $w_1 + w_2 < 1$, then the equilibrium in the subgame after the chance move resulted in $w_3 < 0$ (resp. $w_3 > 0$) involves player 4 choosing $w_4 = 3$ (resp. $w_4 = 2$) and player 3 giving the move to player 4 (resp. ending the game). In that case player 1's (resp. player 2's) payoff function is $3w_1/2$ (resp. $3w_2/2$). Therefore, player 1 (resp. player 2) will choose $w_1 = 1$ (resp. $w_2 = 1$) if she expects player 2 to choose $w_2 < 1$ (resp. if player 1 has chosen $w_1 < 1$).

Now consider the subgame after $w_1 = w_2 = 1$. Then the chance move is degenerate at $w_3 = 0$ and player 4 is indifferent between $w_4 = 2$ and $w_4 = 3$. Denote by $\alpha \in [0,1]$ the probability that player 3 passes the move on to player 4, and by $\beta \in [0,1]$ the probability that player 4 chooses $w_4 = 3$. The equilibria in

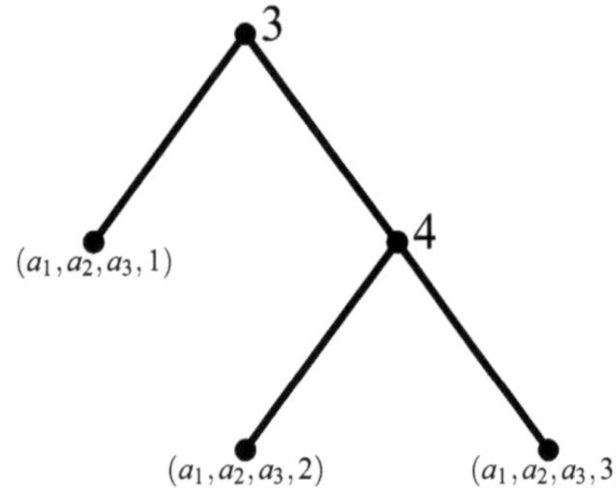

Fig. 7.2 A final subgame of Example 7.5. Player 3 observes $(w_1, w_2, w_3) = (a_1, a_2, a_3)$ (determined by previous players and nature) and chooses between ending the game ($w_4 = 1$) or giving the move to player 4. If called upon to choose, the latter chooses $w_4 = 2$ or $w_4 = 3$

this subgame are given by $\alpha = 0$ together with $\beta \in [0, 1/2]$, by $\alpha \in [0, 1]$ together with $\beta = 1/2$, and by $\alpha = 1$ together with $\beta \in [1/2, 1]$. Accordingly, player 2's (equilibrium) payoffs from choosing $w_2 = 1$ are

$$2(1-\alpha) + \alpha\beta = \begin{cases} 2 & \text{if } \alpha = 0, \beta \in [0, 1/2] \\ 2 - 3\alpha/2 & \text{if } \alpha \in [0, 1], \beta = 1/2 \\ \beta & \text{if } \alpha = 1, \beta \in [1/2, 1] \end{cases}$$

Since player 2 can guarantee herself a payoff slightly below $3/2$ by choosing $w_2 = 1 - \varepsilon$ for some small $\varepsilon > 0$, her optimization problem has a solution only if either $\alpha = 0$ and $\beta \in [0, 1/2]$ or if $\alpha \in [0, 1/3]$ and $\beta = 1/2$. These combinations are thus the only candidates for an equilibrium.

If player 1 expects player 2 to respond to $w_1 = 1$ with $w_2 < 1$, her unique best reply is $w_1 = 1$. But if she expects player 2 to respond with $w_2 = 1$ to $w_1 = 1$, then 1's expected payoff is

$$1 - \alpha + 2\alpha\beta = \begin{cases} 1 & \text{if } \alpha = 0, \beta \in [0, 1/2] \text{ or } \alpha \in [0, 1], \beta = 1/2 \\ 2\beta & \text{if } \alpha = 1, \beta \in [1/2, 1] \end{cases}$$

Since player 1 can also guarantee herself a payoff close to $3/2$, her maximization problem only has a solution if $\alpha = 1$ and $\beta \geq 3/4$. But the two ranges for α and β, which are necessary for solutions of player 1's resp. player 2's optimization problems, are disjoint. Therefore, there exists no subgame perfect equilibrium in this perfect information game with chance.

Intuitively what happens in this example is as follows. Non-degenerate chance moves may induce probability distributions over plays that are infeasible in their absence. For, nothing guarantees that $\phi(S)$ is convex. (In fact, it may not even be a vector space.) If a chance move may become degenerate as a function of what players do, such additional outcomes can suddenly disappear in the limit. This may lead to discontinuities in the players' payoff functions and, a fortiori, to an existence failure of equilibrium. There seems to be no general condition ruling this out, as such a condition would presume knowledge of the equilibrium path beforehand.

Because of examples like this we concentrate on perfect information games without chance moves.

Definition 7.1 A discrete extensive form (DEF) has **perfect information** if all choices are nodes, i.e. $c \in N$, for all $c \in C_i$ and all $i \in I$; in this case, $P(c) = \{p(c)\}$ for each $c \in C_i$ and each $i \in I$.

There are no chance moves, because the set I consists of personal players only. Further, this definition implies that only a single player is active at each move, $J(x) = \{i\}$. In view of (DEF2) from Definition 6.2, it also implies that every finite node is a choice, i.e. $F(N) \setminus \{W\} \subseteq \{c \in C_i \,|\, i \in I\}$. To obtain a game from a DEF preferences for the players need to be added.

Definition 7.2 A **discrete extensive form game** is a pair $((T,C), \precsim)$, where (T,C) is a DEF and $\precsim = (\precsim_i)_{i \in I}$ is an $|I|$-tuple of complete, reflexive, and transitive binary relations $\precsim_i$ on W, the players' preference relations. An extensive form game $((T,C), \precsim)$ has *perfect information* if the DEF (T,C) has perfect information.

A discrete extensive form game with perfect information will be referred to as an EFPI, which is short for "extensive form game with perfect information."

Since in an EFPI all choices are nodes, it is completely determined by a discrete game tree $T = (N, \supseteq)$, a partition $\mathscr{X} = (X_i)_{i \in I}$ of X into the sets of decision points for players $i \in I$, and a preference profile $\precsim = (\precsim_i)_{i \in I}$. Given an EFPI each X_i is determined by $X_i = \{p(c) \,|\, c \in C_i \subseteq N\}$. That is, if $x \in X_i$ then $i \in I$ decides at $x \in X$, choosing among the immediate successors of x. Hence, it is possible to give an alternative, equivalent definition of EFPI without reference to choices.

Definition 7.3 An **extensive form with perfect information** (EFPI) is a tuple $(T, I, \mathscr{X}, \precsim)$, where $T = (N, \supseteq)$ is a discrete game tree, I is the (possibly infinite) set of players, $\mathscr{X} = (X_i)_{i \in I}$ is a partition of the set of moves X of T into the sets of decision points for players $i \in I$, and $\precsim = (\precsim_i)_{i \in I}$ is a profile of complete, reflexive, and transitive binary relations on the set of plays W of T, one *preference relation* $\precsim_i$ for each player $i \in I$.

A *subgame* of an EFPI $((T,C), \precsim)$ is the perfect information game that starts at some move $x \in X$. Under perfect information every move $x \in X$ is the root of a subgame.

A *pure strategy* for player $i \in I$ in an EFPI $((T,C), \precsim)$ is a function $s_i : X_i \to C_i$ such that $s_i \circ p$ is the identity on $s_i(X_i)$.[2] That is, in view of the fact that choices are nodes, a pure strategy for player $i \in I$ in an EFPI is a function $s_i : X_i \to p^{-1}(X_i)$ such that $p(s_i(x)) = x$ for all $x \in X_i$, i.e., a function that specifies which of the immediate successors of each $x \in X_i$ are chosen by player i. The set S_i of all such functions is player i's pure strategy space. A *pure strategy combination* is an $|I|$-tuple $s = (s_i)_{i \in I} \in S \equiv \times_{i \in I} S_i$ of pure strategies, one for each player.

[2] It is not difficult to show that this condition agrees with (4.4) for the case of perfect information.

A (pure) *Nash equilibrium* is a pure strategy combination $s \in S$ such that no player has an incentive to deviate from s. A (pure) *subgame perfect equilibrium* (SPE; Selten 1965) is a pure strategy combination $s \in S = \times_{i \in I} S_i$ that induces a Nash equilibrium in every subgame.

7.3.2 Backwards Induction

Subgame perfect equilibrium is a natural solution concept for extensive form games of perfect information (Selten 1965, 1975). In the finite case subgame perfect equilibria are precisely those delivered by the backwards induction algorithm due to Kuhn (1953). The equivalence is essentially a "one-shot deviation principle" (see, e.g., Perea 2001, Chapter 3, or Osborne and Rubinstein 1994, Lemma 98.2, for the finite case, and Blackwell 1965, for the principle in dynamic programming). In order to check whether a strategy combination satisfies backwards induction the algorithm starts at the end of the tree and proceed backwards, verifying at every decision point (node) that the player controlling it has no profitable deviation. Suppose, however, that a player i plays at least twice, once at a given node x and once further "down" the tree, i.e., at some other node y that comes after his or her first decision. Backwards induction will eventually fix the player's decision at y, and then proceed up until the first node, x, is reached. There, the algorithm will simply check whether player i can profitably deviate *at that node* (x), taking the behavior of all players in the future, including player i at y, as given.

In principle, such a strategy combination may satisfy backwards induction and still fail to constitute a subgame perfect equilibrium, indeed even a Nash equilibrium. For, a priori, even if there is no profitable deviation for i at x, taking i's behavior at y (and later) as given, there may still be *multiple* profitable deviations, where i changes his behavior both at x and at y, and possibly even at further nodes following y. This, of course, cannot happen in finite games, and in that sense the equivalence of backwards induction and subgame perfection amounts to the claim that checking for "one-shot deviations" is enough to establish subgame perfection.

In this section, we prove a one-shot deviation principle for the general case, allowing for extensive form games with perfect information that potentially have an infinite horizon, infinite action spaces, and even infinitely many players. Moreover, we consider ordinal preferences rather than cardinal payoff functions. It turns out that, in general, in infinite games backwards induction may not be equivalent to subgame perfection, but that the equivalence (in the form of a one-shot deviation principle) holds under additional assumptions.

Somewhat surprisingly, those assumptions are topological in nature. The one-shot deviation principle holds as long as all the players' preferences are lower semi-continuous; in particular, full continuity is sufficient. Of course, this begs the question of which topology is used, and the even more primitive question of which space is endowed with a topology. Following the original formulation by von Neumann and Morgenstern (1944), we consider the space of ultimate outcomes (on which preferences are defined) as the appropriate primitive on which a topology is defined. This approach has the convenient feature that all elements of the tree, e.g. the nodes, become sets of outcomes rather than, say, abstract elements of a graph. With this approach, the one-shot deviation principle holds for *any* topology which provides the minimal conditions necessary for players to actually be able to solve optimization problems at every node, in particular that the topology is compact and every node is a closed set of outcomes.[3] Of course, these conditions are trivial in the finite case for the discrete topology (which makes every payoff function continuous), and one obtains the standard equivalence as a straightforward corollary.

We start by formalizing the notion of backwards induction without reference to an endpoint. First, recall that for a fixed move $x \in X$ in a discrete game tree $(N, \supseteq)$ the immediate successor that contains a particular play $w \in x$ is given by $\gamma(x,w) = \cup\{y \in N \,|\, w \in y \subset x\}$, as in (4.7). That $\gamma(x,w)$ is indeed a node follows from Lemma 5.3(c). Second, a discrete game tree is, in particular, up-discrete and regular and, therefore, everywhere playable (see Theorems 6.3, 5.2, and 5.5, and Corollary 5.4). Since this also applies to every subgame, for every $x \in X$ there is a function $\phi_x : S \to W(x)$ which assigns to the subgame starting at $x \in X$ the play $\phi_x(s) \in x$ that the strategy combination $s \in S$ (uniquely) induces in the subgame starting at x. Call a function $f : N \to W$ that satisfies $f(x) \in x$ for all $x \in N$ a *selection* and let $\mathscr{F}$ denote the set of all selections. Then, combining the functions ϕ_x for all $x \in X$, a strategy combination $s \in S$ uniquely defines a selection $f_s \in \mathscr{F}$ by $f_s(x) = \phi_x(s)$ for all $x \in X$ and $f_s(\{w\}) = w$ for all $w \in W$. Call this f_s the selection *induced* by $s \in S$. Conversely, in an EFPI every selection $f \in \mathscr{F}$ uniquely defines a strategy combination $s \in S$ by $x \in X_i \Rightarrow s_i(x) = \gamma(x, f(x))$ for all $x \in X$.

Definition 7.4 For an EFPI $\left((T,C), (\precsim_i)_{i \in I}\right)$ with player set I a strategy combination $s \in S$ satisfies **backwards induction** if it induces a selection $f \in \mathscr{F}$ such that, for all players $i \in I$ and all $x \in X_i$,

(BI1) if $f(p(x)) \in x$ then $f(x) = f(p(x))$, and
(BI2) $f(y) \precsim_i f(x)$ for all $y \in p^{-1}(x)$.

If (BI2) fails, then $f \in \mathscr{F}$ can clearly not be the selection induced by an SPE, because then player $i \in J(x)$ has a profitable deviation at $x \in X_i$. If (BI1) fails, then player $i \in J(p(x))$ expects the play to continue differently than player $j \in J(x)$, which can also not be the case in an SPE. Therefore, every SPE satisfies backwards induction. In a finite game the converse would also hold. In an infinite game, though, a strategy combination may satisfy backwards induction, but not subgame perfection, as the following example illustrates.[4]

---

[3]Those topologies are called "admissible" in Alós-Ferrer and Ritzberger (2016a,b).

[4]The example would be excluded if one insisted on Selten's (1965) original concept of subgame perfection. For, Selten also requires that the subgame perfect equilibria of an infinite game are approximated by subgame perfect equilibria of finite truncations, a property which he calls "asymptotic insensitivity."

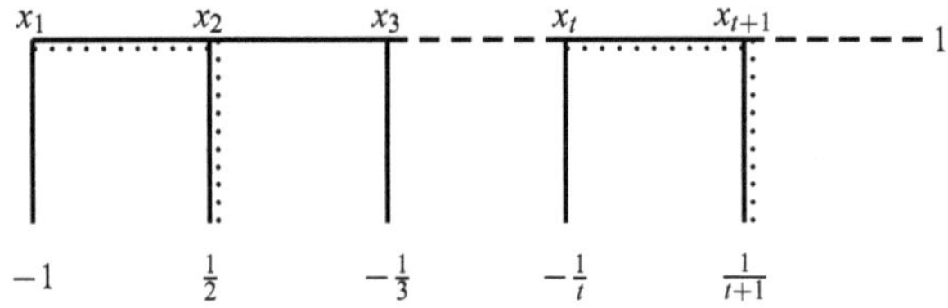

Fig. 7.3 Graphical representation of Example 7.6. Player 1 plays at odd-numbered moves, player 2 at even-numbered ones. Both players have the same payoffs. The *dotted-lines* profile where player 1 always continues and player 2 always stops fulfills backwards induction but is not subgame-perfect

Example 7.6 (Failure of the one-shot deviation principle) Let $W = \{1,2,\ldots,\infty\}$ be the Alexandroff compactified natural numbers with the discrete topology[5] and $N = \{(x_w)_{w\in W}, (\{w\})_{w\in W}\}$, where $x_w = \{w' \in W \,|\, w' \geq w\}$ for all $w \in W$. Then (W,τ) is a compact separated (Hausdorff) space[6] in which all nodes are closed. This tree (which is a complete infinite centipede) and topology will be used in several examples below.

Consider two players, $I = \{1,2\}$, and let $X_1 = \{x_w \in X \,|\, w \text{ is odd}\}$ and $X_2 = \{x_w \in X \,|\, w \text{ is even}\}$. Preferences are represented by the payoff functions $U_i(w) = (-1)^w/w$ for all $w = 1,2,\ldots$ and $U_i(\infty) = 1$ for $i = 1,2$. See Fig. 7.3 for a representation. Consider a strategy combination $s \in S$ such that player 1 always "continues" (chooses x_{w+1} at any $x_w \in X_1$) and player 2 always "stops" (chooses $\{w\} \in E$ at any $x_w \in X_2$). This strategy combination induces the selection $f \in \mathcal{F}$ given by

$$f(x) = \begin{cases} w+1 & \text{if } x = x_w \in X \text{ and } w \text{ is odd} \\ w & \text{if } x = x_w \in X \text{ and } w \text{ is even or } x = \{w\} \end{cases}$$

which satisfies backwards induction, but is not an SPE. To see that backwards induction holds, let first $f(p(x)) \in x \in X$. Then $p(x) = x_w$ with w odd, so that $f(x) = f(x_{w+1}) = w+1 = f(p(x))$. Otherwise, if $f(p(x)) \in \{w\} \in E$, then $p(x) = x_w \in X$ with w even, so that $f(x) = f(\{w\}) = w = f(x_w)$. Therefore, condition (BI1) in the definition of backwards induction holds. To see (BI2), let first $x = x_w \in X$ for w odd. Then $x \in X_1$ and

$$U_1(f(x_w)) = U_1(w+1) = 1/(w+1) = U_1(f(x_{w+1})) > U_1(f(\{w\})) = -1/w.$$

---

[5]The *Alexandroff compactification* of a topological space adds the element ∞ to the space and defines the topology on the space to consist of the open sets of the original space together with all subsets that contain ∞ and are such that their complements (in the original space) are closed and compact (Kelley 1975, p. 150).

[6]A topological space is T_1 (accessible or Frèchet) if, for each pair of distinct points, each has a neighborhood that does not contain the other. It is T_2 (separated or Hausdorff) if any two distinct points can be separated by disjoint neighborhoods. See the Mathematical Appendix A.2.

If, on the other hand, $x = x_w \in X$ for w even, then $x \in X_2$ and

$$U_2(f(x_w)) = U_2(w) = 1/w = U_2(f(\{w\})) > U_2(f(x_{w+1})) = 1/(w+2).$$

Since this completes the verification of (BI2), backwards induction holds. But $s \in S$ does not constitute an SPE. If it would, the resulting payoffs would be $(1/2, 1/2)$. But if player 2 deviates to "always continue" (choosing x_{w+1} at all $x \in X_2$), she obtains $U_2(\infty) = 1$.

Backwards induction fails to detect this deviation, because the deviation is infinite. That 2's infinite deviation is profitable, even though 2 has no profitable finite deviation, is due to a failure of lower semi-continuity of preferences. Payoffs are upper but not lower semi-continuous, and "jump upwards" at ∞.[7]

For finite games a *one-shot deviation principle* holds that backwards induction is equivalent to subgame perfection.[8] The last example shows that, without further assumptions, this is incorrect for large extensive form games. That is, in general backwards induction is necessary, but not sufficient for an SPE. The next Theorem shows that the one-shot deviation principle in infinite games takes lower semi-continuous preferences. The proof requires a lemma.

Lemma 7.1 *Let $(N, \supseteq)$ be a discrete game tree with set W of plays and τ a compact topology on W in which all nodes are closed. If $w \in u \in \tau$, then there is $x \in \uparrow\{w\} \cap F(N)$ such that $x \subseteq u$.*

Proof Let $w \in u \in \tau$. If $\{w\} \in F(N)$, there is nothing to prove. Otherwise, because the tree is discrete, it follows that $\{w\} \in I(N)$. Then by Lemma 4.2

$$\bigcap\{y \mid y \in \uparrow\{w\} \setminus \{\{w\}\}\} = \{w\}.$$

If there is no y such that $\{w\} \subset y \subseteq u$, then for every $y \in \uparrow\{w\} \setminus \{\{w\}\}$ there is $w_y \in y \setminus u$. The set $\{w_y \in W \setminus u \,|\, y \in \uparrow\{w\} \setminus \{\{w\}\}\}$ ordered by $w_y \leq w_z \Leftrightarrow z \subseteq y$ forms a sequence (by discreteness, Theorem 6.1(d)) in $W \setminus u$. Since $W \setminus u$ is closed, hence, compact, it contains a subsequence that converges to some $w' \in W \setminus u$. Since $\uparrow\{w\} \setminus \{\{w\}\}$ is a chain, this subsequence is eventually contained in every $y \in \uparrow\{w\} \setminus \{\{w\}\}$. Hence w' is a cluster point of every such node y. Since nodes are closed by hypothesis, $w' \in y$ for all $y \in \uparrow\{w\} \setminus \{\{w\}\}$. But then $w' \in \bigcap\{y \mid y \in \uparrow\{w\} \setminus \{\{w\}\}\} = \{w\}$ implies $w' = w$, in contradiction to $w' \in W \setminus u$ and $w \in u$. Hence, there is $x \in \uparrow\{w\} \setminus \{\{w\}\} \subseteq F(N)$, the latter by Theorem 6.1(b), such that $x \subseteq u$. This completes the proof. ■

---

[7]A real valued function $f : W \to \mathbb{R}$ is *upper* resp. *lower semi-continuous* (u.s.c. resp. l.s.c.) if for each $r \in \mathbb{R}$ the upper resp. lower contour set $\{w \in W \,|\, f(w) \geq r\}$ resp. $\{w \in W \,|\, f(w) \leq r\}$ is closed. See the Mathematical Appendix A.2.

[8]Hendon et al. (1996) and Perea (2002) consider one-shot deviation principles for finite games with imperfect information.

With this at hand the following result establishes a one-shot-deviation principle under the hypothesis of lower semi-continuous preference relations.[9]

Theorem 7.1 *Let $((T,C),\precsim)$ be an EFPI on a discrete game tree $T=(N,\supseteq)$ with set W of plays and τ a compact topology on W in which all nodes are closed. If $\precsim_i$ is lower semi-continuous with respect to τ for all $i \in I$, then a strategy combination $s \in S$ constitutes an SPE if and only if $s \in S$ satisfies backwards induction.*

Proof By the remarks after Definition 7.4 it is enough to show the if-part. Let $f \in \mathscr{F}$ satisfy (BI1)–(BI2) and $s \in S$ be defined by $x \in X_i \Rightarrow s_i(x) = \gamma(x, f(x))$ for all $x \in X$, so that $s \in S$ satisfies backwards induction. If $s \in S$ is not an SPE, then there is some player $i \in I$ and a move $x \in X_i$ for which there exists a function $g : \downarrow x \to W(x)$ such that $g(y) \in y$, $g(p(y)) \in y \Rightarrow g(y) = g(p(y))$, and $y \notin X_i \Rightarrow \gamma(y, g(y)) = \gamma(y, f(y))$ for all $y \in \downarrow x$, and $f(x) \prec_i g(x)$. The selection g captures the deviation of player i in the subgame starting at x. The play induced by this deviation in the subgame is $g(x)$.

Consider first the case where player i deviates only finitely often along the play $g(x)$, i.e., there are only finitely many $y \in X_i \cap \downarrow x \cup \uparrow \{g(x)\}$ such that $\gamma(y, g(y)) \neq \gamma(y, f(y))$. Let y_0 be the minimum of this chain, i.e., the last deviation by player i along the play $g(x)$. Note that, for every $z \in X_i \cap \downarrow x \cup \uparrow \{g(x)\}$, the play $g(z)$ coincides with the play $g(x)$. By (BI2) $f(y_0) \succsim_i f(\gamma(y_0, g(x))) = g(x)$. Let y_1 be the minimum of the set of $y \in X_i \cap \downarrow x \cup (\uparrow y_0 \setminus \{y_0\})$ such that $\gamma(y, g(y)) \neq \gamma(y, f(y))$; that is, y_1 is the last deviation of player i before y_0. For nodes z such that $y_0 \subset z \subset y_1$ the active player at z does not deviate from the original profile. This implies that for all these nodes $f(z) = f(y_0)$. It now follows, again from (BI2), that $f(y_1) \succsim_i f(\gamma(y_1, g(x))) = f(y_0) \succsim_i g(x)$. Proceeding inductively, it follows that $f(x) \succsim_i g(x)$, contradicting $f(x) \prec_i g(x)$.

Next, consider the case where player i deviates infinitely often along the play $g(x)$, i.e. there are infinitely many $y \in X_i \cap \downarrow x \cup \uparrow \{g(x)\}$ such that $\gamma(y, g(y)) \neq \gamma(y, f(y))$. We claim that then there is another profitable deviation that deviates from f only at a finite chain of moves—which will then lead to a contradiction.

Because $\precsim_i$ is l.s.c. on W, the strict upper contour set $\{w \in x \mid f(x) \prec_i w\}$ is open and $g(x) \in W$ belongs to it. By Lemma 7.1 there is $y' \in \uparrow \{g(x)\} \cap F(N)$ such that $y' \subseteq \{w \in x \mid f(x) \prec_i w\}$. Therefore, $f(x) \prec_i f(z)$ for all $z \in \downarrow y'$. Since player i deviates infinitely often along $g(x)$, there exists $y \in F(N)$ with $y \in \downarrow y' \cap \uparrow \{g(x)\} \cap X_i$ such that $y \subseteq \{w \in x \mid f(x) \prec_i w\}$.

[9]A binary relation $\precsim$ on W is *upper* resp. *lower semi-continuous* (u.s.c. resp. l.s.c.) if for each $w \in W$ the upper resp. lower contour set $\{w' \in W \mid w \precsim w'\}$ resp. $\{w' \in W \mid w' \precsim w\}$ is closed in τ; see the Mathematical Appendix A.2.

Let $w^* = f(\gamma(y, g(y)))$ and define the function $g' : \downarrow x \to W(x)$ by

$$g'(z) = \begin{cases} w^* & \text{if } z \in \downarrow x \cap \uparrow y \\ f(z) & \text{otherwise} \end{cases}$$

Note that $\gamma(z, g'(z)) = \gamma(z, f(z))$ for any player $j \neq i$ and all $z \in X_j$. This is because for any move of player j before y the successor that contains w^* coincides with the one that contains $g(x)$.

Because the tree is discrete and $y \in F(N)$, the chain $\downarrow x \cap \uparrow y$ is finite (Theorem 6.1(c)), thus, g' deviates from f only at finitely many moves. That $\downarrow x \cap \uparrow y$ is a chain and the definition of w^* imply that (BI1) holds. Since $w^* \in y$, it follows that $w^* \in \{w \in x | f(x) \prec_i w\}$. Therefore, g' describes a finite and profitable deviation after x, which according to the previous argument cannot exist. This completes the proof. ■

If, in particular, all preference relations are continuous, then backwards induction is necessary and sufficient for an SPE.

Theorem 7.1 is conceptually related to the literature on dynamic programming, because a dynamic optimization problem is a one-player game. For this case the necessity of appropriate continuity assumptions is known. Subgame perfection corresponds to the concept of policy optimality in that literature, while backwards induction reduces to the statement that a policy cannot be improved through a single deviation; the latter property is sometimes referred to as "unimprovability". Blair (1984, Example 1) provides an example showing that without what amounts to a lower semi-continuity at infinity (his axiom A2') a one-shot deviation principle may *not* hold. That is, backwards induction for the one-player case does not imply optimality. Under A2' (plus a weak axiom of monotonicity), however, the implication does hold (Blair 1984, Theorem 4). Thus, for dynamic (single-player) optimization problems it is known that lower semi-continuity is needed for a one-shot deviation principle to hold. Our result extends this logic in two directions: First, to subgame perfection in multi-player games and, second, to arbitrary (admissible) topologies. The topologies that we consider allow greater flexibility, as will be illustrated now.

It is important to realize that Theorem 7.1 applies for any compact topology on plays in which all nodes are closed. As commented in Sect. 7.2, topological approaches to infinite horizon games have often relied on a product construction, where the set of actions at each step is assumed to be compact and compactness of the space of plays is then derived from Tychonoff's theorem (e.g., Fudenberg and Levine 1983; Harris 1985a; Hellwig and Leininger 1987). This imposes a particular choice of topology, which Theorem 7.1 does not rely on. The following example illustrates that the freedom to choose a natural topology can be advantageous.

Example 7.7 (The Long Interval) Consider the following two-player, infinite-horizon game. The outcome of the game is a real number $w \in [0, 1]$, to be determined jointly by two players in an iterative fashion. In period 1 player 1 can either choose

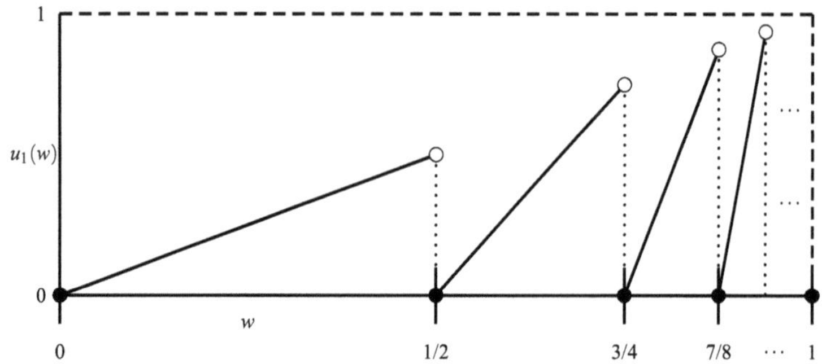

Fig. 7.4 Graphical representation of player 1's payoff in Example 7.7. Horizontal scale is stretched to increase visibility

some $w \in [0,1/2)$ (in which case the game ends), or pass the decision to player 2. In the latter case in period 2 player 2 can either choose some $w \in [1/2,3/4)$ (in which case the game ends), or return the decision to player 1, and so on. That is, player 1 moves at odd periods $t = 1,3,5,\ldots$, and player 2 at even periods, $t = 2,4,6,\ldots$ Let $q_t = 1 - \frac{1}{2^t}$, $t = 0,1,\ldots$, and $I_t = [q_{t-1}, q_t)$, $t = 1,2,\ldots$ At period t, the player who moves can either fix some $w \in I_t$, in which case the game ends with that choice of w, or pass the decision to the other player. If no player ever picks a number, the outcome is $w = 1$. Payoffs are functions $u_i : [0,1] \to \mathbb{R}$, for $i = 1,2$.

Our modeling approach for this game is simple. The natural set of plays is obviously $W = [0,1]$, and the most natural choice of topology is the relative Euclidean topology on this interval. The tree is formed by all singleton nodes $\{w\}$, $w \in W$, and the "continue" nodes $x_t = [q_t, 1]$ for $t \geq 0$. At period t, the active player decides at move x_{t-1} and can choose among the successors $\{w\}$, with $q_{t-1} \leq w < q_t$, and the "continue" successor x_t. Since W is compact and all nodes are closed, the relative Euclidean topology fulfills the required properties.

By contrast, the product construction would go as follows. At period t, let the set of actions be $A_t = [q_{t-1}, q_t]$, where the action q_t stands for "continue". Hence A_t is compact in the relative Euclidean topology and the product set $A = \times_{t=1}^{\infty} A_t$ is compact. As noted in Sect. 7.2, the Tychonoff construction used in the sequence approach blows up the space of plays, in this case generating an infinite-dimensional space even though the set of outcomes is simply a compact interval of real numbers.

Let us complete the description of the game by specifying payoffs. For simplicity, let u_2 be constant, $u_2(w) = k$ for all $w \in [0,1]$. The payoff function for player 1 will be specified in a piecewise way as follows. For each $t = 1,2,\ldots$,

$$u_1(w) = \left(2^t - 1\right)(w - q_{t-1}) \quad \text{for all } w \in I_t.$$

Note that $u_1(q_{t-1}) = 0$ and $\lim_{w \to q_t-} u_1(q_t) = q_t$ for all $t \geq 1$. Further, set $u_1(1) = 0$. Figure 7.4 represents this payoff function.

Consider any payoff π with $0 \leq \pi < 1$. For each $t \geq 0$, there exists a unique $w_t(\pi) \in [q_{t-1}, q_t]$ such that, for $w \in I_t$, $u_1(w) > \pi$ if and only if $w \in (w_t(\pi), q_t)$.

Further, the interval $(w_t(\pi), q_t)$ can be empty ($w_t(\pi) = q_t$) only for finitely many values of t. That is, the strict upper contour set for payoff π is given by

$$U^1(\pi) = \{w \in W | u_1(w) > \pi\} = \bigcup_{t=1}^{\infty}(w_t(\pi), q_t)$$

which is open in the relative Euclidean topology. Further, $U^1(1) = \emptyset$. Hence, the payoff function $u_1 : [0,1] \to \mathbb{R}$ is lower semi-continuous in the relative Euclidean topology, and, since the constant function u_2 is also lower semi-continuous, Theorem 7.1 applies and every strategy profile satisfying backwards induction is a subgame perfect equilibrium.

This game has many SPEs. Consider a strategy profile as follows. Player 1 always continues, i.e. $s_1(x_t) = x_{t+1}$ for all $t \geq 0$. For each $t = 2, 4, \ldots$, choose any w_t such that $u_1(w_t) > q_{t-1}$ (recall Fig. 7.4). Let player 2 stop and choose one such w_t at period t, i.e., $s_2(x_t) = w_t$. Player 2's choice is always optimal, since u_2 is constant. Player 1's choice to continue is always optimal, since player 2 "promises" him a strictly larger payoff in the following period than the supremum of the payoffs he can obtain by stopping. Hence backwards induction holds, and the profile is an SPE by Theorem 7.1. Other, qualitatively different SPE exist, for instance where player 2 also continues for finitely many periods and then behaves as described above.[10]

Analyzing this simple example with the Tychonoff construction gives rise to a number of difficulties. First, the payoff function u_1 is not lower semi-continuous. To see this, recall that numbers $w \in [0,1]$ giving a strictly larger payoff than $\pi \in [0,1)$ are those in $\bigcup_{t=1}^{\infty}(w_t(\pi), q_t)$. If plays are taken to be elements of the infinite product $A = \times_{t=1}^{\infty}[q_{t-1}, q_t]$, essentially the strict upper contour set would map into the product set $\times_{t=1}^{\infty}(w_t(\pi), q_t)$ (see below for the precise argument). But this is not an open set in the product topology. Any open set has to contain a basic open neighborhood of each of its points, where basic open neighborhoods are product sets such that finitely many coordinates are open subsets of the coordinate (action) spaces, but the remaining infinitely many coordinates are "unrestricted", i.e., they contain the whole coordinate space. It follows that, for the strict upper contour set to be open, we would need to have that $[q_{t-1}, q_t] = (w_t(\pi), q_t)$ for all but finitely many coordinates, a contradiction.

The precise argument showing that u_1 is not lower semi-continuous is slightly more involved, due to the technicalities involved in the product approach. Specifically, the function u_1 is not defined on the full product space A, because this space is far too large. Rather, as observed e.g. by Harris (1985a), one needs to reconstruct the set of plays as a strict subset of A. (For the analysis to be fruitful the set of plays also needs to be a closed subset of the product set, as commented in Sect. 7.2, but we will ignore this added difficulty here.) In the present case the set of plays

[10]Hence the two-player game has infinitely many SPEs. It is interesting to note that, if the example is recast as a one-player game (with player 1 making all choices), there exist no subgame perfect equilibrium. The example then becomes a continuous-action version of "Denardo's counterexample" as reported in Sobel (1975, Example 4).

can be represented as a subset of the product space by recalling the convention that "continuing" corresponds to the choice $q_t \in A_t = [q_{t-1}, q_t]$. A play where players continue for $t-1$ periods and then stop by choosing the number $w_t \in I_t$ at period t can be represented by the infinite sequence $(q_1, q_2, \ldots, q_{t-1}, w_t, q_{t+1}, q_{t+2}, \ldots)$. With this convention the set of plays is given by

$$H = \bigcup_{t=1}^{\infty} H_t \quad \text{where } H_t = \left\{ \left(w^t\right)_{t=1}^{\infty} \mid w^t \in A_t,\ w^r = q_r\ \forall r \neq t \right\}.$$

For a payoff $\pi \in [0,1)$, the strict upper contour set is then

$$U^1(\pi) = \bigcup_{t=1}^{\infty} B_t \quad \text{where } B_t = \left\{ \left(w^t\right)_{t=1}^{\infty} \mid w^t \in (w_t(\pi), q_t),\ w^r = q_r\ \forall r \neq t \right\}.$$

If B were open in the relative topology on H inherited from the product topology, there would be a basic open set of the product topology whose intersection with H would be contained in B. This would imply $(w_t(\pi), q_t) = A_t$ for all but finitely many t, a contradiction. Hence, u_1 is not lower semi-continuous if one insists on the product topology. And Theorem 7.1 does not apply, simply because the "wrong" topology has been selected.

7.3.3 Games Where Players Move Finitely Often

Consider the case of finite games. The set of outcomes, W, is finite, and one can simply take the discrete topology on W, which is trivially compact and such that all nodes are closed. All payoff functions are automatically continuous in this topology, and the standard equivalence between backwards induction and subgame perfection follows as a trivial corollary of Theorem 7.1.

Corollary 7.1 *In finite extensive form games with perfect information a strategy combination constitutes an SPE if and only if it satisfies backwards induction.*

In the infinite case, however, the discrete topology will in general not be compact. In Example 7.6 the set W of plays is not compact in the discrete topology. It follows from Theorem 7.1 that in this example there exist no compact topology such that the nodes are closed and the payoff functions are lower semi-continuous.

There are classes of games where the equivalence can be established without explicit reference to a topology. One such class is obtained as follows. Given an EFPI, a player i *plays finitely often* if for every play w, viewed as a maximal chain of nodes $\uparrow\{w\}$, i decides only finitely often along the play, i.e. $\uparrow\{w\} \cap X_i$ is finite for all plays $w \in W$.

The class of extensive form games where all players play finitely often includes many applications. For instance, in every finite-horizon game, all plays are finite, and hence every player can play only finitely often. A different subclass is the one where every player has only finitely many moves, i.e. X_i is finite for each $i \in I$. For instance, games with infinite horizon and infinitely many players where each player only has finitely many decisions belong to this category.[11]

The following result shows that the one-deviation principle holds for this class of games without topological assumptions. The key is that the proof of Theorem 7.1 does not rely on continuity if it is always possible to identify a last decision for every player along any play.

Corollary 7.2 *Let* $(T, I, \mathscr{X}, \precsim)$ *be an EFPI on a discrete game tree* $T = (N, \supseteq)$ *with set* W *of plays such that every player plays finitely often. Then, a strategy combination constitutes an SPE if and only if it satisfies backwards induction.*

Proof Let $s \in S$ satisfy backwards induction. Proceeding as in the proof of Theorem 7.1, there is a player $i \in I$ who can profitably deviate in the subgame starting at some move $x \in X_i$. Since every player plays finitely often, it follows that i deviates only finitely often along the play induced by the deviation. The proof of Theorem 7.1 in that case reaches a contradiction without using any topological assumption. ■

7.3.4 Continuity at Infinity

Theorem 7.1 constitutes a "one-shot deviation principle." Previous results of that sort and beyond the finite case relied on a notion of "continuity at infinity" as introduced by Fudenberg and Levine (1983). For instance, Fudenberg and Tirole (1991, Theorem 4.2) use it to prove a one-stage deviation principle for infinite horizon multi-stage games with observed actions. Hendon et al. (1996, p. 281) also rely on this condition to prove a one-shot deviation principle for games with finite information sets and finite choice sets at each information set, but for an infinite horizon. For the case of perfect information this implies that backwards induction is equivalent to subgame perfection in this particular setting. These results are also conceptually related to the work on consumer patience and myopia as initiated by Koopmans (1960) (see also Brown and Lewis 1981).

Here we will show that the concept of continuity at infinity coincides with full continuity in a particular compact topology in which all nodes are closed. The insight from that is that the assumptions of Theorem 7.1 are strictly weaker than continuity at infinity. Hence, Theorem 7.1 will imply a one-shot deviation principle for games with payoff functions that are continuous at infinity.

We first construct the appropriate topology and demonstrate that it is compact and all nodes are closed. Suppose that at every move only finitely many options are

[11] This does *not* mean that all games in this class are such that every player is active at finitely many moves only. Consider, for example, the classical Stackelberg duopoly (von Stackelberg 1934). The game has a finite horizon and hence every player plays finitely often. However, the follower has a continuum of moves (one for each possible production decision of the leader).

available, that is, $p^{-1}(x)$ is finite for each $x \in X$. For $t \geq 0$ let Y_t again denote the slice of nodes which can be reached after t steps from the root, that is, $x \in Y_t$ if and only if $\uparrow x$ contains exactly $t+1$ nodes. Let A_t be the set containing all possible options available at moves in Y_t plus a dummy option $*$, that is, $A_t = \left(\cup_{x \in Y_t} p^{-1}(x)\right) \cup \{*\}$. The set of plays W can be identified with a subset of the product space $\times_{t \geq 0} A_t$. For instance, a finite play would correspond to an infinite sequence where all entries after some finite t are the dummy option $*$.

Endow each A_t with the discrete topology. Then $A = \times_{t \geq 0} A_t$ can be endowed with the product topology and W inherits the relative topology. Call this topology the *product-discrete topology* on plays. This topology is compact and all nodes are closed sets.

Lemma 7.2 *Consider an EFPI on a discrete game tree T such that every move has finitely many immediate successors. Then the product-discrete topology on plays is compact and all nodes are closed.*

Proof For a given sequence $a \in A$ and a number k define

$$V(a,k) = \left(\times_{t \leq k} \{a(t)\}\right) \times \left(\times_{t \geq k+1} A_t\right). \tag{7.2}$$

This set is open in the product topology on A, because each $\{a(t)\}$ is open in the discrete topology on A_t and only finitely many coordinates are restricted. The set $V(a,k)$ is also closed (i.e., it is clopen) in the product topology, because it is a product of closed sets.

Now, notice that each A_t is compact by virtue of being finite, hence by Tychonoff's Theorem A is compact with the product topology. We now show that W is a closed subset of A, hence compact with the relative topology. Equivalently we show that $A \setminus W$ is open. Let a be a sequence in $A \setminus W$. This is the case if and only if some coordinate $a(t)$ is followed by a coordinate $a(t+1)$ which is incompatible with the tree, that is, one of the following three conditions holds: either

(a) $a(t) \in X$ but $a(t+1) \notin p^{-1}(a(t))$,
(b) $a(t)$ is a terminal node but $a(t+1) \neq *$, or
(c) $a(t) = *$ but $a(t+1) \neq *$.

Let k be the first index at which one of the three violations occurs. In all three cases the open set $V(a,k+1)$ contains a and is contained in $A \setminus W$. That is, every point in $A \setminus W$ has an open neighborhood contained in $A \setminus W$. It follows that $A \setminus W$ is open, therefore W is closed, hence compact.

Furthermore, every node, identified with the corresponding set of plays in $W \subseteq A$, is closed. To see this, notice that a node is the intersection of W with a set of the form $V(a,k)$ for some $a \in A$ and some fixed k. As observed above, the latter set is closed in the product topology of A. Hence, the node is closed in the product-discrete topology on plays. ■

To define continuity at infinity requires representable preferences. Hence, assume that for each players $i \in I$ there is a payoff function v_i that represents i's preferences.

Then, a game is *continuous at infinity* (Fudenberg and Levine 1983) if for each $\varepsilon > 0$ there is a K such that whenever $s, s' \in S$ are identical on all Y_t for any $t \leq K$, then $|v_i(\phi(s)) - v_i(\phi(s'))| < \varepsilon$, for all $i \in I$. Since the payoffs only depend on the induced play $\phi(s)$, this definition can be stated equivalently on plays. That is, the game is continuous at infinity if for each $\varepsilon > 0$ there is a K such that whenever $w, w' \in W$ coincide up to K, then $|v_i(w) - v_i(w')| < \varepsilon$, for all $i \in I$.[12]

It follows from Theorem 7.1 that backwards induction and subgame perfection are equivalent whenever the preferences represented by the payoff functions v_i are lower semi-continuous with respect to some compact topology in which all nodes are closed. Continuity at infinity is a stronger criterion, though. It is equivalent to full continuity with respect to the product-discrete topology on plays.

Lemma 7.3 *Consider an EFPI on a discrete game tree T such that every move has finitely many immediate successors and preferences are representable by payoff functions. It satisfies continuity at infinity if and only if payoff functions are continuous with respect to the product-discrete topology on plays.*

Proof "if:" Let $\varepsilon > 0$ and $w \in W$. By continuity of v_i in the product-discrete topology the preimage $v_i^{-1}((v_i(w) - \varepsilon, v_i(w) + \varepsilon))$ is open. By definition of the product-discrete topology there exists a $t(w, \varepsilon)$ such that

$$w \in V(w, t(w, \varepsilon)) \subseteq v_i^{-1}((v_i(w) - \varepsilon, v_i(w) + \varepsilon))$$

for $V(w, t(w, \varepsilon))$ as in (7.2). That is, $|v_i(w') - v_i(w)| < \varepsilon$ for every $w' \in W$ that coincides with w until $t(w, \varepsilon)$. However, continuity at infinity entails a uniformity in the sense that $t(w, \varepsilon)$ has to be independent of w (see also Fudenberg and Levine 1983, Lemma 3.1). To show this, we use the fact that W is compact by Lemma 7.2.

Fix $\varepsilon > 0$. The collection of sets $V(w, t(w, \varepsilon/2)) \cap W$ for all $w \in W$ form an open covering of W. By compactness there is a finite subcovering with indices $w_1, \ldots, w_L$ for some finite L. Let $\bar{t} = \max_{\ell=1,\ldots,L} t(w_\ell, \varepsilon/2)$ and $w, w' \in W$ be such that they coincide up to $\bar{t}$. There is a w_ℓ such that $w \in V(w_\ell, t(w_\ell, \varepsilon/2))$. Hence w and w_ℓ coincide up to $t(w_\ell, \varepsilon/2)$. By the previous observation this implies that $|v_i(w_\ell) - v_i(w)| < \varepsilon/2$. Since w and w' coincide up to $\bar{t} \geq t(w_\ell, \varepsilon/2)$, it follows that w' also coincides with w_ℓ up to $t(w_\ell, \varepsilon/2)$. Hence, $|v_i(w_\ell) - v_i(w')| < \varepsilon/2$. We conclude that $|v_i(w') - v_i(w)| \leq |v_i(w_\ell) - v_i(w)| + |v_i(w_\ell) - v_i(w')| < \varepsilon$. This proves continuity at infinity.

"only if:" Let (b_0, b_1) be an open interval of real numbers and let $w \in W$ be such that $v_i(w)$ lies in this interval. By continuity at infinity there is a $t \geq 0$ and a set $V(w, t)$ (as in (7.2)), which is open in the product-discrete topology on A, that contains w, and such that every play in this set yields a payoff in (b_0, b_1). This shows that $v_i^{-1}((b_0, b_1))$ is open. It follows that v_i is continuous in the product-discrete topology on plays. ■

[12]The equivalence requires the fact that for every play there is a strategy combination that induces it, see Theorem 5.1.

Putting together the last two lemmata with Theorem 7.1 we obtain the following result.

Proposition 7.2 *In any extensive form game of perfect information where players have finitely many options available at every move and payoff functions are continuous at infinity, a strategy combination constitutes an SPE if and only if it satisfies backwards induction.*

The discussion above shows that Theorem 7.1 is strictly stronger than results based on continuity at infinity, because it relies on weaker assumptions. First, it relies on lower semi-continuity rather than on full continuity. Second, continuity can be with respect to any compact topology for which nodes are closed, rather than a particular one based on a product structure. Third, our result relies on ordinal preferences only, not on cardinal payoff functions.

7.3.5 Well-Behaved Perfect Information Games

Having clarified when backwards induction and subgame perfection agree, we return to when equilibrium exists and when not. Studying equilibrium entails maximization problems. And optimization naturally leads to topological considerations. The existence result later in this chapter will establish existence of subgame perfect equilibria for all (large) perfect information games when the topology renders preferences continuous and satisfies a few additional conditions.

The first condition concerns compactness. If some node were not compact, it would be easy to endow the player active at that node with continuous preferences having no maximum, and hence SPE would not exist in general. Hence, it is natural to work with compact topologies such that all nodes are closed sets. Since closed subsets of compact sets are compact, in such a topology all nodes are compact.

Even for compact topologies where all nodes are closed sets, however, a subtlety can arise that may lead to equilibrium nonexistence. The problem, which the following example illustrates, concerns the assignment $\mathscr{X} = (X_i)_{i \in I}$ of decision points.

Example 7.8 (Compactness and continuity are not enough) Let $W = [0,1]^2$ be the plays,

$$N = \Big\{ W, (\{a\} \times [0,1])_{a \in [0,1]}, (\{w\})_{w \in W} \Big\},$$

$X_1 = \{W\}$, $X_2 = \Big\{ (\{a\} \times [0,1])_{a \in [0,1/2)} \Big\}$, and $X_3 = \Big\{ (\{a\} \times [0,1])_{a \in [1/2,1]} \Big\}$, for three players, $I = \{1,2,3\}$ (see Fig. 7.5). That is, player 1 first chooses from the unit interval. If she chose $a < 1/2$, player 2 gets to choose from the unit interval; otherwise player 3 chooses from the unit interval.

Endow W with the relative Euclidean topology of the plane, making W compact and all nodes closed. Preferences are represented by the continuous payoff functions

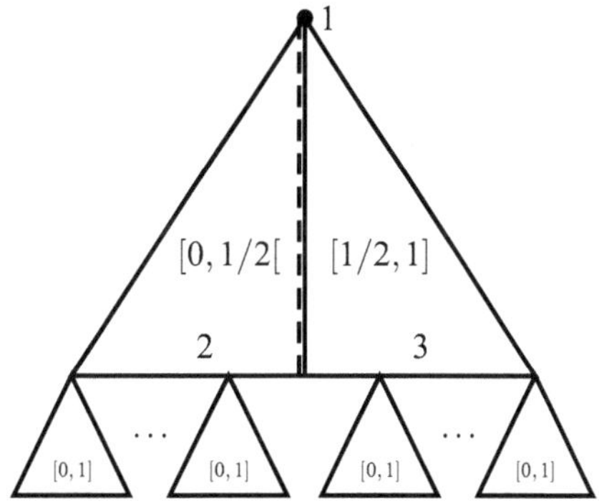

Fig. 7.5 Graphical representation of Example 7.8. Player 1 chooses a number a in $[0,1]$, then either player 2 (if $a < 1/2$) or player 3 (if $a \geq 1/2$) choose another number in $[0,1]$

$U_1(a,b) = a(1-b)$, $U_2(a,b) = -b$, and $U_3(a,b) = b$, where $w = (a,b) \in W$. Player 3 will optimally choose $b = 1$ and player 2 will choose $b = 0$. Thus, player 1's value function yields a if $a < 1/2$ and zero if $a \geq 1/2$ and has no maximum.

In this example the union of the decision points for player 2 in the slice Y_1 is not closed, even though all individual nodes are closed. And this is the reason for why there is no equilibrium. This shows that the assignment of moves to players, $\mathscr{X}$, needs to satisfy a minimal condition with respect to the topological space (W, τ).

Likewise, existence of equilibria will fail if preferences are not continuous. If a single player chooses $a \in [0,1]$ from the unit interval (endowed with the relative Euclidean topology) such that $a \prec_1 a' \Leftrightarrow a < a'$ for all $a, a' \in [0,1)$, but $1 \precsim_1 \bar{a}$ for some $\bar{a} < 1$, then there is no equilibrium. This existence failure is caused not by the assignment of moves to players, but by a failure of continuity for the preference relation $\precsim_1$.

Remark 7.1 For discontinuous payoffs it is often natural to focus on the weaker concept of ε-equilibria. However, Example 3 of Solan and Vieille (2003) shows that a pure subgame perfect ε-equilibrium ($\varepsilon \geq 0$) may not exist if the players' payoff functions are not lower semi-continuous, and Flesch et al. (2010a) show that Borel measurability of payoffs does not suffice for existence of subgame perfect ε-equilibrium in infinite horizon games, even if there are only finitely many available actions at each node. Flesch et al. (2010b) and Purves and Sudderth (2011) analyze ε-equilibrium existence ($\varepsilon > 0$) when only upper or lower semi-continuity is assumed, respectively. (The latter also includes an existence result for $\varepsilon = 0$ for the case of finitely many actions.) Flesch and Predtetchinski (2016) show ε-equilibrium existence ($\varepsilon > 0$) when payoff functions are continuous everywhere except on a countable union of discrete sets.

In summary, the two objects $\mathscr{X}$ and $\precsim$ that make a tree into an EFPI need to satisfy conditions with respect to the topological space (W, τ) in order to admit existence of equilibrium. To isolate these conditions, they are summarized in the next definition of a "well-behaved" perfect information game.

Definition 7.5 An EFPI $((T,C),(\precsim_i)_{i\in I})$ with player set I is **well-behaved** with respect to a topology τ on W if

(WB1) for each $t=0,1,\ldots$ the nonterminal part $Y_t \setminus E$ of the slice Y_t is partitioned into finitely many nonempty cells of the form $Y_{it}=X_i\cap Y_t$ with $i\in I$ such that $W(Y_{it})$ is closed in τ for all $i\in I$;

(WB2) for each player $i\in I$ the preference relation $\precsim_i$ is continuous with respect to the topology τ.[13]

Example 7.8 fails (WB1) while Example 7.2 fulfills it, as the only move other than the root, $\{1\}\times[0,1]\in Y_1$, is a closed set of plays. Example 7.3 also fulfills (WB1), because player 2 decides at all nodes in the intermediate slice Y_1 and $W(Y_1)=W$.

The requirement in (WB1) that $W(Y_{it})$ is closed is a natural one. Since $Y_{it}=Y_t\cap X_i$ denotes i's decision points at slice Y_t, the set $W(Y_{it})$ contains all plays in which player i moves at Y_t. If this set fails to be closed, it is easy to build examples of equilibrium nonexistence where a player is simply unable to maximize in a given slice. Further, note that under (WB1) the number of players who decide at every given slice is finite.

The definition of well-behaved perfect information games rules out misspecified decision points, (WB1), as in Example 7.8, and restricts the analysis to preferences that are continuous with respect to the topology τ, (WB2). Hence, both problems that could surface when the tree is turned into a game, with the preference relations $\precsim$ and with the assignment of decision points $\mathscr{X}$, are taken care of.

The objective in this chapter is to study and characterize the class of topologies on the set of plays of a given tree such that all well-behaved perfect information games defined on that tree admit a subgame perfect equilibrium. This is reflected by the following definition.

Definition 7.6 The topology τ on the set W of plays of a discrete game tree $(N,\supseteq)$ **admits equilibrium analysis** if some well-behaved EFPI can be defined on it and every EFPI that is well-behaved with respect to τ has an SPE.

If a topology τ does not admit equilibrium analysis, then there will be a perfect information game on this tree with this topology that does not have an equilibrium, even though preferences are continuous and decision points are suitably assigned. Thus, Definition 7.6 captures a minimal requirement on a topology. It is also natural to require that a topology only admits equilibrium analysis if some well-behaved EFPI can be defined on it. This is merely a non-triviality requirement. For, otherwise a topology would trivially admit equilibrium analysis when *no* well-behaved EFPI can be defined on it. This is illustrated in the following example.

Example 7.9 (An example without well-behaved EFPI) Let $W=[0,1]^2$ and $(N,\supseteq)$ be given by

$$N=\left\{W,(\{a\}\times[0,1])_{a\in[0,1/2)},(\{w\})_{w\in W}\right\}.$$

[13] The relation $\precsim$ on W is *continuous* if it is both upper and lower semi-continuous.

Denote by $\tau|W(Y_t)$ the topology τ relative to $W(Y_t)$ for all $t = 0,1,\ldots$[14] Here this is the Euclidean topology relative to $W(Y_0) = W(Y_1) = [0,1]^2$ and $W(Y_2) = [0,1/2) \times [0,1]$. Then $W(Y_1 \cap X) = [0,1/2) \times [0,1]$ is not closed relative to $W(Y_1)$. But by (WB1) the moves in this slice, $Y_1 \setminus E = Y_1 \cap X$, must be partitioned into finitely many cells Y_{i1} with $i \in I$ such that each $W(Y_{i1})$ is closed, so that $W(Y_1 \cap X)$ must be closed as a finite union of closed sets—a contradiction.

7.4 A Characterization

The description of an extensive form game entails two levels. The tree provides an "objective" description of the potential events. Choices, the assignment of decision points, and preferences bring in the decision makers and make the tree into a game. Accordingly, an EFPI may not have an SPE for two sets of reasons. As explained above, one has to do with the topology τ on the set W of plays of the tree (the first level). The second set has to do with the game (the second level).

Example 7.1 *is* a well-behaved perfect information game with respect to the relative Euclidean topology. But this topology does not admit equilibrium analysis. This is an instance of the second set of reasons for existence failures. There are, in fact, two conditions that a topology τ on the set W of plays has to respect in order to admit equilibrium analysis. Strikingly, these conditions are actually *necessary and sufficient* for equilibrium analysis. They are neither about preferences $\precsim$ nor about the assignment $\mathscr{X}$ of decision points. Rather, they both reflect the relation between nodes and plays.[15]

As in the last example, denote by $\tau|W(Y_t)$ the relative topology on the plays passing through slice Y_t induced by the topology τ on W, for all $t = 0,1,\ldots$

Theorem 7.2 *A compact perfectly normal topology τ on the set W of plays of a discrete game tree $(N,\supseteq)$ for which there exists a well-behaved EFPI admits equilibrium analysis if and only if τ satisfies*

(CN) $W \setminus x \in \tau$ *for all* $x \in N$;
(OP) *for all* $t = 0,1,\ldots$ *and all sets* $V \subseteq Y_{t+1}$ *of nodes, if* $W(V) \in \tau|W(Y_{t+1})$, *then* $W(p(V)) \in \tau|W(Y_t)$.

[14] That is, for a set $U \subseteq Y_t$ of nodes $W(U) \in \tau|W(Y_t)$ if and only if there is $u \in \tau$ such that $W(U) = u \cap W(Y_t)$.

[15] A topological space is *perfectly normal* if it is T_1 and any two disjoint closed sets can be precisely separated by a continuous function; see the Mathematical Appendix A.2.

A topology τ on the set W of plays of a discrete game tree $(N, \supseteq)$ that satisfies (CN) and (OP) will henceforth be called a *tree topology*. With this terminology Theorem 7.2 states that a compact and perfectly normal topology admits equilibrium analysis if and only if it is a tree topology (provided, of course, that a well-behaved EFPI can be defined).

The "closed nodes" property (CN) requires that all nodes are closed as sets of plays. If it fails, simple counterexamples to existence of SPE can be constructed. For instance, if $W = [0,2]$ were endowed with the relative Euclidean topology and $N = \{W, [0,1), [1,2], (\{w\})_{w \in W}\}$, then a single player maximizing the continuous payoff function $U_1(w) = w$ at the node $[0,1) \in N$ would not be able to do so.

The "open predecessors" condition (OP) requires that if the union of a set of nodes in a slice is open, then the union of its immediate predecessors is also open. That is, if $y \in Y_{t+1}$ is slightly perturbed, this translates into a slight perturbation of $x = p(y) \in Y_t$.[16] This has always been assumed, to the best of our knowledge, in previous proofs of existence of SPE for perfect information games without much ado, even though sometimes in disguised form. For instance, in the formalism of Harris (1985a, p. 618, Assumption 4) it shows up as lower hemi-continuity[17] of the "action correspondence." (Lemma 7.9 below will show that the immediate successors of a move vary lower hemi-continuously in the decision point if and only if (OP) holds.) Harris (1985a, p. 618) calls this property "curious." Apparently it has not been recognized so far that (OP) is necessary for the existence of SPE.

Example 7.10 (Four Intervals revisited) Recall Example 7.3. All nodes except the root are finite, hence (CN) follows immediately since we adopt the (relative) Euclidean topology on $W = [0,1] \times \{1,2,3,4\}$. Consider condition (OP). For Y_1, this condition is trivially fulfilled, since the predecessor of any node in this slice is the root. Hence, (OP) only needs to be verified for Y_3. In this slice, all nodes are terminal, and $W(Y_3) = W(Y_2) = W$. Let V be a set of nodes in Y_3 such that $W(V)$ is open. Condition (OP) requires that the set $W(p(V))$ is also open. Let $(a,n) \in W(p(V))$ with $a < 1$. This implies that $(a,n') \in W(V)$ for some n'. Since $W(V)$ is open in W, the projection of V on $[0,1]$ is an open set, and hence contains an open neighborhood B of (a,n'). Then $B \times \{1,2,3,4\}$ is an open neighborhood of (a,n) in $W(p(V))$. Let now $(1,n) \in W(p(V))$, and suppose $(1,n) \in x_L = \{(1,1),(1,2)\}$ (the case where it is in x_H is analogous). Hence there is $(1,n') \in W(V) \cap x_L$. By an analogous argument as above, the projection of V on $[0,1]$ is an open set, hence it contains an open neighborhood B' of (a,n') in that open set. Then, $(B' \cap [0,1)) \times \{1,2,3,4\} \cup \{(1,1),(1,2)\}$ is an open set contained in $W(p(V))$ that

[16] If τ is a compact separated topology, then a strictly coarser topology cannot be separated and a strictly finer topology cannot be compact (see, e.g., Steen and Seebach 1978, pp. 25–26). Therefore, one cannot simply iteratively delete sets from or add sets to τ in order to enforce (OP).

[17] A correspondence $\Phi : (X, \tau_1) \twoheadrightarrow (Y, \tau_2)$ between topological spaces (X, τ_1) and (Y, τ_2) is *upper* resp. *lower hemi-continuous* (u.h.c. resp. l.h.c.) if for every open set $u \in \tau_2$ in Y the upper preimage $\Phi^+(u) = \{x \in X \,|\, \Phi(x) \subseteq u\}$ resp. the lower preimage $\Phi^-(u) = \{x \in X \,|\, \Phi(x) \cap u \neq \emptyset\}$ is open in X (belongs to τ_1); see the Mathematical Appendix A.2.

contains $(1,n)$. Therefore, $W(p(V))$ is open in $W(Y_2) = W$, completing the proof of (OP).

Condition (OP) is important for pasting together solutions of optimization problems. Imagine that at every $x \in Y_t$ player 1 chooses a $y \in p^{-1}(x) \subseteq Y_{t+1}$, and at every $y \in Y_{t+1}$ player 2 chooses a $z \in p^{-1}(y) \subseteq Y_{t+2}$. If player 2 has continuous preferences, her optimal choice of z will vary upper hemi-continuously in $y \in Y_{t+1}$ (by Berge's maximum theorem). That is, if $y \in Y_{t+1}$ is slightly perturbed, then player 2's maximizers will not "explode." They may still "implode," though.[18] If so, this can lead to a "jump" in player 1's "indirect" preferences (that take into account 2's optimal reaction). This is what fails in Example 7.1 (recall Fig. 7.1).

Example 7.11 (Yes, No, or More continued) Recall Example 7.1. Let the set of plays be $W = ([0,1) \times \{0,1\}) \cup (\{1\} \times [0,1])$ and nodes be given by

$$N = \left\{ W, (\{a\} \times \{0,1\})_{a \in [0,1)}, \{1\} \times [0,1], \{\{w\}\}_{w \in W} \right\}.$$

Endow W with the relative Euclidean topology, so that W is compact. All nodes are closed, i.e. (CN) holds. Yet, (OP) fails. For, let $V \subseteq Y_2$ be the terminal nodes in the intersection of W with an open ε-ball centered at $\bar{w} = (1, 1/2)$ for $\varepsilon \in (0, 1/2)$, so that $W(V) \subset \{1\} \times [0,1] \in Y_1$. Then $W(V) = \{1\} \times (1/2 - \varepsilon, 1/2 + \varepsilon)$ is (relatively) open, but $W(p(V)) = \{1\} \times [0,1]$ is not.

Example 7.12 (Square example and (OP)) Condition (OP) is also what fails in Example 7.2 (recall also Example 7.4) when the unit square $W = [0,1]^2$ carries the relative Euclidean topology. Consider the set $V = \{\{(1,b)\} | \; 1/2 - \varepsilon < b < 1/2 + \varepsilon\} \subseteq Y_2$, for some $\varepsilon \in (0, 1/2)$, of terminal nodes reached after player 2 has decided. The set $W(V) \subseteq W(Y_2)$ is clearly relatively open.[19] Its immediate predecessor set $p(V)$, though, consists of the singleton $\{1\} \times [0,1] \in X$, which is not open. Hence, (OP) fails in this example. This shows that the ultimate reason for the failure of existence in this example is a violation of (OP), that is, a problem with the specification of the tree, and not a failure of compactness in a derived Tychonoff construction.

Condition (OP) has an interesting implication. For, if $V = Y_{t+1}$ then trivially $W(V) = W(Y_{t+1})$ is relatively open and (OP) implies that $W(p(Y_{t+1})) = W(Y_t \cap X)$ is relatively open, for all t. Therefore, $W(Y_t \cap E) = W(Y_t \setminus X)$ is relatively closed. This fails in the following example (for Y_1) that again violates (OP). Once more, the failure of (OP) can be seen directly on the graphical representation (Fig. 7.6) by observing that the successors along a slice of nodes behave in a discontinuous way.

[18]Upper hemi-continuity is sometimes discussed as not allowing "implosions." This corresponds to a terminology that refers to approximations. Ours is one that refers to perturbations. That is, we take the stance that a map is being evaluated at a reference point $y \in Y_{t+1}$ and then implosions or explosions refer to what happens if this reference point is slightly perturbed.

[19]As will be seen below, this implies that V is open in the—natural—quotient topology on Y_2.

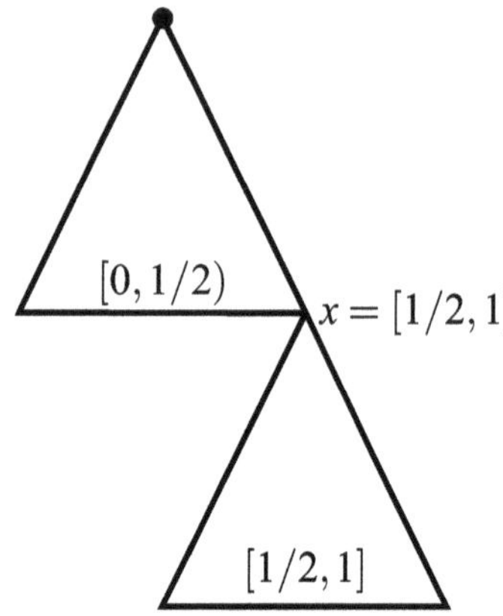

Fig. 7.6 Graphical representation of Example 7.13. Player 1 chooses either a number $a \in [0, 1/2)$, or lets player 2 decide, who then chooses a number in $[1/2, 1]$

Example 7.13 (Split Interval) Let $W = [0,1]$, $x = [1/2,1]$, $N = \{W, x, (\{w\})_{w \in W}\}$, $X_1 = \{W\}$, and $X_2 = \{x\}$ for two players, $I = \{1,2\}$ (see Fig. 7.6). The topology τ on W is the relative Euclidean topology on the unit interval. Then W is compact and all nodes are closed. If preferences are represented by the payoff functions $U_1(w) = w(4/3 - w)$ and $U_2(w) = w$, there is no equilibrium. For, given 2's optimal decision $(w = 1)$ player 1 has no best reply.

The proof of Theorem 7.2 follows from two stronger results in sections below. The first (Theorem 7.3) is a result on necessity that is stronger in the sense that it does not require a compactness hypothesis. The second (Theorem 7.4) is a result on sufficiency, i.e., an SPE existence theorem, that is stronger in the sense that it requires a weaker separation axiom, Hausdorff (T_2) rather than perfect normality.

7.5 Necessary Conditions

The examples in the previous sections are suggestive. The failures of (CN) and (OP) that they illustrate in fact generalize. This gives rise to the following theorem.

Theorem 7.3 *For a discrete game tree $(N, \supseteq)$ with set W of plays every perfectly normal topology τ on W that admits equilibrium analysis is a tree topology.*

Theorem 7.3 does not invoke compactness. What is maintained is the separation axiom that τ is perfectly normal. This is a rather weak assumption. For, in practice one may want to employ a topology that is, for instance, induced by a metric (as e.g. in Hellwig and Leininger 1987) and, therefore, satisfies a much stronger separation axiom. Therefore, this hypothesis captures the relevant cases. In particular, every metric space is perfectly normal.

Of course, counterexamples to existence of SPE are not enough to establish the necessity of a tree topology, because the theorem says nothing about the tree nor about the particular topology (except for perfect normality). Therefore, we devote the following subsection to the proof of Theorem 7.3.

7.5.1 *Proof of Theorem 7.3*

The first lemma combines the observation from Example 7.13 with the hypothesis that the topology admits equilibrium analysis—thus, a well-behaved EFPI can be defined—to deduce that the subset $W(Y_t)$ of W cannot be connected when it contains terminal nodes.[20]

Lemma 7.4 *Let τ be a topology on the set W of plays for a discrete game tree $(N, \supseteq)$ such that a well-behaved EFPI can be defined.*

(a) The sets $W(Y_t)$ and $W(Y_t \cap X)$ are closed in τ, for all $t = 1, 2, \ldots$
(b) If (OP) holds, then the sets $W(Y_t)$, $W(Y_t \cap X)$, and $W(Y_t \cap E)$ are both open and closed in τ, for all $t = 1, 2, \ldots$

Proof (a) Suppose that some well-behaved EFPI with respect to τ can be defined. Then the set $Y_t \setminus E = Y_t \cap X$ is partitioned into finitely many cells of the form $Y_{it} = X_i \cap Y_t$ for $i \in I$. Since nodes in a slice are disjoint, it follows that $W(Y_t \cap X)$ is partitioned into finitely many cells of the form $W(Y_{it})$ for $i \in I$. Further, each $W(Y_{it})$ is closed. Therefore, $W(Y_t \cap X)$ is closed as a finite union of closed sets. That also $W(Y_t)$ is closed in τ follows from $W(Y_0) = W$ and $W(Y_t) = W(Y_{t-1} \cap X)$ for $t = 1, 2, \ldots$

(b) Let $V = Y_{t+1}$. Trivially, $W(V) \in \tau | W(Y_{t+1})$. Note that $p(V) = Y_t \cap X = Y_t \setminus E$. By (OP) it follows that $W(p(V)) \in \tau | W(Y_t)$, that is,

$$\exists u_t \in \tau : W(p(V)) = W(Y_t \cap X) = u_t \cap W(Y_t) \tag{7.3}$$

for all $t = 1, 2, \ldots$ Clearly, $W(Y_1) = W$ implies trivially $W(Y_1) \in \tau$. Proceeding inductively, if $W(Y_t) \in \tau$, then (7.3) implies $W(Y_t \cap X) = u_t \cap W(Y_t) \in \tau$ and $W(Y_{t+1}) = W(Y_t \cap X)$ implies $W(Y_{t+1}) \in \tau$. Hence, $W(Y_t) \in \tau$ for all $t = 0, 1, \ldots$ Together with part (a) this implies that each $W(Y_t)$ and each $W(Y_t \cap X)$ is clopen. Because $W(Y_t \cap E) = W(Y_t) \setminus W(Y_{t+1})$ for all $t = 1, 2, \ldots$, it follows that also each $W(Y_t \cap E)$ is clopen. ■

The first part of Lemma 7.4 states that, for any well-behaved EFPI, the set of plays $W(Y_t)$ is a closed set for every slice Y_t. If additionally (OP) holds, those sets are both open and closed ("clopen"). Further, the terminal parts of all slices are

[20] A topological space is *connected* if its only subsets which are both open and closed are the empty set and the full space. A subset that is a connected space with its relative topology is called *connected*.

also open and closed as sets of plays. The following is a direct consequence of Lemma 7.4(a).

Lemma 7.5 *If a topology* τ *on the set* W *of plays for a discrete game tree* $(N, \supseteq)$ *admits equilibrium analysis, then any EFPI for which all moves in each slice are assigned to the same player,* $Y_t \cap X = Y_{it}$ *for some* $i \in I$ *for all* $t = 0, 1, \ldots$*, satisfies (WB1) from Definition 7.5.*

The next lemma states the obvious, that existence of subgame perfect equilibria for one-player EFPIs boils down to existence of maxima at every node.

Lemma 7.6 *If a topology* τ *on the set* W *of plays for a discrete game tree* $(N, \supseteq)$ *admits equilibrium analysis, then every continuous preference relation* $\precsim$ *on* W *has a maximum at every node.*

Proof By Lemma 7.5 assigning all moves in the tree to a single player, i.e. $X_1 = X$, endowed with continuous preferences $\precsim$ yields a well-behaved EFPI. By hypothesis an SPE s^* exists. Fix a move $x \in X$ and let $w^* = \phi_x(s^*)$ be the play induced by s^* in the subgame starting at x. If $w' \succ w^*$ for some $w' \in x$, there is a strategy that picks $w' \succ w^*$ (since every play can be reached by some strategy profile by Theorem 5.1), a contradiction. ■

7.5.1.1 Necessity of (CN)

Suppose that for some $t = 0, 1, \ldots$ there is $x \in Y_t$ that is not closed. Then there is $w_0 \in W \setminus x$ (a cluster point of x) such that for every neighborhood $u \in \tau$ of $w_0 \in u$ there is $w \in u \cap x$. Fix such a neighborhood $u_0 \in \tau$ of $w_0 \in u_0$. Because (W, τ) is perfectly normal by hypothesis, Urysohn's Lemma (see Aliprantis and Border 1990, p. 45, or Sect. A.2 in the Mathematical Appendix) implies that there is a continuous function $f : W \to [0, 1]$ such that $f^{-1}(0) = W \setminus u_0$ and $f^{-1}(1) = \{w_0\}$, as singletons are closed in a perfectly normal (thus, in particular, T_1) space. We claim that the preferences on W represented by the continuous function $U_1(w) = f(w)$ for all $w \in W$ have no maximum at the node x, in contradiction to Lemma 7.6.

To see the claim it is enough to observe that for every $\lambda \in [0, 1)$ the player can guarantee herself a payoff that is strictly larger than λ. For, the strict upper contour set $u_\lambda = \{w \in W | U_1(w) > \lambda\}$ is an open neighborhood of w_0 by continuity and, therefore, has a nonempty intersection with x. Choosing a play $w \in u_\lambda \cap x$ the player obtains more than λ. But the maximal payoff 1 remains infeasible, because $U_1(w) = 1$ implies $w = w_0 \notin x$. Therefore, the payoff function has no maximum at x.

7.5.1.2 Necessity of Closed Terminal Slices

Before the necessity of (OP) is established, it is shown that if τ admits equilibrium analysis, then the set $W(Y_t \cap E)$ of plays ending at slice Y_t must be closed, even if (OP) does not hold. This implication is used in the next subsection. As

Lemma 7.4(b) deduces the same conclusion from (OP), the latter condition could be weakened so as to apply only to proper subsets ($V \subset Y_{t+1}$) if an independent property, that $W \setminus W(Y_t \cap E) \in \tau$ for all t, were added.

Suppose that for some $t = 0, 1, \ldots$ the set of plays $W(Y_t \cap E)$ passing through the terminal nodes in Y_t is not closed. This implies that $Y_t \cap X \neq \emptyset$, because otherwise $W(Y_t) = W(Y_t \cap E)$ would be closed by Lemma 7.4(a). It follows that there is a cluster point w_0 of $W(Y_t \cap X)$ which does not belong to $W(Y_t \cap X)$. Since $W(Y_t \cap E) \subseteq W(Y_t)$ and the latter is closed by Lemma 7.4(a), it follows that $w_0 \in W(Y_t)$, hence $w_0 \in W(Y_t \cap X)$. Since w_0 is a cluster point of $W(Y_t \cap E)$, for every neighborhood $u \in \tau$ of $w_0 \in u$ there is $w \in u \cap W(Y_t \cap E)$. Fix such a neighborhood $u_0 \in \tau$ of $w_0 \in u_0$. Because (W, τ) is perfectly normal, Urysohn's lemma again implies that there is a continuous function $f : W \to [0,1]$ such that $f^{-1}(0) = W \setminus u_0$ and $f^{-1}(1) = \{w_0\}$.

Assign all moves before Y_t to player 1, $X_1 = \cup_{k=0}^{t-1}(Y_k \cap X)$, and all moves from Y_t onwards to player 2, $X_2 = \cup_{k \geq t}(Y_k \cap X)$. Preferences are represented by the continuous functions

$$U_1(w) = f(w) \text{ and } U_2(w) = -f(w) \text{ for all } w \in W.$$

By continuity of U_1 and U_2 and Lemma 7.5 this is a well-behaved EFPI.

For every $\lambda \in [0,1)$ player 1 can guarantee herself a payoff that is strictly larger than λ. For, the strict upper contour set $u_\lambda = \{w \in W \,|\, U_1(w) > \lambda\}$ is an open neighborhood of w_0 by continuity and, hence, there is $w' \in u_\lambda \cap W(Y_t \cap E)$, that is, there is $\{w'\} \in Y_t \cap E$ such that $U_1(w') > \lambda$. If player 1 picks $w' \in u_\lambda \cap W(Y_t \cap E)$ she obtains more than λ, since player 2 does not get to choose. Therefore, player 1 must obtain payoff $U_1(w) = 1$ in any Nash equilibrium. But $U_1(w) = 1$ implies $w = w_0$, yet $\{w_0\} \notin Y_t \cap E$. Hence, $w_0 \in x_0$ for some $x_0 \in Y_t \cap X = Y_t \cap X_2$. Since $x_0 \in X$ and $E = \{\{w\} \mid w \in W\}$, there is $w' \in x_0 \setminus \{w_0\}$ such that $U_2(w') > -1 = U_2(w_0)$. Because player 2 controls all moves from Y_t onwards, picking $w' \in x_0$ constitutes a profitable deviation for player 2. Therefore, there is no Nash equilibrium and in particular no SPE.

7.5.1.3 Necessity of (OP)

If (OP) fails, then for some $t = 0, 1, \ldots$ there is a set $V \subseteq Y_{t+1}$ of nodes such that $W(V) \in \tau|W(Y_{t+1})$, but $W(p(V)) \notin \tau|W(Y_t)$, where $p(V) = \{p(y) \,|\, y \in V\}$. It follows that the set $W(\{y \in Y_{t+1} \,|\, p(y) \notin p(V)\})$ is not closed in $\tau|W(Y_{t+1})$. For, if it were, then its complement would be open in $\tau|W(Y_{t+1})$, i.e., there would be $v \in \tau$ with $W(\{y \in Y_{t+1} \,|\, p(y) \in p(V)\}) = v \cap W(Y_{t+1})$. Since $W(Y_t \cap E)$ is closed by the argument in the previous subsection, $v' = v \setminus W(Y_t \cap E) \in \tau$ as the complement of a closed set, hence,

$$W(p(V)) = W(\{y \in Y_{t+1} \,|\, p(y) \in p(V)\}) = v \cap W(Y_{t+1}) = v' \cap W(Y_t)$$

would be open in $\tau|W(Y_t)$, in contradiction to the hypothesis that $W(p(V)) \notin \tau|W(Y_t)$.

That $W(\{y \in Y_{t+1} | p(y) \notin p(V)\})$ is not closed implies that there exists some $w_0 \in W(\{y \in Y_{t+1} | p(y) \in p(V)\})$ such that for every neighborhood $\hat{v} \in \tau$ of $w_0 \in \hat{v}$ there is some $w \in \hat{v} \setminus \{w_0\}$ with $w \in W(\{y \in Y_{t+1} | p(y) \notin p(V)\})$. Let $y_0 = \text{proj}_{t+1}(w_0) \in Y_{t+1}$ be the unique node in Y_{t+1} such that $w_0 \in y_0$. Then y_0 does not belong to V, i.e. $y_0 \notin V$. For, if it did, then, with $\hat{v} \in \tau$ such that $\hat{v} \cap W(Y_{t+1}) = W(V)$, that $w_0 \in y_0 \subseteq \hat{v}$ would imply that there is some $w' \in \hat{v} \setminus \{w_0\}$ with $p(\text{proj}_{t+1}(w')) \notin p(V)$ even though $w' \in \hat{v}$ implies $\text{proj}_{t+1}(w') \in V$, a contradiction. On the other hand, $w_0 \in W(\{y \in Y_{t+1} | p(y) \in p(V)\})$ implies that $p(y_0) \in p(V)$, and the latter implies that there is some $y_1 \in V$, $y_1 \neq y_0$, such that $p(y_1) = p(y_0)$. Let $\bar{x} = p(y_0) = p(y_1) \in Y_t$.

Choose $w_1 \in y_1 \in p^{-1}(\bar{x}) \cap V$ and recall that $w_0 \in y_0 \in p^{-1}(\bar{x}) \setminus V$. Because (W, τ) is separated, there are $u_0, \hat{u}_1 \in \tau$ such that $w_0 \in u_0$, $w_1 \in \hat{u}_1$, and $u_0 \cap \hat{u}_1 = \emptyset$, and the singletons $\{w_0\}$ and $\{w_1\}$ are closed. Setting $u_1 = \hat{u}_1 \cap v \in \tau$ it follows that $u_1 \cap W(Y_{t+1}) \subseteq W(p(V))$. Since $W \setminus u_0$ and $W \setminus u_1$ are closed as complements of open sets, again by Urysohn's lemma there are continuous functions $f_1 : W \to [0,1]$ and $f_2 : W \to [0,1]$ such that $f_1^{-1}(1) = \{w_0\}$, $f_1^{-1}(0) = W \setminus u_0$, $f_2^{-1}(1) = \{w_1\}$, and $f_2^{-1}(0) = W \setminus u_1$, as (W, τ) is perfectly normal by hypothesis.

Assign all moves before Y_t to player 1, $X_1 = \cup_{k=0}^{t-1}(Y_k \cap X)$, and all moves from Y_t onwards to player 2, $X_2 = \cup_{k \geq t}(Y_k \cap X)$, so that $Y_t \setminus E \subseteq X_2$ but $Y_{t-1} \setminus E \subseteq X_1$. Preferences for the two players $i = 1, 2$ are represented by the continuous functions

$$U_1(w) = f_1(w) \text{ and } U_2(w) = \frac{1}{2} f_1(w) + f_2(w) \text{ for all } w \in W.$$

By Lemma 7.5 and continuity of U_i $(i = 1, 2)$ this yields a well-behaved EFPI.

In an SPE player 1 can guarantee herself a payoff that strictly exceeds λ for any $\lambda \in [0, 1)$. For, the strict upper contour set $u_\lambda = \{w \in W | U_1(w) > \lambda\}$ is open in τ by continuity of U_1 and a neighborhood of $w_0 \in u_\lambda$. Thus, there is $w_\lambda \in u_\lambda \setminus \{w_0\}$ such that $w_\lambda \in W(\{y \in Y_{t+1} | p(y) \notin p(V)\})$. Hence, there is $y_\lambda = \text{proj}_{t+1}(w_\lambda) \in Y_{t+1}$ such that $p(y_\lambda) \equiv x_\lambda \in Y_t \setminus p(V)$. Since $u_1 \cap W(Y_{t+1}) \subseteq W(p(V))$ implies $W(Y_t \setminus p(V)) = W(Y_t) \setminus W(p(V)) \subseteq W \setminus u_1$ and $U_1(w) = 2U_2(w)$ for all $w \notin u_1$, if player 1 chooses x_λ, then (by subgame perfection) player 2 will choose $w \in \arg\max_{w' \in x_\lambda} U_2(w') = \arg\max_{w' \in x_\lambda} U_1(w')$, which yields player 1 strictly more than λ, because $w_\lambda \in y_\lambda \subset x_\lambda \Rightarrow w_\lambda \in x_\lambda \cap u_\lambda$.

Therefore, player 1 must obtain payoff $U_1(w) = 1$ in each SPE. But $U_1(w) = 1$ implies $w = w_0 \in \bar{x} \in Y_{2t}$. Consider the subgame starting at $\bar{x} \in Y_{2t}$. Since player 2 controls all moves from Y_t onwards, she can in particular choose $w_1 \in y_1 \in V$. This gives her a payoff of at least 1 which is strictly more than what she can obtain outside of u_1, that is, $\arg\max_{w' \in \bar{x}} U_2(w') \subseteq u_1 \cap W(Y_{t+1})$. But then player 1's best payoff from choosing $\bar{x} \in Y_t$ is zero. Therefore, player 1's value function in an SPE has no maximum and no SPE exists. This completes the proof of Theorem 7.3.

7.6 Sufficient Conditions

This section establishes the sufficiency part of the characterization. That is, the necessary conditions (CN) and (OP) are also sufficient for a topology to admit equilibrium analysis. Theorem 7.4 below takes a weaker separation axiom and is hence stronger than the if-part of Theorem 7.2: that τ is separated (Hausdorff or T_2). Assuming a weaker separation property makes the sufficiency result more useful.

Theorem 7.4 *For a discrete game tree* $(N, \supseteq)$ *with set* W *of plays every compact separated tree topology* τ *on* W *for which there exists a well-behaved EFPI admits equilibrium analysis.*

In other words, what is needed for existence of SPE in well-behaved EFPIs is a compact separated topology which satisfies (CN) and (OP). Earlier approaches (e.g., Fudenberg and Levine 1983; Harris 1985a; Hellwig and Leininger 1987) have imposed compactness directly on the set of choices ("action sets") at each move (plus extra conditions on the set of histories). They then derive compactness of the space of plays from Tychonoff's theorem. As discussed in Sect. 7.2 above, this leads to difficulties because, first, the product space might be much larger than the relevant space of outcomes, and, second, the latter space needs to be recovered as a subset of the product space, but compactness of that subset is not guaranteed. In contrast, we impose compactness on the set of outcomes W, which is the natural space on which preferences are defined.

Kuhn's (1953) existence proof of an SPE for finite EFPIs used an algorithm. In the infinite case a direct generalization is, of course, infeasible, because there may be no "end" where the backwards induction procedure can begin. Consequently, earlier proofs of existence for the infinite case (in particular Harris 1985a; Hellwig and Leininger 1987; Hellwig et al. 1990) have used a "finite truncation" approach. This technique truncates the game at some finite stage (slice), then lengthens the horizon of the truncation, and finally looks at a limit. The present proof, on the other hand, provides an explicit algorithm that is *not* based on finite truncations.

The idea of the algorithm is as follows. Players start naively, that is, when they are called upon to choose they pick a play, as if they had full control of all other players moving afterwards. This is the first step. In the second step players develop some anticipation and reoptimize, now under the constraints generated by what later players have done in the first step. That is, in the second step players have become "smarter" and anticipate the choices (from the first round) of later players. This is repeated in the third step. Players now reoptimize under the constraints generated by the choices of later players in the second step. For this procedure an appropriate limit contains an SPE.

In what follows we develop the algorithm and demonstrate that any compact separated tree topology for a discrete game tree admits equilibrium analysis. That is, if the game is well-behaved with respect to a compact separated tree topology, then there exists an SPE.

7.6.1 Topology on Nodes

The construction of the algorithm will take a topology on the set of nodes, at least for a given slice Y_t. Since the set Y_t of nodes (i.e., the t-th slice) forms a partition of the set $W(Y_t)$ of plays, the projection $\text{proj}_t : W(Y_t) \to Y_t$ is uniquely defined by $w \in \text{proj}_t(w)$ for all $w \in W(Y_t)$. The *quotient topology* on Y_t (induced by τ), denoted τ_t^q, is the finest topology that makes the projection continuous (when $W(Y_t)$ is endowed with the relative topology induced by τ). We begin with a characterization of open sets of nodes (in a given slice) in the quotient topology: A set of nodes in a slice is open if and only if the set of plays passing through them is open. This follows directly from the definition of the quotient topology (e.g. Kelley 1975, p. 94) and the fact that $\text{proj}_t^{-1}(U) = W(U)$ for any $U \in \tau_t^q$.

Lemma 7.7 *The set $U \subseteq Y_t$ of nodes is open in τ_t^q if and only if $W(U)$ is open in τ relative to $W(Y_t)$, for all $t = 0, 1, \ldots$*

The second observation concerns a property of well-behaved EFPIs (see also Lemma 7.5). If (WB1) holds for Y_t, then Y_t is compact.

Lemma 7.8 *If (W, τ) is compact and $W(Y_t)$ is closed in τ, then Y_t is compact in τ_t^q, for all $t = 0, 1, \ldots$*

Proof If $W(Y_t)$ is closed, then it is compact in τ, since (W, τ) is compact. Because τ_t^q makes the surjection proj_t continuous, and the continuous image of a compact space is compact, the statement follows. ■

The third observation concerns lower hemi-continuity of the correspondence $p^{-1} : Y_t \setminus E \twoheadrightarrow Y_{t+1}$ that assigns to each move $x \in Y_t$ the set of its immediate successors, $p^{-1}(x) = \{y \in Y_{t+1} | p(y) = x\}$.

Property (OP) may be restated, in terms of the quotient topologies on slices, as the immediate predecessor function $p : Y_{t+1} \to Y_t$ from (7.1) being an open map, i.e. $V \in \tau_{t+1}^q \Rightarrow p(V) \in \tau_t^q$ (by Lemma 7.7). Applying a well-known result (see, e.g., Aliprantis and Border 1990, Theorem 17.7(2), p. 560) yields:

Lemma 7.9 *The immediate successor correspondence $p^{-1} : Y_t \setminus E \twoheadrightarrow Y_{t+1}$ is l.h.c. if and only if (OP) holds, when slices are endowed with the quotient topologies, for all $t = 0, 1, \ldots$*

Since the sufficient conditions entail only (OP) and (CN), the present theorem does not require an assumption on upper hemi-continuity of successors. Lower hemi-continuity, on the other hand, ensures that the optimization problems of different players can be pasted together.

The next results concern the function γ from (4.7). This is defined on the graph of the correspondence $W(\cdot)$, denoted $\mathscr{G} = \{(x, w) \in N \times W | w \in x\}$, and maps to N. For a fixed move x this function is continuous in $w \in x$.

Lemma 7.10 *For each $x \in X \cap Y_t$ the evaluation map $\gamma_x : x \to p^{-1}(x)$, defined by $\gamma_x(w) = \gamma(x,w)$ for all $w \in x$, is continuous in $w \in x$, when x is endowed with τ relative to x and $p^{-1}(x)$ with τ^q_{t+1} relative to $p^{-1}(x)$, for all $t = 0,1,\ldots$*

Proof Let $x \in X \cap Y_t$ so that $p^{-1}(x) \subseteq Y_{t+1}$. Then the function γ_x coincides with the map from $W(p^{-1}(x))$ to the partition $p^{-1}(x)$ of $x = W(p^{-1}(x))$ that assigns to each $w \in x$ the node $\text{proj}_{t+1}(w) \cap x = \text{proj}_{t+1}(w) \in Y_{t+1}$ by Proposition 6.1(a). Since τ^q_{t+1} relative to $p^{-1}(x)$ makes proj_{t+1} continuous, γ_x is also continuous. ■

Lemma 7.10 only concerns continuity of γ in plays for a fixed move. It does not say that γ is continuous in both arguments. To achieve that takes condition (OP). This property makes γ continuous in both arguments, when it operates from one slice to the next. Formally, let $\mathcal{G}_t = \{(x,w) \in (Y_t \setminus E) \times W(Y_t \setminus E) \,|\, w \in x\}$ denote the graph of $W(\cdot)$ restricted to the non-terminal part of a slice Y_t, for $t = 0,1,\ldots$

Lemma 7.11 *If (OP) holds, then the function $\gamma_t : \mathcal{G}_t \to Y_{t+1}$, defined by $\gamma_t(x,w) = \gamma(x,w)$ for all $(x,w) \in \mathcal{G}_t$, is continuous, when $\mathcal{G}_t$ is endowed with the product topology of τ^q_t and τ relative to $\mathcal{G}_t$ and Y_{t+1} is endowed with τ^q_{t+1}, for all $t = 0,1,\ldots$*

Proof Fix $V \in \tau^q_{t+1}$ and $(x,w) \in \gamma_t^{-1}(V) = \{(x',w') \in \mathcal{G}_t \,|\, \gamma_t(x',w') \in V\}$. By property (OP), $p(V) = \{p(y) \in Y_t \,|\, y \in V\} \in \tau^q_t$. By Lemma 7.10, the set $\gamma_x^{-1}(V) = \{w \in x \mid \gamma(x,w) \in V\}$ is (nonempty and) open in τ relative to $x \in Y_t$. That is, there exists $u_x \in \tau$ with $\gamma_x^{-1}(V) = u_x \cap x$.

As $V \in \tau^q_{t+1}$, by Lemma 7.7 there is $\bar{v} \in \tau$ such that $W(V) = \bar{v} \cap W(Y_{t+1})$. Let $\bar{u}_x = u_x \cap \bar{v} \in \tau$. Since $w \in \gamma(x,w) \in V$, on the one hand $x \in p(V)$; on the other hand, $w \in W(V) \subseteq \bar{v}$, which together with $w \in u_x$ implies $w \in \bar{u}_x$. In summary, $(x,w) \in p(V) \times \bar{u}_x$. Therefore $A = (p(V) \times \bar{u}_x) \cap \mathcal{G}_t$ is an open set in the product topology of τ^q_t and τ relative to $\mathcal{G}_t$ such that $(x,w) \in A$.

We claim that $A \subseteq \gamma_t^{-1}(V)$. To see this, let $(y,w') \in A$. Since $A \subseteq \mathcal{G}_t$, $w' \in y$. Since $y \in p(V)$, y is not a terminal node. Therefore, $\gamma(y,w') \in Y_{t+1}$, implying that $w' \in W(Y_{t+1})$. This together with $w' \in \bar{u}_x \subseteq \bar{v}$ yields that $w' \in W(V)$. That nodes in Y_{t+1} are pairwise disjoint implies that $\gamma(y,w') \in V$, hence $(y,w') \in \gamma_t^{-1}(V)$. It follows that A is an open neighborhood of (x,w) that is contained in $\gamma_t^{-1}(V)$. This completes the proof. ■

7.6.2 The Algorithm

With this preparation in place, we turn to the construction of the algorithm. For all that follows let a compact and separated tree topology τ on the set W of plays of a discrete game tree $(N, \supseteq)$ be given and consider an EFPI that is well-behaved with respect to τ. First we seek to define, for each $k = 1,2,\ldots$, three correspondences $\underline{\xi}_k, \xi_k, \overline{\xi}_k : N \twoheadrightarrow W$ by the following recursive procedure. For terminal nodes $\{w\} \in E$ set $\overline{\xi}_k(\{w\}) = \xi_k(\{w\}) = \underline{\xi}_k(\{w\}) = w$ for all $k = 0,1,\ldots$ For moves $x \in Y_t \cap X$

initially define $\xi_0(x) = \{w \in x | w' \precsim_i w \; \forall w' \in x, \; \{i\} = J(x)\}$ and

$$\overline{\xi}_0(x) = \{w \in x | (x,w) \in \mathrm{cl}(\{(y,v) \in Y_t \times W(Y_t) | v \in \xi_0|_{Y_t}(y)\})\} \\ = \{w \in x | (x,w) \in \mathrm{cl}(\mathrm{graph}(\xi_0|_{Y_t}))\},$$

where the closure is taken in the product topology of τ_t^q and τ. Since $x \in Y_t$ is closed by (CN), thus compact (as (W,τ) is compact), $\xi_0(x)$ is nonempty, closed (by continuity of preferences, (WB2)), and contained in x. The graph of $\overline{\xi}_0\big|_{Y_t}$ is closed by construction, and its range, $W(Y_t)$ is compact (by Lemma 7.4(a) and (WB1)) and separated. Hence, by the Closed Graph Theorem (Aliprantis and Border 1990, Theorem 17.11, p. 561), $\overline{\xi}_0\big|_{Y_t}$ is u.h.c. with nonempty closed values. For $k = 1,2,\dots$ assume that ξ_{k-1} and $\overline{\xi}_{k-1}$ are already defined such that $\emptyset \neq \xi_{k-1}(x) \subseteq x$ for all $x \in N$ and the correspondences $\overline{\xi}_{k-1}\big|_{Y_t}$ are u.h.c. with closed values. Let again $x \in Y_t \cap X$ so that $p^{-1}(x) \subseteq Y_{t+1}$ for $t = 0,1,\dots$ Define

$$\underline{\xi}_k(x) = \left\{w \in x \middle| w \in \overline{\xi}_{k-1}(\gamma(x,w))\right\} \tag{7.4}$$

$$= \left\{w \in x \middle| w \in \overline{\xi}_{k-1}(y), p(y) = x\right\} = x \cap \mathrm{proj}_W\left(\mathrm{graph}\left(\overline{\xi}_{k-1}\big|_{Y_{t+1}}\right)\right).$$

Then $\underline{\xi}_k(x)$ is closed in τ, hence compact, because it is the intersection of two closed sets. The first is a node, hence closed by (CN). The second is the projection of the graph of an u.h.c. correspondence (which has a closed graph, because its range is regular; see Klein and Thompson 1984, Theorem 7.1.15, p. 78) onto the compact space (W,τ). It follows that

$$\xi_k(x) = \left\{w \in \underline{\xi}_k(x) \middle| w' \precsim_i w \; \forall w' \in \underline{\xi}_k(x), \; \{i\} = J(x)\right\} \tag{7.5}$$

is nonempty, closed, and contained in $x \in Y_t$. Define $\underline{\xi}_k$ and ξ_k in this way for all $x \in Y_t$ and all $t = 0,1,\dots$ Finally, define for each $x \in Y_t \cap X$ and all $t = 0,1,\dots$

$$\overline{\xi}_k(x) = \{w \in x | (x,w) \in \mathrm{cl}(\{(y,v) \in Y_t \times W(Y_t) | v \in \xi_k|_{Y_t}(y)\})\} \tag{7.6}$$
$$= \{w \in x | (x,w) \in \mathrm{cl}(\mathrm{graph}(\xi_k|_{Y_t}))\},$$

where the closure is taken in the product topology of τ_t^q and τ. Because the range is compact, this construction guarantees that each $\overline{\xi}_k\big|_{Y_t}$ is u.h.c. with nonempty closed values (see Klein and Thompson 1984, Theorem 7.1.16, p. 78). This completes the construction of $\underline{\xi}_k, \xi_k, \overline{\xi}_k : N \twoheadrightarrow W$ for all $k = 1,2,\dots$ by induction.

The intuition of this construction is as follows. The correspondence $\underline{\xi}_k$ in (7.4) serves as the constraint generated from the previous step. The correspondence ξ_k

in (7.5) captures the optimization under the constraints from the previous iteration round. Finally, the correspondence $\overline{\xi}_k$ in (7.6) generates the input for the next step by taking a closure. Why the closure operation is required is nicely explained by an example from Hellwig and Leininger (1987, pp. 62), reproduced below.

Example 7.14 (Hellwig and Leininger's example) Let $W = [0,1]^3$ and the tree be given by

$$N = \left\{ W, \left(\{x\} \times [0,1]^2\right)_{x \in [0,1]}, (\{(x,y)\} \times [0,1])_{(x,y) \in [0,1]^2}, (\{w\})_{w \in W} \right\}.$$

There are three players. Player 1 chooses from $C_1 = \left\{ \{x\} \times [0,1]^2 \in N \,|\, x \in [0,1] \right\}$, player 2 from $C_2 = \left\{ (x,y) \times [0,1] \in N \,\middle|\, (x,y) \in [0,1]^2 \right\}$, and player 3 from terminal nodes, $C_3 = \{\{w\} \in N \,|\, w \in W\}$. Payoffs are

$$u_1(x,y,z) = -x - z,\ u_2(x,y,z) = y/2 + z, \text{ and } u_3(x,y,z) = z(z - 1/2 - x - y).$$

Player 3's payoff is strictly convex in $z \in [0,1]$, so that her best replies are given by

$$z \in \begin{cases} \{0\} & \text{if } x + y > 1/2 \\ \{0,1\} & \text{if } x + y = 1/2 \\ \{1\} & \text{if } x + y < 1/2 \end{cases}$$

Substituting this into player 2's payoff function yields a piecewise linear function (correspondence) of $y \in [0,1]$ that equals $1 + y/2$ if $y \le 1/2 - x$ and equals $y/2$ if $y \ge 1/2 - x$. Player 2, of course, always prefers $z = 1$ to $z = 0$. So, when she selects a play from $\underline{\xi}_3\left(\{x\} \times [0,1]^2\right)$, she chooses $y = 1/2 - x$ whenever $x \le 1/2$. Only for $x > 1/2$ player 2 will choose $y = 1$. Therefore, after $x \in [0,1]$,

$$\xi_3\left(\{x\} \times [0,1]^2\right) = \begin{cases} (1,0) & \text{if } x > 1/2 \\ (1/2 - x, 1) & \text{if } x \le 1/2 \end{cases}$$

The point $(x,y,z) = (1/2,1,0)$ does not belong to the graph of $\xi_3\left(\{x\} \times [0,1]^2\right)$, but only to its closure. This explains the construction of $\overline{\xi}_3\left(\{x\} \times [0,1]^2\right)$.

By choosing x close to but above $1/2$ player 1 can always force the continuation $(y,z) = (1,0)$ and thereby attain a payoff close to (but slightly below) $-1/2$. In order for this to be equilibrium behavior player 2 has to be prevented from choosing her preferred continuation $(y,z) = (0,1)$. This is achieved by breaking player 3's indifference in favor of $z = 0$ after $x + y = 1$. This restricts player 2's choice set after $x = 1/2$ to the single line segment $y/2$ and induces player 2 to choose the continuation $(y,z) = (1,0)$. The ultimate equilibrium then involves the choices $(x,y,z) = (1/2,1,0)$ and yields player 1 a payoff $-1/2$.

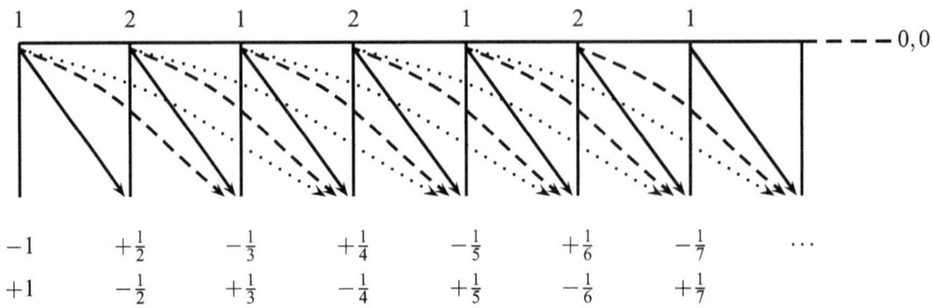

Fig. 7.7 Graphical representation of Example 7.15. Players 1 and 2 alternate moves. Solid, dashed, and dotted arrows represent the 1st, 2nd, and 3rd iteration of the algorithm, respectively

The next example illustrates how the algorithm described above works in practice.

Example 7.15 (A centipede game with a unique SPE) Take the tree from Example 7.6. Consider again two players, $I = \{1,2\}$, and let $X_1 = \{x_w \in X \,|\, w \text{ is odd}\}$ and $X_2 = \{x_w \in X \,|\, w \text{ is even}\}$. Preferences are represented by the continuous payoff functions $U_1(w) = (-1)^w/w$ and $U_2(w) = -U_1(w)$ for all $w = 1,2,\ldots,$ and $U_i(\infty) = 0$ for $i = 1,2$. See Fig. 7.7 for a representation. That (CN) and (OP) hold has been verified in Example 7.6.

In the first iteration of the algorithm each player maximizes her payoff by continuing once and stop at the next node, i.e. $\xi_0(x_w) = \{w+1\}$ for all w. In the second iteration players take into account the decision of the next player, and payoffs are maximized by continuing and letting the next player stop one node further away, i.e. $\xi_1(x_w) = \{w+2\}$ for all w. In the third iteration players anticipate this new pattern for the later players, and payoffs are maximized by continuing and letting the next player let the following player stop a further node away, i.e. $\xi_2(x_w) = \{w+3\}$. Iterating, the process converges to the strategy profile where every player always continues, $s_i(x_w) = x_{w+1}$, which is indeed the SPE of the game.

7.6.3 *Proof of Theorem 7.4*

The proof of Theorem 7.4 from here onwards proceeds via five lemmata and one final proposition. First, it is shown that at each iteration round the players' expectations about the continuation are generated from the previous round. Second, at each move a punishment for deviations from the equilibrium path is identified that is consistent with the players' anticipations at the current iteration round. Once this is done for all $k = 1,2,\ldots$ an appropriate limit of the correspondence $\overline{\xi}_k$ is taken. The fourth and fifth step show, first, that now expectations are correct and, second, that punishments for deviations are still feasible. Finally, these properties are combined with Theorem 7.1 and a "forward induction" argument to deduce Theorem 7.4.

The first step shows that the iteration procedure defined in (7.4), (7.5), and (7.6) uses the anticipations of later players' behavior as generated from the previous iteration step.

Lemma 7.12 *If $w \in \overline{\xi}_k(x)$, then $w \in \overline{\xi}_{k-1}(\gamma(x,w))$ for all $x \in Y_t \cap X$ and all $k =$* $1,2,\ldots$

Proof If $w \in \xi_k(x)$ for $x \in Y_t \cap X$, then $\xi_k(x) \subseteq \underline{\xi}_k(x) = \cup_{y \in p^{-1}(x)} \overline{\xi}_{k-1}(y)$ implies that there is $y \in p^{-1}(x)$ such that $w \in \overline{\xi}_{k-1}(y)$ by Proposition 6.1(a), and $\overline{\xi}_{k-1}(y) \subseteq y$ implies $y = \gamma(x,w)$. If $w \in \overline{\xi}_k(x) \setminus \xi_k(x)$, then, as $\text{graph}\left(\overline{\xi}_k\big|_{Y_t}\right) =$ $\text{cl}\left(\text{graph}\left(\xi_k|_{Y_t}\right)\right)$, there is a net

$$\left\{\left(x^d, w^d\right) \in Y_t \times W(Y_t) \,|\, d \in D\right\} \subseteq \text{graph}\left(\xi_k|_{Y_t}\right),$$

where $(D, \geq)$ is a directed set,[21] such that $\left(x^d, w^d\right) \to (x,w)$ (see Aliprantis and Border 1990, Theorem 2.14, p. 30).[22] Since $w^d \in \xi_k\left(x^d\right)$ for all $d \in D$, it follows that $w^d \in \underline{\xi}_{k-1}\left(x^d\right) = \cup_{z^d \in p^{-1}(x^d)} \overline{\xi}_{k-1}\left(z^d\right)$ for all $d \in D$. Therefore, for all $d \in D$ there is $y^d \in p^{-1}\left(x^d\right)$ with $w^d \in \overline{\xi}_{k-1}\left(y^d\right)$. Since the net $\left\{y^d \,|\, d \in D\right\}$ is contained in Y_{t+1} which is compact by Lemma 7.8, there is a subnet[23] that converges to some $y \in Y_{t+1}$ (see Aliprantis and Border 1990, Theorem 2.31, p. 39). Take the corresponding subnet from $\left\{w^d \,|\, d \in D\right\}$ so that $y^d \to y$, $w^d \to w$, and $w^d \in \overline{\xi}_{k-1}\left(y^d\right)$. Then, that $\overline{\xi}_{k-1}$ is u.h.c. on Y_{t+1} implies that $w \in \overline{\xi}_{k-1}(y) \subseteq y$ (see Aliprantis and Border 1990, Corollary 17.17, p. 564; or Kelley 1975, Theorem 20, p. 148). Because $w \in \overline{\xi}_k(x) \subseteq x$, that $w \in x \cap y$ implies by Trivial Intersection (2.4) either $y \subset x$ or $x \subseteq y$. Since the latter contradicts $x \in Y_t$ and $y \in Y_{t+1}$, it follows that $p(y) = x$, which implies $y = \gamma(x,w)$, as required. ■

The second step is to show that the iteration procedure also uses punishments (for deviations) that are consistent with the later players' behavior from the previous iteration step.

Lemma 7.13 *For all $t = 0, 1, \ldots$, all $x \in X_i \cap Y_t$, all $w_x \in \overline{\xi}_k(x)$, and all $y \in p^{-1}(x)$ there is $w_y \in \overline{\xi}_{k-1}(y)$ such that $w_y \precsim_i w_x$ for all $k = 1, 2, \ldots$*

---

[21] A *directed set* $(D, \geq)$ is a set D together with a reflexive transitive binary relation $\geq$ on D such that for all $i, j \in D$ there is $k \in D$ with $k \geq i$ and $k \geq j$. A *net* in a set is an indexed subset, where the index set is directed; see the Mathematical Appendix A.1.2.

[22] A net $\left\{\left(x^d, w^d\right) \in Y_t \times W \,|\, d \in D\right\}$ *converges* to (x,w), denoted $\left(x^d, w^d\right) \to (x,w)$, if for all neighborhoods of (x,w) there is $d \in D$ such that $\left(x^j, w^j\right)$ belongs to the neighborhood for all $j \geq d$. It *accumulates* at (x,w) if for all neighborhoods of (x,w) and all $d \in D$ there is $j \geq d$ such that $\left(x^j, w^j\right)$ belongs to the neighborhood.

[23] A *subnet* of a net $\left\{y^d \,|\, d \in D\right\}$ is a set $\left\{z^d \,|\, d \in D'\right\}$, where $(D', \geq')$ is a directed set, such that for all $d \in D$ there is $d' \in D'$ with $\left\{z^k \,|\, k \geq' d', k \in D'\right\} \subseteq \left\{y^k \,|\, k \geq d, k \in D\right\}$. See the Mathematical Appendix A.2.

Proof Let $x \in Y_{ti}$, $w_x \in \overline{\xi}_k(x)$, and $y \in p^{-1}(x)$. If $w_x \in \xi_k(x)$, the statement is true by the construction of ξ_k. Thus, suppose that $w_x \in \overline{\xi}_k(x) \setminus \xi_k(x)$. Then there is a net $\{(x^d, w^d) \in Y_t \times W(Y_t) \,|\, d \in D\} \subseteq \text{graph}(\xi_k|_{Y_t})$, where D is directed, such that $(x^d, w^d) \to (x, w_x)$.

We claim that there is a subnet $\{(x^d, w^d) \in Y_t \times W(Y_t) \,|\, x^d \in X_i, \forall d \in D'\}$ with D' directed. If not, then there is some $d_0 \in D$ such that $x^d \notin Y_{ti}$ for all $d \geq d_0$. On the other hand, $x^d \to x \in Y_{ti}$, even though $\{x^d \,|\, d \geq d_0\} \subseteq \cup_{j \neq i} Y_{tj}$. But this contradicts the fact that $\cup_{j \neq i} Y_{tj}$ is closed by property (WB1) of a well-behaved perfect information game (Definition 7.5) as a finite union of closed sets. Therefore, we can work with a subnet such that $x^d \in Y_{ti}$ for all $d \in D'$.

Since $(x^d, w^d) \in \text{graph}(\xi_k|_{Y_t})$ for all $d \in D'$, $w^d \in \xi_k(x^d)$ implies $w \precsim_i w^d$ for all $w \in \underline{\xi}_k(x^d)$, where $\{i\} = J(x) = J(x^d)$, for all $d \in D'$. Because $x^d \to x$ and $y \in p^{-1}(x)$, (OP) implies via the l.h.c. property from Lemma 7.9 that there is a subnet $\{x^d \in Y_{ti} \,|\, d \in D''\}$, with D'' directed, and a net $\{y^d \in Y_{t+1} \,|\, d \in D''\}$ such that $y^d \in p^{-1}(x^d)$ for all $d \in D''$ and $y^d \to y$ (see Aliprantis and Border 1990, Theorem 17.19, pp. 564). Choose $v^d \in \overline{\xi}_{k-1}(y^d)$ for all $d \in D''$. Since $\overline{\xi}_{k-1}(y^d) \subseteq \underline{\xi}_k(x^d)$, it follows that $v^d \in \underline{\xi}_k(x^d)$ for all $d \in D''$. From $w^d \in \xi_k(x^d)$ it follows that $v^d \precsim_i w^d$, where $\{i\} = J(x) = J(x^d)$ for all $d \in D''$.

Due to property (WB1) of a well-behaved perfect information game (Definition 7.5) $W(Y_{t+1})$ is closed by Lemma 7.5, thus, compact. Therefore, the net $\{v^d \,|\, d \in D''\}$ contains a subnet that converges to some $v_y \in W(Y_{t+1})$ (see Aliprantis and Border 1990, Theorem 2.31, p. 39). Take the corresponding subnet $\{y^d \,|\, d \in D'''\}$ such that $(y^d, v^d) \to (y, v_y)$. Then, that $\overline{\xi}_{k-1}|_{Y_t}$ has a closed graph (since the range is regular; see Klein and Thompson 1984, Theorem 7.1.15, p. 78) implies that $v_y \in \overline{\xi}_{k-1}(y)$ (see Aliprantis and Border 1990, Theorem 2.14, p. 30).

We claim that $(w^d, v^d) \to (w_x, v_y)$ and $v^d \precsim_i w^d$ for all $d \in D'''$ imply that $v_y \precsim_i w_x$. For suppose otherwise, i.e. $w_x \prec_i v_y$. Then there are two possibilities. First, if there is $w^* \in W(Y_t)$ such that $w_x \prec_i w^* \prec_i v_y$, then the set $u = \{w \in W(Y_t) \,|\, w \prec_i w^*\} \times \{w \in W(Y_t) \,|\, w^* \prec_i w\}$ is a neighborhood of (w_x, v_y) by continuity of preferences.[24] Therefore, there is $d_0 \in D'''$ such that $(w^d, v^d) \in u$ for all $d \geq d_0$, which implies $w^d \prec_i w^* \prec_i v^d$ for all $d \geq d_0$, in contradiction to $v^d \precsim_i w^d$ for all $d \in D'''$. Otherwise, if there is no $w^* \in W(y_t)$ such that $w_x \prec_i w^* \prec_i v_y$, then $u = \{w \in W(Y_t) \,|\, w \prec_i v_y\} \times \{w \in W(Y_t) \,|\, w_x \prec_i w\}$ is a neighborhood of (w_x, v_y). Hence, there is $d_0 \in D'''$ such that $(w^d, v^d) \in u$ for all $d \geq d_0$. But then $v^d \precsim_i w^d$ implies by transitivity of preferences that $w_x \prec_i v^d \precsim_i w^d \prec_i v_y$ for all $d \geq d_0$, in contradiction to the hypothesis that there is no w^* with $w_x \prec_i w^* \prec_i v_y$. Therefore, $v_y \precsim_i w_x$, as required. ■

[24]Note that this step requires that $\precsim_i$ is both u.s.c. and l.s.c.

7.6.3.1 The Limit Correspondence

Once the sequence of u.h.c. correspondences $\left\{\bar{\xi}_k\right\}_{k=0,1,\ldots}$ (with nonempty closed values) is constructed, one can take an appropriate limit. In particular, define the correspondence $\bar{\xi} : N \twoheadrightarrow W$ by the (topological) limit superior of the graphs, i.e.,

$$\bar{\xi}(x) = \left\{ w \in x \,\middle|\, (x,w) \in \cap_{t=1}^{\infty} \operatorname{cl}\left(\cup_{k=t}^{\infty} \operatorname{graph}\left(\bar{\xi}_k \Big|_{Y_t} \right)\right)\right\} \tag{7.7}$$

for all $x \in Y_t \cap X$ and all $t = 0,1,\ldots$ That is, $w \in \bar{\xi}(x)$ if and only if for all neighborhoods $\hat{u}$ of (x,w) in the product topology of τ_t^q and $\tau|\, W(Y_t)$ there are infinitely many k with $\hat{u} \cap \operatorname{graph}\left(\bar{\xi}_k\Big|_{Y_t}\right) \neq \emptyset$. By construction the restriction $\bar{\xi}\big|_{Y_t}$ is u.h.c., because the graph is closed and the range is compact and separated.

Lemma 7.14 *The correspondence $\bar{\xi} : N \twoheadrightarrow W$ has nonempty closed values.*

Proof Since $\cup_{k=t}^{\infty} \operatorname{graph}\left(\bar{\xi}_k\Big|_{Y_t}\right) \subseteq \cup_{k=t'}^{\infty} \operatorname{graph}\left(\bar{\xi}_k\Big|_{Y_t}\right)$ for $t \geq t'$ implies

$$\operatorname{cl}\left(\cup_{k=t}^{\infty} \operatorname{graph}\left(\bar{\xi}_k\Big|_{Y_t}\right)\right) \subseteq \operatorname{cl}\left(\cup_{k=t'}^{\infty} \operatorname{graph}\left(\bar{\xi}_k\Big|_{Y_t}\right)\right),$$

every finite subfamily $\left\{\left\{\operatorname{cl}\left(\cup_{k=t}^{\infty} \operatorname{graph}\left(\bar{\xi}_k\Big|_{Y_t}\right)\right)\right\} \,\middle|\, t = t_1,\ldots,t_n,\, n \in \mathbb{N}\right\}$ has a non-empty intersection. Since $x \in X$ is compact as a closed subset of the compact set W, by the finite intersection property (Aliprantis and Border 1990, Theorem 2.31, p. 39) the infinite intersection $\bar{\xi}(x)$ is nonempty for all $x \in X$. Since $\bar{\xi}$ has a closed graph, $\bar{\xi}(x)$ is closed. ∎

For the following two proofs abbreviate notation by denoting

$$\Gamma_k^t = \operatorname{graph}\left(\bar{\xi}_k\Big|_{Y_t}\right) \quad \text{for all } k,t = 0,1\ldots$$

Next, we verify that the properties identified in Lemmata 7.12 and 7.13 also hold for the limit correspondence $\bar{\xi}$. As the fourth step consider the "rational expectations" property.

Lemma 7.15 *If $w \in \bar{\xi}(x)$, then $w \in \bar{\xi}(\gamma(x,w))$ for all $x \in X$.*

Proof Let $w \in \bar{\xi}(x)$ for $x \in Y_t \cap X$ for some $t = 0,1,\ldots$ For any open neighborhood G of (x,w) in the product topology of τ_t^q and $\tau|\, W(Y_t)$ and any $n = 1,2,\ldots$ there is some $\left(x_{(G,n)}, w_{(G,n)}\right) \in G \cap \left(\cup_{k\geq n}\Gamma_k^t\right) \neq \emptyset$. The product D of all open neighborhoods of (x,w) with $\{1,2,\ldots\}$, endowed with the order $(\subseteq,\geq)$, is a directed set. Thus, $\left\{\left(x_{(G,n)}, w_{(G,n)}\right) \middle| (G,n) \in D\right\}$ is a net. By construction $\left(x_{(G,n)}, w_{(G,n)}\right) \to (x,w)$.

That $\left(x_{(G,n)}, w_{(G,n)}\right) \in \cup_{k\geq n}\Gamma_k^t$ for all $(G,n) \in D$ implies that there is some $K(G,n) \geq n$ such that $w_{(G,n)} \in \overline{\xi}_{K(G,n)}\left(x_{(G,n)}\right) \subseteq x_{(G,n)}$. Thus, by Lemma 7.12, $w_{(G,n)} \in \overline{\xi}_{K(G,n)-1}\left(y_{(G,n)}\right)$, where $y_{(G,n)} = \gamma\left(x_{(G,n)}, w_{(G,n)}\right)$, hence, $\left(y_{(G,n)}, w_{(G,n)}\right) \in \Gamma_{K(G,n)-1}^{t+1}$. Because the set $\left\{\left(y_{(G,n)}, w_{(G,n)}\right) | (G,n) \in D\right\}$ is a net and the node $y_{(G,n)} = \gamma\left(x_{(G,n)}, w_{(G,n)}\right)$ converges by continuity of γ (Lemma 7.11) to $\gamma(x,w) = y \in p^{-1}(x)$, it follows that $\left(y_{(G,n)}, w_{(G,n)}\right) \to (y,w)$.

Let $\tilde{G}$ be any open neighborhood of (y,w) in the product topology of τ_t^q and $\tau | W(Y_{t+1})$. Because $\left(y_{(G,n)}, w_{(G,n)}\right) \to (y,w)$ there is some $(G',n') \in D$ such that $\left(y_{(G,n)}, w_{(G,n)}\right) \in \tilde{G}$ for all $(G,n) \in D$ greater than (G',n'). Since also $\left(y_{(G,n)}, w_{(G,n)}\right) \in \Gamma_{K(G,n)-1}^{t+1}$, it follows that for all $(G,n) \in D$ greater than $(G',n') \in D$ the intersection $\tilde{G} \cap \Gamma_{K(G,n)-1}^{t+1}$ is nonempty. In particular, for all open neighborhoods $\tilde{G}$ of (y,w) there are infinitely many k such that $\tilde{G} \cap \Gamma_k^{t+1}$ is nonempty. Therefore, $(y,w) \in \cap_{t=1}^{\infty} \mathrm{cl}\left(\cup_{k=t}^{\infty}\Gamma_k^{t+1}\right)$ implies that $w \in \bar{\xi}(y) = \bar{\xi}(\gamma(x,w))$. ■

Next, consider punishments for deviations. The fifth step shows that the possibility to punish is also preserved in the limit, now based on correct anticipations.

Lemma 7.16 *For all $w_x \in \bar{\xi}(x)$ and all $y \in p^{-1}(x)$ there is $w_y \in \bar{\xi}(y)$ such that $w_y \precsim_i w_x$, for all $x \in X_i$ and all $i \in I$.*

Proof Let $w \in \bar{\xi}(x)$ for $x \in Y_{ti}$ and $y \in p^{-1}(x) \subseteq Y_{t+1}$. Then for any open neighborhood G of (x,w) in the product topology of τ_t^q and $\tau | W(Y_t)$ and any $n = 1,2,\ldots$ there is some $\left(x_{(G,n)}, w_{(G,n)}\right) \in G \cap \left(\cup_{k\geq n}\Gamma_k^t\right) \neq \emptyset$. In particular, there is some $K(G,n) \geq n$ such that $\left(x_{(G,n)}, w_{(G,n)}\right) \in G \cap \left(\Gamma_{K(G,n)}^t\right)$. Again $\left\{\left(x_{(G,n)}, w_{(G,n)}\right) | (G,n) \in D\right\}$ is a net, when D (the product of the open neighborhoods of (x,w) with the integers) is ordered by $(\subseteq, \geq)$, that converges to (x,w) by construction. Like in the proof of Lemma 7.13 it can be assumed without loss of generality that $x_{(G,n)} \in Y_{ti}$ for all $(G,n) \in D$.

Let $y_{(G,n)} = \gamma\left(x_{(G,n)}, w_{(G,n)}\right)$ for all $(G,n) \in D$. Then by continuity of γ (Lemma 7.11) $y_{(G,n)} \to \gamma(x,w) = y \in p^{-1}(x)$. By Lemma 7.13 for all $(G,n) \in D$ there is some $v_{(G,n)} \in \overline{\xi}_{K(G,n)-1}\left(y_{(G,n)}\right) \subseteq y_{(G,n)}$ such that $v_{(G,n)} \precsim_i w_{(G,n)}$. Since $W(Y_{t+1})$ is compact, the net $\left\{v_{(G,n)} | (G,n) \in D\right\}$ contains a subnet that converges to $w_y \in y = \gamma(x,w)$. Because $w_{(G,n)} \to w$, continuity of preferences (property (WB2) of a well-behaved perfect information game) implies that $w_y \precsim_i w$ by a similar argument as in the proof of Lemma 7.13.[25] ■

[25]Once again, note that the last step requires that $\precsim_i$ is both u.s.c. and l.s.c.

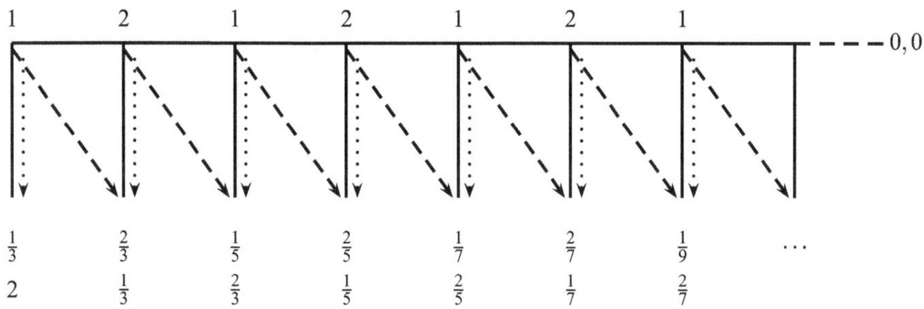

Fig. 7.8 Graphical representation of Example 7.16. Players 1 and 2 alternate moves. Dashed arrows represent iterations number 1, 3, 5, and so on. Dotted arrows represent iterations number 2, 4, 8, and so on

7.6.3.2 Forward Induction

With this hard work out of the way we are ready to prove Theorem 7.4. The (easy) construction of the proof combines the "one-shot deviation principle" from Theorem 7.1 with a "forward induction" argument.

Proposition 7.3 *For each $w \in \bar{\xi}(W)$, there exists an SPE inducing the play w.*

Proof Construct a selection $f \in \mathscr{F}$ from $\bar{\xi}$ as follows. At the root choose $f(W) = w \in \bar{\xi}(W)$. Proceeding inductively, assume that the selection f from $\bar{\xi}$ has been determined for all $x \in Y_t \cap X$ for $t = 0, 1, \ldots, T-1$. For $x \in Y_T \cap X$ choose $f(x) = f(p(x))$ if $f(p(x)) \in x$ so that $f(x) \in \bar{\xi}(x)$ by Lemma 7.15; otherwise set $f(x) = w_x \in \bar{\xi}(x)$ such that $w_x \precsim_i f(p(x))$ whenever $J(x) = \{i\}$, which is feasible by Lemma 7.16. Appealing to induction completes the construction of f. Since f satisfies (BI1) and (BI2) from the definition of backwards induction (Definition 7.4), constructing strategies by $x \in X_i \Rightarrow s_i(x) = \gamma(x, f(x))$ for all $i \in I$ yields an SPE by Theorem 7.1. ■

This completes the proof of Theorem 7.4. The following example clarifies in which sense the algorithm "converges." It has the added virtue of presenting an example for multiplicity of SPE in a generic perfect information game, which is not possible in the finite case.

Example 7.16 (A centipede game with multiple SPE) The game form is an Alexandroff-compactified infinite centipede with two alternating players, as in Examples 7.6 and 7.15. Preferences are represented by the payoff functions

$$U_1(w) = \begin{cases} 1/(w+2) & \text{if } w \text{ is odd} \\ 2/(w+1) & \text{if } w \text{ is even} \end{cases} \qquad U_2(w) = \begin{cases} 2/w & \text{if } w \text{ is odd} \\ 1/(w+1) & \text{if } w \text{ is even} \end{cases}$$

See Fig. 7.8 for a representation. Again, it is easy to verify that payoff functions are continuous. The characteristics of the example arise from the ordinal representation of preferences and not from payoff ties. That is, this is a generic example.

In the first iteration of the algorithm each player maximizes payoff by continuing once and stopping at the next node, i.e. $\xi_0(x_w) = \{w+1\}$ for all w. In the second iteration players foresee the decision of the next player and payoffs are maximized by stopping right away, i.e. $\xi_1(x_w) = \{w\}$ for all w. In the third iteration players anticipate the stop-decision of the next player and payoffs are maximized by continuing and letting the next player stop, i.e. $\xi_2(x_w) = \{w+1\}$. Iterating, the process alternates between always continuing and always stopping.

At first glance this does not look like "convergence." Yet, the algorithm actually identifies *sets* of plays for each node. Indeed, the algorithm "converges" very quickly to $\bar{\xi}(x_w) = \{w, w+1\}$ for all w. The actual statement, once the algorithm has delivered $\bar{\xi}$, is that every play in $\bar{\xi}(W) = \{1,2\}$ is the equilibrium path of an SPE, which can be constructed by appropriately selecting choices from $\bar{\xi}(x_w)$, as in Proposition 7.3. There are in fact two SPE, $s^1, s^2 \in S$, as follows:

$$s_1^1(x_w) = w \text{ for all } w \text{ odd}, \quad s_2^1(x_w) = x_{w+1} \text{ for all } w \text{ even}$$

$$s_1^2(x_w) = x_{w+1} \text{ for all } w \text{ odd}, \quad s_2^2(x_w) = w \text{ for all } w \text{ even.}$$

In the first player 1 always stops and player 2 always continues. Hence, the equilibrium play is $w = 1$. In the second player 1 always continues and player 2 always stops. Hence, the equilibrium play is $w = 2$.

7.7 Discussion

In this Section we discuss a few fine points. The first establishes by means of an example that Theorem 7.4 is more general than prior proofs for the existence of subgame perfect equilibrium in large perfect information games. In particular, the immediate successor correspondence need not be upper hemi-continuous, nor does the topology on nodes in a slice have to be separated. The second concerns topologies that are so coarse that they are not perfectly normal and admit equilibrium analysis even in the absence of (OP). The third point concerns the hypothesis of a compact topology in Theorem 7.2, and the fourth derived topologies on strategies. The last point takes up again the issue about upper hemi-continuity of the immediate successor correspondence and proposes a way how this can be added.

7.7.1 On Generality

Theorem 7.4 does not employ an assumption that makes the immediate successor correspondence upper hemi-continuous with respect to the quotient topologies on nodes in a slice. It also does not assume any separation property for these topologies.

This makes Theorem 7.4 the most general existence theorem for perfect information games in the literature. To see that this is the case an example is helpful.

Example 7.17 (Cube) Two players have to determine the values of three quantities $a,b,c \in [0,1]$. Which player gets to choose $b \in [0,1]$ is endogenous. Player 1 can either pick some $a < 1$, in which case player 2 will decide on b and c upon having seen $a < 1$. Alternatively, player 1 may set $a = 1$, in which case she also gets to choose the value of b. In this case, player 2 will subsequently decide on c.

The set of plays is the unit cube $W = [0,1]^3$ endowed with the (relative) Euclidean topology. The set N of nodes is given by

$$N = \left\{ W, \left(\{a\} \times [0,1]^2\right)_{a \in [0,1)}, (\{(1,b)\} \times [0,1])_{b \in [0,1]}, (\{w\})_{w \in W} \right\}.$$

Player 1 decides at the root, $Y_0 = X_1 = \{W\}$. The last slice consists of the singletons, $Y_2 = E = \{\{(a,b,c)\} \mid (a,b,c) \in [0,1]^3\}$. Player 2 decides at the intermediate Y_1, made up of all the "vertical squares" where player 1 has decided an $a < 1$ and all the "horizontal segments" where player 1 has fixed $a = 1$ and chosen a value of b. That is,

$$X_2 = \{\{a\} \times [0,1]^2 \mid 0 \le a < 1\} \cup \{\{(1,b)\} \times [0,1] \mid 0 \le b \le 1\}$$

(see Fig. 7.9). With the Euclidean topology W is compact and all nodes are closed sets. The Euclidean topology is also separated. Condition (WB1) is trivially fulfilled, because at every slice one and only one player decides. Thus, for any specification of continuous preferences on $[0,1]^3$, the game is well-behaved.

It remains to show that (OP) holds. Since Y_0 contains only the root, (OP) holds trivially for $t = 0$. Hence, it is enough to verify it for Y_1. Let $V \subseteq Y_2$ be a subset of terminal nodes such that $W(V)$ is open. Since terminal nodes are singletons, this is simply any open set in the Euclidean topology on $[0,1]^3$. To see that $W(p(V))$ is also open, let $(a,b,c) \in W(p(V))$ and distinguish two cases.

If $(a,b,c) \in W(p(V))$ and $a < 1$, then $(a,b',c') \in W(V)$ for some $(b',c') \in [0,1]^2$. Since $W(V)$ is open, there exists some $\varepsilon > 0$ such that $(a-\varepsilon,a+\varepsilon) \times (b'-\varepsilon,b'+\varepsilon) \times (c'-\varepsilon,c'+\varepsilon) \subseteq W(V)$. By the construction this implies that $(a-\varepsilon,a+\varepsilon) \times [0,1]^2 \subseteq W(p(V))$. This latter set is an open neighborhood of (a,b,c).

If $(a,b,c) \in W(p(V))$ and $a = 1$, then $(1,b,c') \in W(V)$ for some $c' \in [0,1]$. Since $W(V)$ is open, there exists $\varepsilon > 0$ such that $(1-\varepsilon,1] \times (b-\varepsilon,b+\varepsilon) \times (c'-\varepsilon,c'+\varepsilon) \subseteq W(V)$. By construction this implies that $\left((1-\varepsilon,1) \times [0,1]^2\right) \cup (\{1\} \times (b-\varepsilon,b+\varepsilon) \times [0,1]) \subseteq W(p(V))$. In particular, the product set $(1-\varepsilon,1] \times (b-\varepsilon,b+\varepsilon) \times [0,1]$ is contained in $W(p(V))$. This latter set is an open neighborhood of $(1,b,c)$.

This argument shows that $W(p(V))$ contains an open neighborhood for each of its elements and is, therefore, an open set. Thus, (OP) holds. It follows from Theorem 7.4 that, for any specification of continuous preferences, the resulting game has an SPE.

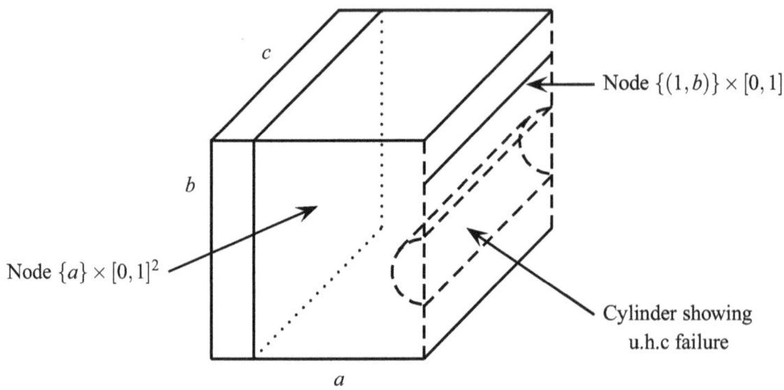

Fig. 7.9 Graphical representation of the set of plays of Example 7.17. Player 1 either chooses $a<1$ or fixes $a=1$ and chooses b. In the first case, player 2 chooses b and c. In the second case, player 2 chooses only c

Example 7.17 illustrates three differences of the present approach with earlier ones, which are now examined in detail.

7.7.1.1 Topologies on Slices May Not Be Separated

Existence theorems for large games often begin by specifying nicely behaved topologies on the slices (or stages), e.g., compact and Hausdorff, and then rely on a Tychonoff product to construct the (compact and Hausdorff) topology on the set of plays. By contrast, the present approach starts with an arbitrary compact and separated topology on the set of plays. This does *not* necessarily imply that the (quotient) topologies at the slices behave nicely. In Example 7.17 the induced topology on the set of nodes Y_1 (the relevant intermediate slice) is *not* Hausdorff.

To see this, consider the nodes $y_1=\{(1,0)\}\times[0,1]$ and $y_2=\{(1,1)\}\times[0,1]$. Suppose these two moves can be separated by open sets in the topology on the set of nodes Y_1. This means that there exist two sets of nodes U_1 and U_2 in Y_1 such that $W(U_1)$ and $W(U_2)$ are open in the Euclidean topology, that $y_1\in U_1$ and $y_2\in U_2$, and that $W(U_1)\cap W(U_2)=\emptyset$. It follows that there exist some $\varepsilon>0$ such that $(1-\varepsilon,0,0)\in W(U_1)$ and $(1-\varepsilon,1,0)\in W(U_2)$. Hence, the node $\{1-\varepsilon\}\times[0,1]^2$ must be contained in both U_1 and U_2—a contradiction with the fact that $W(U_1)$ and $W(U_2)$ must be disjoint.

This argument shows that starting with conditions at the level of plays allows for cases that would be excluded if analogous conditions were imposed directly on the slices of the tree.

7.7.1.2 The Successor Correspondence May Not Be Upper Hemi-Continuous

Some of the literature uses an "action correspondence," which in the present setting corresponds to the mapping that assigns immediate successors to moves. Lemma 7.9 shows that condition (OP) is equivalent to lower hemi-continuity of the immediate successor correspondence. The present approach, however, does *not* require upper hemi-continuity. This is in stark contrast with, for instance, Harris (1985a), who states that his assumptions imply full continuity ("the overall continuity requirement [...] is quite substantial"; Harris 1985a, p. 618).

To see that upper hemi-continuity is not needed, return to Example 7.17. Consider the immediate successor correspondence p^{-1} defined from the intermediate slice Y_1 to the terminal slice made up of the singletons. Let D be the two-dimensional open half-disk of radius ε centered on $(a,b) = (1,1/2)$, and define the open half-cylinder $B = D \times [0,1] \subseteq W$ (see Fig. 7.9). Clearly, the set of terminal nodes $V = \{\{(a,b,c)\} \in N \mid (a,b,c) \in B\}$ is open in the quotient topology on the final slice. Upper hemi-continuity requires the upper preimage

$$\left(p^{-1}\right)^{+}(V) = \left\{x \in Y_1 \;\middle|\; p^{-1}(x) \subseteq V\right\}.$$

to be open. Yet, by construction of the game no node of the form $\{a\} \times [0,1]^2$ with $a < 1$ can have *all* its successors contained in V. In fact, $W\left(\left(p^{-1}\right)^{+}(V)\right) = \{1\} \times (1/2-\varepsilon, 1/2+\varepsilon) \times [0,1]$ and this set is *not* open in the Euclidean topology. The conclusion is that, even though Theorem 7.4 covers Example 7.17, the immediate successor correspondence may not be upper hemi-continuous.

7.7.1.3 Topologies on Plays Allow Additional Flexibility

In Example 7.17 it is natural to conceive of the set of outcomes as the unit cube and use the three-dimensional Euclidean topology. Constructions based on a product approach, on the other hand, may *not* end up with this choice of set and topology. To illustrate this point, we consider the approach of Harris (1985a) applied to Example 7.17.

Harris (1985a) examines each stage of the game and endows each acting player with a fixed universal set of actions for that stage. This is without loss of generality in his approach, because a later restriction reduces the resulting product set to the subset of plays, incorporating all restrictions given by the game. In Example 7.17 the natural choice of action set for player 2 is $A_2 = [0,1]^2$, since this player may (but need not) choose both b and c. Yet, player 1 can also choose a and b. Thus, the natural choice for this player is also $A_1 = [0,1]^2$. This results in the product space $A = A_1 \times A_2 = [0,1]^4$. That is, an extra dimension has slipped in through the back door. To recover the structure of the game, Harris (1985a) then defines the set of plays H as a subset of A. In the present case this must include all plays where player 1 has fixed $a = 1$ and chosen b, and subsequently player 2 can only choose c. This requires the subset $H_1 = \{((1,b_1),(b_2,c)) \mid b_1 = b_2\}$ to be one part of the set of plays. In H_1 the fact that player 2 does not decide on b is incorporated by the restriction that her decision after $(1,b_1)$ has to be of the form $(b_2,c) = (b_1,c)$.

The set H must also include all plays such that player 1 chooses an $a < 1$ and lets player 2 decide both on b and c. This gives the set

$$H_2 = \left\{((a,0),(b,c)) \;\middle|\; 0 \leq a < 1, (b,c) \in [0,1]^2 \right\}.$$

The coordinate 0 in $(a,0)$ is just an arbitrary marker reflecting that player 1 does not actually decide on b. Any other marker would also do, but it would be incorrect to write (a,b), because if $a < 1$ player 2's choice of b is unconstrained by player 1's decision.

The set of plays finally is $H = H_1 \cup H_2$. Harris (1985a) assumes that the action sets A_i are compact and separated, which is unproblematic with this construction and the (natural) Euclidean topology. Yet, his next assumption requires the set H to be a closed subset of A. Consider the sequence given by $w_n = ((a_n,0),(1/2,1/2))$ with $0 \leq a_n < 1$ for all $n = 1,2,\ldots$ and $a_n \to 1$. Then, $w_n \in H_1 \subseteq H$ for all n, but $w_n \to ((1,0),(1/2,1/2))$, which is not an element of H. Thus, H is not a closed subset of A and the approach in Harris (1985a) does not work. The reason is that it relies on a Tychonoff construction, which first identifies the action sets at each stage, then imposes topological constraints on those, and finally constructs the product set. While this approach is natural if games have a clear stage structure, in other games (as in Example 7.17) it is far less natural.

Perfect information games with a stage structure are special cases of the present setting. The first argument in this section showed that existence of equilibrium can be established without imposing topological constraints on the stages. The second argument showed that upper hemi-continuity of an action correspondence is not needed for equilibrium existence. The third showed that the freedom to choose a topology on the set of plays allows for cases that are excluded by other approaches.

7.7.2 *The Fort Example*

Theorem 7.3 states that any topology τ on the set W of plays that admits equilibrium analysis must be a tree topology. The result employs the hypothesis that (W,τ) is perfectly normal, though. Therefore, there could be very coarse topologies that admit equilibrium analysis which are not tree topologies. Yet, if that is the case, it may severely constrain the set of preferences that are still continuous with respect to such a coarse topology. The following example is constructed to fail perfect normality is such a way that (OP) fails, but the topology still admits equilibrium analysis. The price paid for this is that only very few functions, namely those with countably many values, are continuous with respect to this topology.

Example 7.18 (Fort Example) Let $W = [-2,-1] \cup [1,2]$ and nodes be given by $N = \{W, (\{-w,w\})_{w \in W}, (\{w\})_{w \in W}\}$. For any set of players I let $1 \in I$ denote the player who chooses at the root. The topology is the Fort topology, i.e., the open sets are those whose complements are either finite or contain the point $1 \in W$. (That is, a closed set is either a finite set or contains $1 \in W$.) This is a compact Hausdorff space that is not perfectly normal (see Steen and Seebach 1978, pp. 52).

To see (CN), note that all nodes except the root are finite sets, hence closed, and the root contains $1 \in W$. The Fort topology fails (OP), though. For, consider the terminal node (singleton set) $x = \{-1\}$, which is open by $1 \notin x$. Its immediate predecessor is $p(x) = \{-1, 1\}$, which contains $1 \in W$ but is not cofinite, and thereby not open. That is, while $W(\{x\})$ is open, $W(p(\{x\}))$ is not, and (OP) fails.

Any continuous preference relation $\precsim_i$ on W satisfies the following. First, for every $w \in W$ with $1 \prec_i w$, the upper contour set $\{w' \in W | w \precsim_i w'\}$ is finite, because it must be closed, but cannot contain 1. Analogously, for every $w \in W$ with $w \prec_i 1$, the lower contour set $\{w' \in W | w' \precsim_i w\}$ is finite. Hence, the indifference sets $\{w' \in W | w' \sim_i w\}$ are also finite for any $w \in W$ with $w \prec_i 1$ or $1 \prec_i w$. This implies that the strict upper contour set $\{w' \in W | 1 \prec_i w'\}$ is countable. To see this, consider the quotient space

$$\Phi_1 \equiv \{w \in W | 1 \prec_i w\} / \sim_i .$$

This set can be enumerated (mapped one-to-one to a subset of the natural numbers) by assigning to each class $V \in \Phi_1$ the (finite) number of elements of W which are weakly preferred to the elements of V. Hence, Φ_1 is countable. The strict upper contour set $\{w' \in W | 1 \prec_i w'\}$ is the union of all elements of Φ_1, hence a countable union of finite sets and, therefore, countable. Analogously, the strict lower contour set $\{w' \in W | w' \prec_i 1\}$ is countable.

Let continuous preference relations $\precsim_i$ for the players $i \in I$ be given. At each move $\{-w,w\}$ there is a best element for player $i \in J(\{-w,w\})$, who controls it, say, $m(w)$, by finiteness. We claim that $\precsim_1$ is maximized on the set $M = \{m(w) | w \neq 1\}$. Proceeding indirectly, suppose that there is no best element with respect to $\precsim_1$ on M. It follows that $w \precsim_1 1$ for all $w \in M$, for otherwise taking any $w \in M$ with $1 \prec_1 w$ that $\{w' \in W | w \precsim_1 w'\}$ is finite would imply the existence of a best element. If further $w \prec_1 1$ would hold for all $w \in M$, then the uncountable set M were a subset of the countable set $\{w' \in W | w' \prec_1 1\}$. Therefore, there must be $w \in M$ with $w \sim_1 1$; but then $w \in M$ is a best element with respect to $\precsim_1$, a contradiction.

Let w^* be a best element with respect to $\precsim_1$ in M. Then $\precsim_1$ is also maximized on $M \cup \{m(1)\}$, either at $m(1)$ or at $w^* = m(w^*)$. There is an SPE where each player at $\{-w,w\}$ chooses $m(w)$ and player 1 chooses $\{-1,1\}$ if $w^* \prec_1 m(1)$ or $\{-w^*, w^*\}$ otherwise. It follows that the Fort topology admits equilibrium analysis, but violates (OP).

If the set $W = [-2,-1] \cup [1,2]$ of plays from this example were endowed with the more natural relative Euclidean topology, then (OP) would hold. That is, the failure of (OP) is due to the choice of a coarse topology. (The coarseness of the Fort topology is also responsible for the fact that so few preference relations are continuous with respect to this topology.) This illustrates why Theorem 7.3 is indeed stronger than analogous necessity results would be that only assumed T_1 or T_2. With a stronger separation axiom it shows necessity for cases where more preference relations are continuous.

7.7.3 Is Compactness Necessary?

In Theorem 7.4, compactness is one of the sufficient conditions for equilibrium existence. Of course, compactness is a mild assumption which is usually included in existence theorems involving continuity of preferences without much ado. For, a priori existence of maxima for continuous preferences may fail in non-compact spaces. Hence, it is natural to ask whether compactness can also be obtained as a *necessary* condition in Theorem 7.3.

The answer to this question is surprising. It is closely related to a class of topological spaces called "pseudocompact." A pseudocompact topological space is one where each real-valued continuous function is bounded. It is easy to show that this is equivalent to the statement that every real-valued continuous function has maxima and minima. It is known that every countably compact topological space is pseudocompact.

By Lemma 7.6 it follows that if a topological space admits equilibrium analysis, it must be pseudocompact. If pseudocompactness would imply compactness, then we would conclude that compactness is necessary for equilibrium analysis.

Fedorchuck (1976) and Ostaszewski (1976) constructed (in the set-theoretic sense) examples of countably compact (hence pseudocompact) and perfectly normal spaces that are not compact. Hence, not every pseudocompact space is compact. Further, for this space every real-valued continuous function has maxima. Consider a tree defined on this space (as set of plays) which includes only the root and the singletons of plays. The only possible assignment of players to nodes is a single player, who directly chooses a final outcome. It follows that, for any continuous payoff function, this game has an SPE, even though the topology is not compact.[26] Hence, compactness is not a necessary condition for equilibrium analysis.

Alas, the constructions in Fedorchuck (1976) and Ostaszewski (1976) rely on more than just the standard Zermelo-Fraenkel-Choice (ZFC) axioms, on which standard set theory is based. They make use of an additional axiom known as Jensen's Combinatorial Principle $\diamondsuit$, which is a consequence of Gödel's Axiom of

[26] In order not to (further) complicate the discussion, we focus on equilibrium existence for continuous payoff functions, rather than for continuous preference relations. Technically speaking, the implication discussed here would remain open if we defined and focused on "ordinally pseudocompact spaces."

Constructibility ($V = L$) and is stronger than the Continuum Hypothesis. This axiom is independent of ZFC. Hence, what the result actually means is that there exists an extension of ZFC such that compactness is not a necessary condition for equilibrium analysis.

Weiss (1978) showed that, if one assumes that the Continuum Hypothesis is *false* (which is perfectly justified, as the Continuum Hypothesis is independent of ZFC) and adds a further axiom to ZFC, known as Martin's Axiom, we obtain an alternative extension of the standard mathematical system where a remarkable result (known as Weiss' Theorem) can be shown: every pseudocompact perfectly normal topological space is actually compact. Martin's Axiom is known to be implied by the Continuum Hypothesis, but it is also consistent with its negation. Hence, we obtain a new extension of ZFC in which, as argued above, whenever a topological space admits equilibrium analysis, it must be compact.

In short, there exist extensions of the standard ZFC axiom system where compactness is necessary for equilibrium analysis—and extensions where it is not. That is, the question of whether compactness can be added as a necessary condition to our set of properties has neither a negative nor a positive answer under ZFC. It is *undecidable* (neither provable nor refutable) in the sense of Gödel (1931) (or, in more modern terminology, independent of ZFC). Given this state of affairs, we are content with adding compactness as a hypothesis for Theorem 7.4.

7.7.4 Topologies on Strategies

In the above approach the topological structure is imposed on the set W of plays for a discrete game tree $(N, \supseteq)$. There is a way, of course, to translate this into a topology on the strategy space S. In particular, the *weak* (or initial) topology on S is the weakest (or coarsest) topology on S that makes the function $\phi : S \to W$, as identified by Theorems 5.2 and 5.5, continuous when W is endowed with a tree topology τ. This is the topology generated by the collection of sets $\{\phi^{-1}(u) \,|\, u \in \tau\}$ as a subbasis. A basis for this topology is the collection of finite intersections of the form $\cap_{k=1}^{n} \phi^{-1}(u^k)$ with $u^k \in \tau$ for all $k = 1, \ldots, n \in \mathbb{N}$. Interestingly, topologies on strategies as generated by topologies on the set of plays have been productively used in the previous literature, e.g. by Fudenberg and Levine (1983) and Harris (1985b).

7.7.5 Continuous Game Trees

In a tree topology the immediate successors $y \in p^{-1}(x)$ of a move $x \in X$ vary lower hemi-continuously in the move by Lemma 7.9. For applications it may often be desirable to add *upper* hemi-continuity, yielding a tree topology in which immediate successors vary continuously in the move. That is, applications may call for a *continuous tree* in the sense that successors vary both lower and upper hemi-continuously in the moves. To achieve that takes an additional condition on the topology. One way to do this is to require that (the restriction of) the immediate predecessor function $p : Y_{t+1} \to Y_t$ is a closed map in the quotient topologies τ_t^q on the slices. That is, for all $t = 0, 1, \ldots$ and all sets $V \subseteq Y_{t+1}$ of nodes,

$$\text{if } W \setminus W(V) \in \tau | W(Y_{t+1}) \text{ , then } W \setminus W(p(V)) \in \tau | W(Y_t) \, . \tag{7.8}$$

This condition is equivalent to the immediate successor correspondence $p^{-1} : Y_t \twoheadrightarrow Y_{t+1}$ being u.h.c. by Theorem 17.7(1) of Aliprantis and Border (1990, p. 560).

An alternative (stronger) condition is to assume that, for each slice Y_t, the nodes in Y_t form an *upper semi-continuous decomposition* of the set $W(Y_t)$ of plays (Kelley 1975, p. 99). This is the content of the next result. As for notation, for each subset $Y \subseteq N$ of nodes let $A(Y)$ denote the set of all unions of nodes in Y, that is, the algebra generated by Y.

Proposition 7.4 *Let τ be a tree topology τ on the set W of plays of a discrete game tree $(N, \supseteq)$. If*

$$\begin{array}{c} \textit{for all } x \in Y_t \textit{ and all } u \in \tau \textit{ there is} \\ v \in \tau \textit{ with } x \subseteq v \subseteq u \textit{ such that } v \cap W(Y_t) \in A(Y_t) \end{array} \tag{7.9}$$

for all $t = 0, 1, \ldots$, then the immediate successor correspondence $p^{-1} : Y_t \twoheadrightarrow Y_{t+1}$ is u.h.c. in the quotient topologies τ_t^q on the slices, for all $t = 0, 1, \ldots$

The proof of this proposition relies on the coincidence of the quotient topologies τ_t^q with the relative upper Vietoris topology on the same slice under (7.9). Recall that the correspondence $W : N \twoheadrightarrow W$, defined by $W(x) = x$ for all $x \in N$, assigns to every node, viewed as an element of the tree, the set of its constituent plays, that is, the node itself viewed as a subset of the set of plays. The coarsest topology on N that makes the correspondence $W(.)$ u.h.c. is the *upper Vietoris topology* (see Klein and Thompson 1984), denoted τ^+. This topology is generated by the sets

$$W^+(u) = \downarrow u = \{x \in N \,|\, x \subseteq u\}$$

for all $u \in \tau$, as a subbasis. In fact, they form a basis. Call sets of the form $W^+(u)$ for $u \in \tau$ *upper Vietoris sets*. For each $t = 0, 1, \ldots$ denote by $\tau^+ | Y_t$ the relative upper Vietoris topology on the slice Y_t. The relative upper Vietoris topology $\tau^+ | Y_t$ on a slice Y_t will in general be finer than the quotient topology τ_t^q on the slice. The essence of condition (7.9) is that these two topologies agree on all slices.

Lemma 7.17 *If (W, τ) satisfies (7.9), then for all $t = 0, 1, \ldots$ and each set $U \subseteq Y_t$ of nodes the following three statements are equivalent:*

(a) $U \in \tau^+ | Y_t$;
(b) *there is $u \in \tau$ such that $W(U) = u \cap W(Y_t)$;*
(c) *there is $u \in \tau$ such that $U = W^+(u) \cap Y_t$.*

Proof "(a) implies (b):" If $U \in \tau^+ |Y_t$, then there is $U' \in \tau^+$ such that $U = U' \cap Y_t$ by the definition of the relative topology. Because upper Vietoris sets form a basis for τ^+, there is some index set Λ (with arbitrary cardinality) such that $U' = \cup_{\lambda \in \Lambda} W^+(u_\lambda)$ with $u_\lambda \in \tau$ for all $\lambda \in \Lambda$. Therefore, $U = \cup_{\lambda \in \Lambda} W^+(u_\lambda) \cap Y_t$. It follows that for all $x \in U$ there is some $\lambda(x) \in \Lambda$ such that $x \subseteq u_{\lambda(x)} \in \tau$. By (7.9) there is some $u'_{\lambda(x)} \in \tau$ such that $x \subseteq u'_{\lambda(x)} \subseteq u_{\lambda(x)}$ with $u'_{\lambda(x)} \cap W(Y_t) \in A(Y_t)$, that is, $u'_{\lambda(x)} \cap W(Y_t) = W(Z_x)$ for some $Z_x \subseteq Y_t$. Since $z \subseteq u'_{\lambda(x)}$ for all $z \in Z_x$ and nodes in Y_t are pairwise disjoint,

$$x \in Z_x \subseteq W^+\left(u'_{\lambda(x)}\right) \subseteq W^+\left(u_{\lambda(x)}\right) \subseteq U.$$

Let $u = \cup_{x \in U} u'_{\lambda(x)} \in \tau$. Then by the above $u \supseteq W(U)$. We claim that also $u \cap W(Y_t) \subseteq W(U)$. For, if $w \in u \cap W(Y_t)$, then

$$w \in \left(\cup_{x \in U} u'_{\lambda(x)}\right) \cap W(Y_t) = \left(\cup_{x \in U} \cup_{z \in Z_x} z\right) \cap \left(\cup_{y \in Y_t} y\right),$$

so that there are $x \in U$ and $z \in Z_x$ such that $w \in z$. But then $Z_x \subseteq U$ implies $w \in W(U)$, as required. With $u \supseteq W(U)$ and $W(U) \subseteq W(Y_t)$ it follows that $u \cap W(Y_t) = W(U)$.

"(b) implies (c):" If $W(U) = u \cap W(Y_t)$ for some $u \in \tau$, then set $U' = W^+(u) \cap Y_t \in \tau^+ |Y_t$. We claim that $U' = U$. For, if $x \in U \subseteq Y_t$, then $x \subseteq W(U)$ implies $x \subseteq u$, thus, $x \in U'$. Conversely, if $x \in U' \subseteq Y_t$, then $u \supseteq x \in Y_t$ implies $x \subseteq W(U)$, thus, $x \in U$, because nodes in Y_t are pairwise disjoint.

"(c) implies (a)" follows from the definition of $\tau^+ |Y_t$. ■

Without condition (7.9) "(b) implies (a)" in Lemma 7.17 would go through, but not "(a) implies (b)." In that case $\tau^+ |F_t$ could be finer than the quotient topology.

Corollary 7.3 *If (W, τ) satisfies (7.9), then for all $t = 0, 1, \ldots$ the topology $\tau^+ |Y_t$ coincides with the quotient topology τ_t^q on the slice Y_t.*

Proof If $U \subseteq Y_t$ is open in $\tau^+ |Y_t$, then $\mathrm{proj}_t^{-1}(U) = W(U)$ is open in τ relative to $W(Y_t)$ by "(a) implies (b)" from Lemma 7.17. Hence, proj_t is continuous, when Y_t is endowed with $\tau^+ |Y_t$. For any finer topology there is a set $U \subseteq Y_t$ that is open in the finer topology with $U \notin \tau^+ |Y_t$. But then $\mathrm{proj}_t^{-1}(U) = W(U)$ cannot be open in τ relative to $W(Y_t)$ by "(b) implies (a)" from Lemma 7.17 , so that proj_t cannot be continuous. Hence, $\tau^+ |Y_t$ is precisely the quotient topology τ_t^q. ■

7.7.5.1 Proof of Proposition 7.4

Let $V \in \tau^+ |Y_{t+1}$ and consider any $x \in Y_t$ such that $p^{-1}(x) \subseteq V$. By Lemma 7.17(b) there is $v \in \tau$ such that $W(V) = v \cap W(Y_{t+1})$. Since $\cup_{y \in p^{-1}(x)} y = x$ by Proposition 6.2(a), it follows that $x \subseteq v$. Conversely, if $v \supseteq x \in Y_t$ then $\cup_{y \in p^{-1}(x)} y = x$ implies

$y \subseteq v$ for all $y \in p^{-1}(x)$, thus, $p^{-1}(x) \subseteq V$. Therefore,

$$\{x \in Y_t \,|p^{-1}(x) \subseteq V\} = \{x \in Y_t \,|x \subseteq v\} = W^+(v) \cap Y_t \in \tau^+ |Y_t$$

and this shows that p^{-1} is u.h.c. This completes the proof of Proposition 7.4.

7.8 Summary

The results in this chapter characterize the topologies on the plays of a potentially large discrete game tree such that every well-behaved perfect information game defined on this tree has an equilibrium. A full characterization is obtained for the class of compact and perfectly normal topologies (Theorem 7.2). This characterization rests on two stronger results. By Theorem 7.3 the conditions (CN) and (OP) are necessary under perfect normality, but compactness is not needed. By Theorem 7.4 the same conditions are sufficient for compact topologies, even if those are only separated (Hausdorff). Hence, Theorem 7.2 is the most general existence theorem for subgame prefect equilibrium in perfect information games. As the discussion in Sect. 7.7 showed, all that is left is to tinker with separation axioms and extensions of ZFC.

This result also has significance beyond perfect information games. After all, perfect information is a special case of imperfect information. If a topology on the plays of a given game tree does not admit an equilibrium under perfect information, then existence already fails in this simple case. In this sense a tree topology is the minimal structure that needs to be added when studying large games.

Infinite games are the rule in economics, rather than the exception. To cope with that, one typically uses some stationary structure to reduce the problem, for instance that the "action sets" remain constant across "periods." Our framework requires no such stationarity assumptions and hence opens the door to a potentially large number of applications. For instance, many examples of infinite perfect information games are typically solved under stationarity assumptions, a classical example being bilateral bargaining games (Rubinstein 1982). The present result shows that this is not necessary. Another potentially large field of application includes recent developments in dynamic decision theory, because a one-player dynamic optimization problem is a perfect information game. The results in this paper imply that such a problem has a solution even if there is no stationary structure, provided the defining conditions of a tree topology hold. Potentially, this allows for generalizations of recent contributions as e.g. Gul and Pesendorfer (2004) or Fudenberg and Levine (2006, 2012). Applications in this field might be particularly interesting, because nothing in the formalism relies on specific representations of preferences. The framework is capable of accommodating any preference relation that is continuous with respect to the tree topology.

These and other potential applications arise because the conditions identified in this chapter provide a general existence theorem for large extensive form games. The result, however, goes beyond the existing literature, because those conditions are also necessary. That is, they provide an answer to the question "How general can a topological existence theorem for large extensive form games become?"

References

Aliprantis CD, Border KC (1990) Infinite dimensional analysis: a Hitchhiker's guide. Springer, Berlin/Heidelberg

Alós-Ferrer C, Ritzberger K (2015) Characterizing existence of equilibrium for large extensive form games: a necessity result. Econ Theory. doi:10.1007/s00199-015-0937-0

Alós-Ferrer C, Ritzberger K (2016a) Equilibrium existence for large perfect information games. J Math Econ 62:5–18

Alós-Ferrer C, Ritzberger K (2016b) Does backwards induction imply subgame perfection? Games Econ Behav. doi:10.1016/j.geb.2016.02.005

Bertrand J (1883) Théorie mathématique de la richesse sociale. J des Savants 67:499–508

Blackwell D (1965) Discounted dynamic programming. Ann Math Stat 36:226–235

Blair CE (1984) Axioms and examples related to ordinal dynamic programming. Math Oper Res 9:345–347

Brown DJ, Lewis LM (1981) Myopic economic agents. Econometrica 49(2):359–368

Cournot AA (1838) Recherches sur les principes mathématiques de la théorie des richesses. Hachette, Paris

Fedorchuck V (1976) Fully closed mappings and the compatibility of some theorems in general topology with the axioms of set theory. Matematicheskii Sbornik 99:3–33

Flesch J, Predtetchinski A (2016) Subgame-perfect ε-equilibria in perfect information games with sigma-discrete discontinuities. Econ Theory 61:479–495

Flesch J, Kuipers J, Mashiah-Yaakovi A, Schoenmakers G, Shmaya E, Solan E, Vrieze K (2010a) Non-existence of subgame-perfect ε-equilibrium in perfect information games with infinite horizon. Math Oper Res 35(4):742–755

Flesch J, Kuipers J, Mashiah-Yaakovi A, Schoenmakers G, Solan E, Vrieze K (2010b) Perfect-information games with lower-semicontinuous payoffs. Math Oper Res 35(4):742–755

Fudenberg D, Levine DK (1983) Subgame-perfect equilibria of finite and infinite horizon games. J Econ Theory 31(2):251–268

Fudenberg D, Levine DK (2006) A dual-self model of impulse control. Am Econ Rev 96(5):1449–1476

Fudenberg D, Levine DK (2012) Timing and self-control. Econometrica 80(1):1–42

Fudenberg D, Tirole J (1991) Game theory. MIT, Cambridge

Gödel K (1931) über formal unentscheidbare sätze der principia mathematica und verwandter systeme, i. Monatshefte für Mathematik und Physik 38:173–198

Gul F, Pesendorfer W (2004) Self-control and the theory of consumption. Econometrica 72:119–158

Harris C (1985a) Existence and characterization of perfect equilibrium in games of perfect information. Econometrica 53:613–628

Harris C (1985b) A characterization of the perfect equilibria of infinite horizon games. J Econ Theory 33:461–481

Hellwig M, Leininger W (1987) On the existence of subgame-perfect equilibrium in infinite-action games of perfect information. J Econ Theory 43:55–75

Hellwig M, Leininger W, Reny PJ, Robson A (1990) Subgame perfect equilibrium in continuous games of perfect information: an elementary approach to existence and approximation by discrete games. J Econ Theory 52:406–422

Hendon E, Jacobsen HJ, Sloth B (1996) The one-shot-deviation principle for sequential rationality. Games Econ Behav 12:274–282

Kelley J (1975) General topology. Graduate texts in mathematics, vol 27. Springer, Berlin/Heidelberg/New York

Klein E, Thompson AC (1984) Theory of correspondences. Wiley, New York

Koopmans TC (1960) Stationary ordinal utility and impatience. Econometrica 28(2):287–309

Kuhn H (1953) Extensive games and the problem of information. In: Kuhn H, Tucker A (eds) Contributions to the theory of games, vol II. Annals of mathematics study, vol 28. Princeton University Press, Princeton, pp 217–243

Luttmer EGJ, Mariotti T (2003) The existence of subgame-perfect equilibrium in continuous games with almost perfect information: a comment. Econometrica 71:1909–1911

Osborne MJ, Rubinstein A (1994) A course in game theory. MIT, Cambridge

Ostaszewski AJ (1976) On countably compact, perfectly normal spaces. J Lond Math Soc 2:505–516

Perea A (2001) Rationality in extensive form games. Theory and decision library, series C, vol 29. Kluwer Academic, Boston/Dordrecht/London

Perea A (2002) A note on the one-deviation property in extensive games. Games Econ Behav 40:322–338

Purves RA, Sudderth WD (2011) Perfect-information games with upper-semicontinuous payoffs. Math Oper Res 36(3):468–473

Rubinstein A (1982) Perfect equilibrium in a bargaining model. Econometrica 50:97–109

Selten R (1965) Spieltheoretische Behandlung eines Oligopolmodells mit Nachfrageträgheit. Zeitschrift für die gesamte Staatswissenschaft 121:301–324 and 667–689

Selten R (1975) Reexamination of the perfectness concept for equilibrium points in extensive games. Int J Game Theory 4:25–55

Shapley L (1953) Stochastic games. Proc Natl Acad Sci 39:1095–1100

Sobel MJ (1975) Ordinal dynamic programming. Manag Sci 21:967–975

Solan E, Vieille N (2003) Deterministic multi-player Dynkin games. J Math Econ 1097:1–19

Steen LA, Seebach JA Jr (1978) Counterexamples in topology, 2nd edn. Springer, Berlin/Heidelberg/New York

von Neumann J, Morgenstern O (1944) Theory of games and economic behavior. Princeton University Press, Princeton

von Stackelberg H (1934) Marktform und Gleichgewicht. Springer, Heidelberg

Weiss W (1978) Countably compact spaces and Martin's axiom. Can J Math 30:243–249

Appendix A
Mathematical Appendix

This appendix explains very briefly some of the mathematical terms used in the text. There is no claim of completeness and the presentation is sketchy. It cannot substitute for a text on mathematical methods used in game theory (e.g. Aliprantis and Border 1990). But it can serve as a reminder for the reader who does not have the relevant definitions at hand.

A.1 Sets, Relations, and Functions

To begin with we review some of the basic definitions of set theory, binary relations, and functions and correspondences. Most of the material in this section is elementary.

A.1.1 Sets

Intuitively a *set* is a list of objects, called the *elements* of the set. In fact, the elements of a set may themselves be sets. The expression $x \in X$ means that x is an element of the set X, and $x \notin X$ means that it is not. Two sets are equal if they have the same elements. The symbol $\emptyset$ denotes the empty set, the set with no elements. The expression $X \setminus A$ denotes the elements of X that do not belong to the set A, $X \setminus A = \{x \in X \mid x \notin A\}$, the *complement* of A in (or relative to) X.

The notation $A \subseteq B$ or $B \supseteq A$ means that the set A is a *subset* of the set A or that B is a *superset* of A, that is, $x \in A$ implies $x \in B$. In particular, this allows for equality, $A = B$. If this is excluded, we write $A \subset B$ or $B \supset A$ and refer to A as a *proper* subset of B or to B as a proper superset of A.

C. Alós-Ferrer, K. Ritzberger, *The Theory of Extensive Form Games*, Springer Series in Game Theory, DOI 10.1007/978-3-662-49944-3

The set of all elements that belong to A or B is the *union* of A and B, denoted $A \cup B$. The set of elements that belong both to A and to B is the *intersection* of A and B, denoted $A \cap B$.

A set of sets $\{A_\lambda\}_{\lambda \in \Lambda}$ may be indexed by "names" or "labels" $\lambda \in \Lambda$. In that case $\bigcup_{\lambda \in \Lambda} A_\lambda$ denotes the union of all indexed sets, that is, the elements that belong to A_λ for some $\lambda \in \Lambda$; and $\bigcap_{\lambda \in \Lambda} A_\lambda$ denotes the intersection of all indexed sets.

A familiar set of sets is the *power set* of a given set X which is denoted by 2^X. This is the collection of all subsets of X (including the empty set). Nonempty subsets of 2^X are called families of sets. For an indexed family of subsets of a given set X the following useful identities, known as *de Morgan's laws*, hold:

$$X \setminus \left(\bigcup_{\lambda \in \Lambda} A_\lambda \right) = \bigcap_{\lambda \in \Lambda} (X \setminus A_\lambda) \quad \text{and} \quad X \setminus \left(\bigcap_{\lambda \in \Lambda} A_\lambda \right) = \bigcup_{\lambda \in \Lambda} (X \setminus A_\lambda). \tag{A.1}$$

The *Cartesian product* $\times_{\lambda \in \Lambda} A_\lambda$ of a family $\{A_\lambda\}_{\lambda \in \Lambda}$ of sets is the collection of all tuples $\{x_\lambda\}_{\lambda \in \Lambda}$ with $x_\lambda \in A_\lambda$ for all $\lambda \in \Lambda$. Each set A_λ is a *factor* in the product.

Set theory can be based on axioms. The most popular axioms are the eight known as the *Zermelo-Fraenkel* (ZF) axioms. They are designed so as to guarantee that operations on sets, like unions or power sets, give again sets (and to avoid Russell's paradox). It is often useful to complement the ZF axioms with a ninth axiom, which is both consistent with and independent of ZF set theory proper:

Axiom of Choice If $\{A_\lambda \mid \lambda \in \Lambda\}$ is a nonempty set of nonempty sets, then there is a function $f : \Lambda \to \cup_{\lambda \in \Lambda} A_\lambda$ such that $f(\lambda) \in A_\lambda$ for each $\lambda \in \Lambda$.

In other words, the Cartesian product of a nonempty set of nonempty sets is itself a nonempty set. Though apparently innocuous, this axiom has a lot of hidden power. The axiom system emerging from adding the Axiom of Choice to ZF is known as Zermelo-Fraenkel cum Axiom of Choice (ZFC). There are a few equivalent formulations of the Axiom of Choice, two of which are mentioned below.

A.1.2 Binary Relations

A *binary relation* $\geq$ on a nonempty set A assigns to each ordered pair $(a_1, a_2) \subset A \times A$ exactly one of the two statements "a_1 is in relation $\geq$ to a_2" or "a_1 is not in relation $\geq$ to a_2." Since all ordered pairs $(a_1, a_2) \in A \times A$ that satisfy the first statement constitute a subset of the Cartesian product $A \times A$, the relation $\geq$ can be identified with the set of all such pairs. Hence, a binary relation may be thought of as a subset of $A \times A$.

There are many important properties that a binary relation might satisfy. A binary relation $\geq$ on a set A is

- *reflexive* if $a \geq a$ for all $a \in A$;
- *irreflexive* if there is no $a \in A$ such that $a \geq a$;

- *symmetric* if, for all $a_1, a_2 \in A$, $a_1 \geq a_2$ implies $a_2 \geq a_1$;
- *asymmetric* if there are no $a_1, a_2 \in A$ such that $a_1 \geq a_2$ and $a_2 \geq a_1$;
- *antisymmetric* if, for all $a_1, a_2 \in A$, $a_1 \geq a_2$ and $a_2 \geq a_1$ imply $a_1 = a_2$;
- *transitive* if, given $a_1, a_2, a_3 \in A$, $a_1 \geq a_2$ and $a_2 \geq a_3$ imply $a_1 \geq a_3$;
- *complete* (or connected) if, for all $a_1, a_2 \in A$, either $a_1 \geq a_2$ or $a_2 \geq a_1$ (or both);
- *total* (or weakly connected) if $a_1 \neq a_2$ implies either $a_1 \geq a_2$ or $a_2 \geq a_1$ or both, for all $a_1, a_2 \in A$.

A complete relation is necessarily reflexive and total, but a total relation may or may not be reflexive. Note also that an antisymmetric relation may or may not be reflexive.

In the following a few prominent classes of binary relations (defined by the combination or properties from the list above that they fulfill) are discussed.

A.1.2.1 Partial Orders

A binary relation $\geq$ on a set A is a *partial order* if it is reflexive, transitive, and antisymmetric. For a partial order $\geq$ the notation $>$ is often used to refer to its asymmetric part, that is, $a_1 > a_2$ if and only if $a_1 \geq a_2$ and $a_1 \neq a_2$ for $a_1, a_2 \in A$. The set A endowed with a partial order $\geq$ is denoted $(A, \geq)$ and referred to as a *partially ordered set* or a *poset*. A relation $\geq$ on A is a *linear order* if it is total, transitive, and antisymmetric. If $\geq$ is a linear order, then for every pair $(a_1, a_2) \in A \times A$ exactly one of the statements $a_1 > a_2$, $a_2 > a_1$, or $a_1 = a_2$ holds. A *preorder* (or quasiorder) is reflexive and transitive. An antisymmetric preorder is a partial order. A set endowed with a preorder $\geq$ is referred to as a *preordered set.*.

A *chain* in a poset $(A, \geq)$ is a subset that is totally ordered. Hence, any two elements of a chain are comparable. For a subset $X \subseteq A$ an *upper bound* resp. *lower bound* is an element $a \in A$ such that $a \geq x$ resp. $x \geq a$ for all $x \in X$. A *greatest* resp. *least* element of X is an $x \in X$ such that $x \geq x'$ resp. $x' \geq x$ for all $x' \in X$. An element $a \in A$ is a *maximal* resp. *minimal* element of A if there is no $a' \in A$ such that $a' > a$ resp. $a > a'$. A nonempty subset X of A has at most one greatest resp. least element and, if it exists, it is maximal resp. minimal.

For a subset $X \subseteq A$ of a poset $(A, \geq)$ its *supremum* resp. *infimum* is its least upper bound resp. its greatest lower bound. That is, $a \in A$ is the supremum of $X \subseteq A$ if and only if $a \geq x$ for all $x \in X$ and $a' \geq x$ for all $x \in X$ implies $a' \geq a$ and analogously for the infimum. Note that supremum and infimum need not exist.

The following two assertions are equivalent to the Axiom of Choice. That is, both are theorems if the axiom of choice is assumed; but if one of them is taken as an axiom, then the Axiom of Choice becomes a theorem.

Zorn's Lemma. If every chain in a poset A has an upper bound, then A has a maximal element.
Hausdorff Maximality Principle. If c is a chain in a poset A, then there is a maximal chain in A that contains c.

As an example, the Hausdorff Maximality Principle states that plays exist for a tree, because plays are maximal chains.

A.1.2.2 Lattices and Directed Sets

A poset $(A, \geq)$ is a *lattice* if every pair of elements $a, a' \in A$ has a supremum and an infimum. A *sublattice* of a lattice is a subset that contains the suprema and infima for all its pairs of elements. A sublattice of a lattice $(A, \geq)$ is necessarily a lattice under the restricted partial order, but the converse statement is not true. A lattice is *complete* if every nonempty subset has a supremum and an infimum.

A *direction* $\geq$ on a set A is a reflexive and transitive binary relation such that each pair has an upper bound. That is, for each $a, b \in A$ there is some $c \in A$ such that $c \geq a$ and $c \geq b$. (Beware: A direction need not be antisymmetric.) A *directed set* $(A, \geq)$ is any set A endowed with a direction $\geq$. In Chap. 3, a subset A of a poset P was called directed if every finite subset of A had an upper bound in A; this is equivalent to stating that the partial order restricted to A is a direction. If $(A_\lambda, \geq_\lambda)$ is a family of directed sets, for some index set Λ, then their Cartesian product $\times_{\lambda \in \Lambda} A_\lambda$ is also a directed set under the *product direction*, defined by $(a_\lambda)_{\lambda \in \Lambda} \geq (b_\lambda)_{\lambda \in \Lambda}$ whenever $a_\lambda \geq_\lambda b_\lambda$ for each $\lambda \in \Lambda$.

A.1.2.3 Equivalence Relations

An *equivalence relation* is a reflexive, symmetric, and transitive binary relation. An equivalence relation on a set A is often denoted by $\sim$. Given an equivalence relation $\sim$ on A define the equivalence class $[a]$ of $a \in A$ by $[a] = \{a' \in A \mid a' \sim a\}$. A *partition* $\{A_k\}_{k \in K}$ (for some index set K) of a set A is a collection of nonempty subsets of A such that $A_i \cap A_j \neq \emptyset$ implies $A_i = A_j$ and $\cup_{k \in K} A_k = A$. The $\sim$-equivalence classes partition A. Conversely, every partition of A defines an equivalence relation on A by $a_1 \sim a_2$ if $a_1, a_2 \in A_k$ for some $k \in K$. The collection of $\sim$-equivalence classes of A is called the *quotient of A modulo* $\sim$ and denoted $A/\sim$.

A.1.3 Functions and Correspondences

A relation between elements of a nonempty set A and elements of another nonempty set B is a subset of the Cartesian product $A \times B$. Almost synonymous for relation is the notion of a correspondence, but its connotation is different. A *correspondence* φ from A to B associates to each $a \in A$ a subset $\varphi(a)$ of B. This is denoted by $\varphi : A \twoheadrightarrow B$. The *graph* of φ is the subset $\operatorname{graph}(\varphi) = \{(a,b) \in A \times B \mid b \in \varphi(a)\}$ of $A \times B$. The space A is the *domain* of the correspondence and B is its *codomain*. For a subset $A' \subseteq A$ the *image* $\varphi(A')$ of A' is the subset $\varphi(A') = \cup\{\varphi(a) \mid a \in A'\}$ of B. The *range* of φ is the image of A.

A relation $f \subseteq A \times B$ between two nonempty sets A and B is a *function* if $(a,b) \in f$ and $(a,c) \in f$ imply $b = c$ for all $a \in A$ and all $b,c \in B$. Another way of thinking about functions is as single-valued correspondences, where the set brackets embracing the values are dropped. That is, a function $f : A \to B$ from (domain) A to (codomain) B associates to each $a \in A$ an element $f(a)$ of B. The *graph* of f is the subset $\text{graph}(f) = \{(a,b) \in A \times B \mid b = f(a)\}$ of the Cartesian product $A \times B$. Again, for a subset $A' \subseteq A$ the *image* of A' under f is the subset $f(A') = \{f(a) \in B \mid a \in A'\}$ of B. The *range* of f is the image of A. Note that a function f and the single-valued correspondence $\varphi(a) = \{f(a)\}$ represent the same relation, but their values are not exactly the same objects.

For a nonempty subset $A' \subseteq A$ the *restriction* $f|_{A'}$ of a function $f : A \to B$ to A' is the function $f|_{A'} : A' \to B$ defined by $f|_{A'}(a) = f(a)$ for all $a \in A'$. We also say that f is an *extension* of $f|_{A'}$.

If A, B, and C are nonempty sets and $f : A \to B$ and $g : B \to C$ are functions, the *composition* of f and g, denoted $g \circ f$, is the function $g \circ f : A \to C$ given by $(g \circ f)(a) = g(f(a))$ for all $a \in A$.

A function $f : A \to B$ is *one-to-one* or *injective* if $f(a) = f(a')$ implies $a = a'$, for all $a, a' \in A$. It maps *onto* B, or it is *surjective*, if for every $b \in B$ there is some $a \in A$ such that $f(a) = b$. A *bijection* is a function that is both injective and surjective. For a subset $B' \subseteq B$ the *preimage* of B' under a function $f : A \to B$ is the subset $f^{-1}(B') = \{a \in A \mid f(a) \in B'\}$ of A. This notation is also used for singletons, suppressing the set brackets, though. That is, given $b \in B, f^{-1}(b) = \{a \in A \mid f(a) = b\}$

For a correspondence $\varphi : A \twoheadrightarrow B$ the notion of a preimage is less clear. Indeed there are two possibilities. The *upper* (or strong) inverse image of a set $B' \subseteq B$ is given by $\varphi^{+}(B') = \{a \in A \mid \varphi(a) \subseteq B'\}$. The *lower* (or weak) preimage of a subset B' of B is given by $\varphi^{-}(B') = \{a \in A \mid \varphi(a) \cap B' \neq \emptyset\}$. Clearly, $\varphi^{+}(B') \subseteq \varphi^{-}(B')$ for all $B' \subseteq B$. If the correspondence φ is single-valued, the two definitions agree.

A.2 Topology

When functions and correspondences are considered, it is often helpful to approximate. This requires a notion of what is close. A natural way to develop such a notion is to start with the concept of an open set, that is, a set that is a neighborhood of all of its elements. The list of all open sets is known as a topology.

A *topology* τ on a set W is a collection of subsets of W such that (a) $W, \emptyset \in \tau$, (b) if $x, y \in \tau$, then $x \cap y \in \tau$, and (c) if $x_\lambda \in \tau$ for all $\lambda \in \Lambda$, then $\cup_{\lambda \in \Lambda} x_\lambda \in \tau$, where Λ is an arbitrary index set. The elements of τ (subsets of W) are called *open sets*. The complement of an open set $x \in \tau$, i.e. $W \setminus x$, is a *closed* set.

Closed sets satisfy the dual properties of open sets. That is, (a) both W and $\emptyset$ are closed, (b) a finite union of closed sets is closed, and (c) an arbitrary intersection of closed sets is closed.

A nonempty set W endowed with a topology τ is a *topological space*, denoted (W, τ). A subset of a topological space (W, τ) may be open, closed, neither, or both (clopen). For instance, the power set 2^W constitutes a topology, the *discrete topology*. In the discrete topology every subset of W is clopen. The discrete topology is a natural choice on any finite set; but not on an infinite set. The *indiscrete topology* consists only of W and $\emptyset$.

Topologies can be made up. For, the intersection of a family of topologies on a set W is again a topology. Now, if $\mathscr{W}$ is an arbitrary collection of subsets of a set W, then there is a smallest (with respect to set inclusion) topology that contains $\mathscr{W}$, namely the intersection of all topologies that include $\mathscr{W}$. This is the topology *generated by* $\mathscr{W}$. It consists of W, $\emptyset$, and all sets of the form $\cup_\lambda V_\lambda$, where each V_λ is a finite intersection of sets from $\mathscr{W}$.

For instance, a topology may emerge from a metric. A *semimetric* on a set A is a real-valued function $d : A \times A \to \mathbb{R}_+$ such that, for all $a,b,c \in A$, (a) $d(a,b) = d(b,a)$, (b) $d(a,a) = 0$, and (c) $d(a,c) \leq d(a,b) + d(b,c)$. A *metric* is a semimetric which, in addition, satisfies that $d(a,b) = 0$ implies $a = b$. The collection of open balls $B_\varepsilon(a) = \{b \in A | d(a,b) < \varepsilon\}$, for $\varepsilon > 0$, with respect to the metric d generates a topology. A topological space with a metric is called a *metric space*.

A *basis* for a topology τ is a subfamily $\mathscr{B} \subseteq \tau$ such that each open set in τ is a union of sets in $\mathscr{B}$. A *subbasis* for a topology τ is a subfamily $\mathscr{S} \subseteq \tau$ such that the collection of all finite intersections of sets in $\mathscr{S}$ is a basis for τ.

Topologies on a fixed set W can be partially ordered by set inclusion. Say that a topology τ is *finer* (or stronger) than another topology τ' if every τ'-open set is also τ-open. Alternatively, we say that τ' is then *coarser* (or weaker) than τ.

If V is a subset of W and W comes with a topology τ, then the collection $\tau_V = \{x \cap V | x \in \tau\}$ is a topology on V, called the *relative topology* (or the topology induced by τ on V). The space (V, τ_V) is then a topological *subspace* of (W, τ). A set in τ_V is called *relatively open* in V.

Note that a countably infinite intersection of open sets may not be open. Likewise, a countably infinite union of closed sets need not be closed. Such sets have names, in fact. A countable intersection of open sets is called a $\mathscr{G}_\delta$-set, and a countable union of closed sets is called an $\mathscr{F}_\sigma$-set.

Consider a subset $V \subseteq W$ of a topological space (W, τ). The *interior* of V is the largest (with respect to set inclusion) open set contained in V, that is, the union of all open sets contained in V. The interior of V may be empty. The *closure* of V is the smallest closed set that contains V, that is, the intersection of all closed sets that contain V. If one subset of W contains another, then set inclusion is preserved by taking interior or closure.

A *neighborhood* of a point $w \in W$ is a subset $u \subseteq W$ that contains w in its interior. A neighborhood may or may not be open. Mostly we work with open neighborhoods, though. Obviously, a set is open if and only if it is a neighborhood of each of its elements. A *neighborhood basis* at $w \in W$ is a family of neighborhoods of w such that any neighborhood of w contains a neighborhood from the neighborhood basis.

A point w is a *closure point* of the set $V \subseteq W$ if every neighborhood of w has a nonempty intersection with V. A point $w \in W$ is a *cluster point* (or accumulation point, or limit point) of V if for every neighborhood u of w the intersection $(u \setminus \{w\}) \cap V$ is nonempty. Finally, a point w is a *boundary point* of V if for every neighborhood u of w both $u \cap V \neq \emptyset$ and $u \cap (W \setminus V) \neq \emptyset$. Cluster points and boundary points belong to the closure of V. In fact, the closure of V is the union of its interior and its boundary points. A set V is closed if and only if it contains all its cluster points.

A.2.1 Separation Properties

On top of what its definition implies there are several additional properties that a topological space (A, τ) may satisfy. The most important ones are separation properties.

A topological space (W, τ) is T_0 or *Kolmogorov* if, for every pair of distinct points $w, w' \in W$ with $w \neq w'$, there is either an open set $u \in \tau$ such that $w \in u$ but $w' \notin u$ or an open set $u' \in \tau$ such that $w' \in u'$ but $w \notin u'$. That is, in a T_0 space any two distinct points are topologically distinguishable. It is T_1, *accessible*, or *Fréchet* if, for every pair of distinct points $w, w' \in W$ with $w \neq w'$, there are $u, u' \in \tau$ such that $w \in u$, $w' \notin u$, $w' \in u'$, and $w \notin u'$. This is equivalent to all singletons being closed. Every T_1 space is T_0. Both T_0 and T_1 are very weak axioms—too weak for many applications indeed.

A topology τ on W is T_2, *separated*, or *Hausdorff* if, for any pair of distinct points $w, w' \in W$, there are $u, u' \in \tau$ such that $w \in u$, $w' \in u'$, and $u \cap u' = \emptyset$. That is, a space is separated if any two distinct points can be separated by disjoint open neighborhoods. Every Hausdorff space is T_1.

A topological space is *regular* if, for every nonempty closed set $V \subseteq W$ and every point $w \in W \setminus V$ outside of V, there are $u, u' \in \tau$ such that $w \in u$, $V \subseteq u'$, and $u \cap u' = \emptyset$. It is *normal* if, for every pair $V, V' \in 2^W$ of disjoint nonempty closed sets ($W \setminus V \in \tau$, $W \setminus V' \in \tau$, and $V \cap V' = \emptyset$), there are $u, u' \in \tau$ such that $V \subseteq u$, $V' \subseteq u'$, and $u \cap u' = \emptyset$. A regular or normal space need not be separated (T_2). Yet, every normal T_1-space is indeed separated.

A key property of normal spaces and a further separation property are discussed in Sect. A.2.5 below.

A.2.2 Sequences and Nets

A *sequence* in W is a function from the natural numbers into W. One may also think of a sequence as a subset of W indexed by the natural numbers. A net generalizes sequences by replacing the natural numbers with a directed set. That is, a *net* in a set W is a function $a : D \to W$, where $(D, \geq)$ is a directed set, called the index set. Nets are denoted by $\{w_d\}_{d \in D}$ on the understanding that $w_d = a(d)$ for each $d \in D$. In particular every sequence is a net.

Let (W,τ) be a topological space and $\{w_d\}_{d\in D}$ a net in W. A point $w \in W$ is a *cluster point* (or limit point) of the net $\{w_d\}$ if for each neighborhood u of w and each $d_0 \in D$ there exists some $d \geq d_0$ such that $w_d \in u$. Note that cluster points of nets and cluster points of sets are different (but conceptually related) objects.

A net $\{w_d\}$ in a topological space (W,τ) *converges* to a point $w \in W$ if for each neighborhood u of w there is some $d_u \in D$ such that $w_d \in u$ for all $d \geq d_u$. The point w is then a *limit* of the net. If the topological space is Hausdorff, limits are unique. Convergence is denoted by $w_d \to w$.

A *subnet* $\{v_\lambda\}_{\lambda\in\Lambda}$ of a net $\{w_d\}_{d\in D}$ is a net such that there is a function $f:\Lambda \to D$ that satisfies (a) $v_\lambda = w_{f(\lambda)}$ for each $\lambda \in \Lambda$ and (b) for each $d_0 \in D$ there is some $\lambda_0 \in \Lambda$ such that $\lambda \geq \lambda_0$ implies $f(\lambda) \geq d_0$. In a topological space, a point is a cluster point of a net if and only if it is a limit of some subnet. A net converges to a point if and only if every subnet converges to that same point.

A.2.3 Compactness

The underlying set W of a topological space (W,τ) may well be infinite. (In fact, this is the only interesting case.) A natural substitute for finiteness is the property of compactness, as introduced next.

An *open covering* of a subset $V \subseteq W$ is a family of open sets whose union contains V. A subset V of a topological space is *compact* if every open covering contains a finite subcovering. That is, V is compact if every family $\{u_\lambda \in \tau | \lambda \in \Lambda\}$, for some index set Λ, that satisfies $V \subseteq \cup_{\lambda\in\Lambda} u_\lambda$ contains a finite subfamily $\{u_i \in \tau | i = 1,\ldots,n\}$ such that $V \subseteq \cup_{i=1}^n u_i$. A subset of a topological space is *relatively compact* if its closure is compact.

A topological space (W,τ) is a compact space if W is a compact set. A topological space (W,τ) is compact if and only if every family of closed subsets of W has a nonempty intersection whenever each of its finite subfamilies has a nonempty intersection. (The latter is the *finite intersection property*.) Equivalently, it is compact if and only if every net in W has a cluster point.

Here are a few facts about compact sets. All finite subsets of a topological space are compact. Finite unions of compact sets are compact. Closed subsets of compact sets are compact. If $V \subseteq V' \subseteq W$, then V is a compact subset of W if and only if V is a compact subset of V' in the relative topology. Compact subsets of separated topological spaces are closed. Every compact separated (T_2) space is normal.

Let $\{(W_\lambda,\tau_\lambda)\}_{\lambda\in\Lambda}$ be a family of topological spaces, for some index set Λ, and let $W = \times_{\lambda\in\Lambda} W_\lambda$ denote its Cartesian product, with typical element $w = (w_\lambda)_{\lambda\in\Lambda}$. For each $\lambda \in \Lambda$ the projection $\text{proj}_\lambda : W \to W_\lambda$ is defined by $\text{proj}_\lambda(w) = w_\lambda$. The *product topology* on W is the coarsest topology τ such that each projection proj_λ is continuous (i.e. $\text{proj}_\lambda^{-1}(u_\lambda) \in \tau$ for each $u_\lambda \in \tau_\lambda$). A basis for the product topology consists of all sets of the form $V = \times_{\lambda\in\Lambda} V_\lambda$, where $V_\lambda \in \tau_\lambda$ and $V_\lambda = W_\lambda$ for all but finitely many $\lambda \in \Lambda$. The celebrated Tychonoff Product Theorem states that the product of a family of topological spaces is compact in the product topology if and only if each factor is compact.

A.2.4 Continuity

A major goal of topology is to study continuous functions. A function $f : A \to B$ between two topological spaces (A, τ_A) and (B, τ_B) is *continuous at the point* $a \in A$ if for every open neighborhood $u \in \tau_B$ of $f(a)$ the preimage $f^{-1}(u)$ is a neighborhood of a. It is continuous if it is continuous at every point. Equivalently, a function $f : A \to B$ between two topological spaces is *continuous* if $f^{-1}(u) \in \tau_A$ for each $u \in \tau_B$.

Here are a few facts about continuous functions. The preimage under a continuous function of a closed subset of B is closed in A. The image under a continuous function of the closure of a subset of A is contained in the closure of the image of the subset. If a net in A converges to a point $a \in A$, then the images converge to $f(a)$. The composition of continuous functions between topological spaces is continuous. A function from a topological space into a compact separated topological space is continuous if and only if its graph is closed—the Closed Graph Theorem. A continuous function between two topological spaces carries compact sets to compact sets.

The last property is reminiscent of, but different from the following definitions. A function $f : A \to B$ between topological spaces (A, τ_A) and (B, τ_B) is *open* if it carries open sets to open sets (i.e. $u \in \tau_A$ implies $f(u) \in \tau_B$). It is *closed* if it carries closed sets to closed sets (i.e. $W \setminus u \in \tau_A$ implies $W \setminus f(u) \in \tau_B$).

For real-valued functions $f : A \to \mathbb{R}$, where $\mathbb{R}$ is endowed with the standard Euclidean topology, there are "one-directional" continuity notions. In particular, a real-valued function f on a topological space is *upper semi-continuous* if for each $r \in \mathbb{R}$ the upper contour set $\{a \in A | f(a) \geq r\}$ is closed (or equivalently the set $\{a \in A | f(a) < r\}$ is open). A real-valued function f on a topological space is *lower semi-continuous* if for each $r \in \mathbb{R}$ the lower contour set $\{a \in A | f(a) \leq r\}$ is closed (or equivalently the set $\{a \in A | f(a) > r\}$ is open). Clearly, a real-valued function is continuous if and only if it is both upper and lower semi-continuous.

An important consequence of semi-continuity is a generalization of Weierstrass' Theorem. An upper (resp. lower) semi-continuous function on a compact set attains a maximum (resp. minimum) value, and the nonempty set of maximizers (resp. minimizers) is compact.

Semi-continuity generalizes to complete, reflexive, and transitive binary relations $\precsim$. Such a relation on a topological space (W, τ) is *upper semi-continuous* if for every $w \in W$ the upper contour set $\{w' \in W | w \precsim w'\}$ is closed. It is *lower semi-continuous* if for every $w \in W$ the lower contour set $\{w' \in W | w' \precsim w\}$ is closed. Consequently, an upper semi-continuous total preorder on a compact space has a greatest element.

For correspondences the definition of continuity is more involved, because the preimage of an open set may have two meanings (see above). A correspondence $\varphi : A \twoheadrightarrow B$ between two topological spaces (A, τ_A) and (B, τ_B) is *upper hemi-continuous* (u.h.c.) if for every open set $u \in \tau_B$ the upper preimage $\varphi^+(u)$ is open in A, i.e. $\varphi^+(u) \in \tau_A$. It is *lower hemi-continuous* (l.h.c.) if for every open set $u \in \tau_B$ the lower preimage $\varphi^-(u)$ is open in A.

A correspondence with a closed graph, $\text{graph}(\varphi) = \{(a,b) \in A \times B | b \in \varphi(a)\}$, is always closed-valued. (Beware: The converse is false.) If $f : A \to B$ is a function, then its inverse f^{-1} (which is a correspondence) satisfies: f^{-1} is u.h.c. if and only if f is closed; and f^{-1} is l.h.c. if and only if f is open.

Here are a few facts about correspondences. The image of a compact set under a compact-valued u.h.c. correspondence is compact. A correspondence with compact separated range has a closed graph (in the product topology) if and only if it is u.h.c. and closed-valued. A compact-valued correspondence φ between topological spaces is u.h.c. if and only if for every net $\{(a_d, b_d)\}_{d \in D}$ in the graph of φ that satisfies $a_d \to a \in A$ the net $\{b_d\}_{d \in D}$ has a cluster point in $\varphi(a)$.

A.2.5 Separation by Continuous Functions

Normal spaces (defined above) are important because of Urysohn's Lemma. This result states that a topological space is normal if and only if for every pair of disjoint nonempty closed sets there is a continuous function from W to the unit interval $[0,1]$ that assumes the value 0 on one set and the value 1 on the other. That is, Urysohn's Lemma asserts that a space is normal if and only if any two disjoint closed sets can be *separated by a continuous function*. Unfortunately, normality does not guarantee that the preimages of 0 and 1 (under the continuous function) are precisely the two closed sets that are separated by a Urysohn function. The latter is only true in a perfectly normal space.

A topological space (W, τ) is *perfectly normal* if it is T_1 and, for any two disjoint nonempty closed subsets $V, V' \in 2^W$, there is a continuous function $f : W \to [0,1]$ such that $V = f^{-1}(1)$ and $V' = f^{-1}(0)$. Every perfectly normal space is normal. And every metric space is perfectly normal.

Bibliography

Aliprantis CD, Border KC (1990) Infinite dimensional analysis: a Hitchhiker's guide. Springer, Berlin/Heidelberg

Alós-Ferrer C (1999) Dynamical systems with a continuum of randomly matched agents. J Econ Theory 86:245–267

Alós-Ferrer C (2002) Individual randomness with a continuum of agents. Mimeo, Vienna

Alós-Ferrer C, Kern J (2015) Repeated games in continuous time as extensive form games. J Math Econ 61:34–57

Alós-Ferrer C, Ritzberger K (2005a) Trees and decisions. Econ Theory 25(4):763–798

Alós-Ferrer C, Ritzberger K (2005b) Some remarks on pseudotrees. Order 22(1):1–9

Alós-Ferrer C, Ritzberger K (2008) Trees and extensive forms. J Econ Theory 43(1):216–250

Alós-Ferrer C, Ritzberger K (2013) Large extensive form games. Econ Theory 52(1):75–102

Alós-Ferrer C, Ritzberger K (2015) Characterizing existence of equilibrium for large extensive form games: a necessity result. Econ Theory. doi:10.1007/s00199-015-0937-0

Alós-Ferrer C, Ritzberger K (2016a) Equilibrium existence for large perfect information games. J Math Econ 62:5–18

Alós-Ferrer C, Ritzberger K (2016b) Does backwards induction imply subgame perfection? Games Econ Behav. doi:10.1016/j.geb.2016.02.005

Alós-Ferrer C, Ritzberger K (2016c) Characterizations of perfect recall. Int J Game Theory. doi:10.1007/s00182-016-0534-x

Alós-Ferrer C, Kern J, Ritzberger K (2011) Comment on 'Trees and extensive forms'. J Econ Theory 146(5):2165–2168

Anderson RM (1984) Quick-response equilibrium. Working papers in economic theory and econometrics, #IP-323, C.R.M, University of California at Berkeley

Anscombe F, Aumann RJ (1963) A definition of subjective probability. Ann Math Stat 34:199–205

Aumann RJ (1961) Borel structures for function spaces. Ill J Math 5:614–630

Aumann RJ (1963) On choosing a function at random. In: Wright, FB (ed) Ergodic theory. Academic Press, New York, pp 1–20

Aumann RJ (1964) Mixed and behavior strategies in infinites extensive games. In: Advances in game theory, Annals of mathematics study, vol 52. Princeton University Press, Princeton, pp 627–650

Aumann RJ (1987) Game theory. In: Eatwell J, Milgate M, Newman P (eds) The new Palgrave, a dictionary of economics, vol 2. Macmillan, London, pp 460–482

Aumann RJ, Hart S (2003) Long cheap talk. Econometrica 71:1619–1660

Baur L (2000) Cardinal functions on initial chain algebras on pseudotrees. Order 17:1–21

Baur L, Heindorf L (1997) Initial chain algebras on pseudotrees. Order 14:21–38

C. Alós-Ferrer, K. Ritzberger, *The Theory of Extensive Form Games*,
Springer Series in Game Theory, DOI 10.1007/978-3-662-49944-3

Bergin J, MacLeod WB (1993) Continuous time repeated games. Int Econ Rev 34(1):21–37

Bertrand J (1883) Théorie mathématique de la richesse sociale. J des Savants 67:499–508

Billingsley P (1986) Probability and measure. Wiley, New York

Birkhoff G (1973) Lattice theory, vol XXV, 3rd edn. American Mathematical Society Colloquium Publications, Providence

Blackwell D (1965) Discounted dynamic programming. Ann Math Stat 36:226–235

Blair CE (1984) Axioms and examples related to ordinal dynamic programming. Math Oper Res 9:345–347

Blume LE, Brandenburger A, Dekel E (1991a) Lexicographic probabilities and choice under uncertainty. Econometrica 59:61–79

Blume LE, Brandenburger A, Dekel E (1991b) Lexicographic probabilities and equilibrium refinements. Econometrica 59:81–98

Brown DJ, Lewis LM (1981) Myopic economic agents. Econometrica 49(2):359–368

Carmona G (2009) A remark on the measurability of large games. Econ Theory 39:491–494

Cournot AA (1838) Recherches sur les principes mathématiques de la théorie des richesses. Hachette, Paris

Dalkey N (1953) Equivalence of information patterns and essentially determinate games. In: Kuhn H, Tucker A (eds) Contributions to the theory of games, volume II. Annals of mathematics study, vol 28. Princeton University Press, Princeton, pp 217–243

Davey B, Priestley H (1990) Introduction to lattices and order. Cambridge University Press, Cambridge

Davidson R, Harris R (1981) Nonconvexities in continuous-time investment. Rev Econ Stud 48:235–253

Debreu G, Scarf H (1963) A limit theorem on the core of an economy. Int Econ Rev 4:236–246

Dekel E, Lipman B, Rustichini A (2001) Representing preferences with a unique subjective state space. Econometrica 69:891–934

Dockner E, Jörgensen S, Long NV, Sorger G (2000) Differential games in economics and management science. Cambridge University Press, Cambridge

Elmes S, Reny P (1994) On the strategic equivalence of extensive form games. J Econ Theory 62:1–23

Fedorchuck V (1976) Fully closed mappings and the compatibility of some theorems in general topology with the axioms of set theory. Matematicheskii Sbornik 99:3–33

Feldman M, Gilles C (1985) An expository note on individual risk without aggregate uncertainty. J Econ Theory 35:26–32

Flesch J, Predtetchinski A (2016) Subgame-perfect ε-equilibria in perfect information games with sigma-discrete discontinuities. Econ Theory 61:479–495

Flesch J, Kuipers J, Mashiah-Yaakovi A, Schoenmakers G, Shmaya E, Solan E, Vrieze K (2010a) Non-existence of subgame-perfect ε-equilibrium in perfect information games with infinite horizon. Math Oper Res 35(4):742–755

Flesch J, Kuipers J, Mashiah-Yaakovi A, Schoenmakers G, Solan E, Vrieze K (2010b) Perfect-information games with lower-semicontinuous payoffs. Math Oper Res 35(4):742–755

Friedman JW (1990) Game theory with applications to economics, 2nd edn. Oxford University Press, Oxford

Friedman A (1994) Differential games. In: Aumann RJ, Hart S (eds) Handbook of game theory, vol 2. Elsevier, Amsterdam/Princeton, pp 781–799

Fudenberg D, Levine DK (1983) Subgame-perfect equilibria of finite and infinite horizon games. J Econ Theory 31(2):251–268

Fudenberg D, Levine D (1986) Limit games and limit equilibria. J Econ Theory 38:261–279

Fudenberg D, Levine DK (2006) A dual-self model of impulse control. Am Econ Rev 96(5):1449–1476

Fudenberg D, Levine DK (2012) Timing and self-control. Econometrica 80(1):1–42

Fudenberg D, Tirole J (1991) Game theory. MIT, Cambridge

Gödel K (1931) über formal unentscheidbare sätze der principia mathematica und verwandter systeme, i. Monatshefte für Mathematik und Physik 38:173–198

Gul F, Pesendorfer W (2001) Temptation and self-control. Econometrica 69:1403–1435
Gul F, Pesendorfer W (2004) Self-control and the theory of consumption. Econometrica 72:119–158
Harris C (1985a) Existence and characterization of perfect equilibrium in games of perfect information. Econometrica 53:613–628
Harris C (1985b) A characterization of the perfect equilibria of infinite horizon games. J Econ Theory 33:461–481
Harsanyi JC (1967–1968) Games of incomplete information played by Bayesian players, I, II, and III. Manag Sci 14:159–182; 320–3341; 486–502
Hellwig M, Leininger W (1987) On the existence of subgame-perfect equilibrium in infinite-action games of perfect information. J Econ Theory 43:55–75
Harsanyi JC, Selten R (1988) A general theory of equilibrium selection in games. MIT, Cambridge
Hellwig M, Leininger W, Reny PJ, Robson A (1990) Subgame perfect equilibrium in continuous games of perfect information: an elementary approach to existence and approximation by discrete games. J Econ Theory 52:406–422
Hendon E, Jacobsen HJ, Sloth B (1996) The one-shot-deviation principle for sequential rationality. Games Econ Behav 12:274–282
Hewitt E, Stromberg K (1965) Real and abstract analysis, graduate texts in mathematics, vol 25. Springer, Berlin/Heidelberg/New York
Isbell JR (1957) Finitary games. In: Dresher D, Tucker AW, Wolfe P (eds) Contributions to the theory of games, volume III. Annals of mathematics studies, vol 39. Princeton University Press, Princeton, pp 79–96
Judd K (1985) The law of large numbers with a continuum of i.i.d random variables. J Econ Theory 35:19–25
Kelley J (1975) General topology. Graduate texts in mathematics, vol 27. Springer, Berlin/Heidelberg/New York
Khan MA, Sun Y (1999) Non-cooperative games on hyperfinite loeb spaces. J Math Econ 31:455–492
Klein E, Thompson AC (1984) Theory of correspondences. Wiley, New York
Koopmans TC (1960) Stationary ordinal utility and impatience. Econometrica 28(2):287–309
Koppelberg S (1989) General theory of Boolean algebras. In: Monk J, Bonnet R (eds) Handbook of Boolean algebras, vol 1. Elsevier, Amsterdam/New York, pp 741–773
Koppelberg S, Monk J (1992) Pseudo-trees and Boolean algebras. Order 8:359–577
Kreps DM (1979) A representation theorem for "preference for flexibility." Econometrica 47:565–577
Kreps D, Wilson R (1982) Sequential equilibria. Econometrica 50:863–894
Kuhn H (1953) Extensive games and the problem of information. In: Kuhn H, Tucker A (eds) Contributions to the theory of games, vol. II. Annals of mathematics study, vol 28. Princeton University Press, Princeton, pp 217–243
Luttmer EGJ, Mariotti T (2003) The existence of subgame-perfect equilibrium in continuous games with almost perfect information: a comment. Econometrica 71:1909–1911
Markowsky G (1976) Chain-complete posets and directed sets with applications. Algebra Universalis 6:53–68
Mas-Colell A (1984) On a theorem of Schmeidler. J Math Econ 13:201–206
Mas-Colell A, Whinston M, Green J (1995) Microeconomic theory. Oxford University Press, Oxford
Nash JF (1950) Equilibrium points in n-person games. Proc Natl Acad Sci 36:48–49
Osborne MJ, Rubinstein A (1994) A course in game theory. MIT, Cambridge
Ostaszewski AJ (1976) On countably compact, perfectly normal spaces. J Lond Math Soc 2:505–516
Perea A (2001) Rationality in extensive form games. Theory and decision library, series C, vol 29. Kluwer Academic, Boston/Dordrecht/London
Perea A (2002) A note on the one-deviation property in extensive games. Games Econ Behav 40:322–338

Piccione M, Rubinstein A (1997) On the interpretation of decision problems with imperfect recall. Games Econ Behav 20:3–24

Podczeck K (2010) On the existence of rich Fubini extensions. Econ Theory 45:1–22

Podczeck K, Puzzello D (2012) Independent random matching. Econ Theory 50:1–29

Purves RA, Sudderth WD (2011) Perfect-information games with upper-semicontinuous payoffs. Math Oper Res 36(3):468–473

Rao BV (1971) Borel structures for function spaces. Colloq Math 23:33–38

Ritzberger K (1999) Recall in extensive form games. Int J Game Theory 20:69–87

Ritzberger K (2002) Foundations of non-cooperative game theory. Oxford University Press, Oxford

Rubinstein A (1982) Perfect equilibrium in a bargaining model. Econometrica 50:97–109

Savage L (1954) The foundations of statistics. Wiley, Hoboken

Schmeidler D (1973) Equilibrium points of nonatomic games. J Stat Phys 7:295–300

Schwarz G (1974a) Ways of randomizing and the problem of their equivalence. Isr J Math 17:1–10

Schwarz G (1974b) Randomizing when time is not well-ordered. Isr J Math 19:241–245

Selten R (1965) Spieltheoretische Behandlung eines Oligopolmodells mit Nachfrageträgheit. Zeitschrift für die gesamte Staatswissenschaft 121:301–324; 667–689

Selten R (1975) Reexamination of the perfectness concept for equilibrium points in extensive games. Int J Game Theory 4:25–55

Selten R (1998) Multistage game models and delay supergames. Theory Decis 44:1–36

Shapley L (1953) Stochastic games. Proc Natl Acad Sci 39:1095–1100

Simon L, Stinchcombe MB (1989) Extensive form games in continuous time, Part I: pure strategies. Econometrica 57:1171–1214

Sobel MJ (1975) Ordinal dynamic programming. Manag Sci 21:967–975

Solan E, Vieille N (2003) Deterministic multi-player Dynkin games. J Math Econ 1097:1–19

Spence AM (1973) Job market signaling. Q J Econ 83:355–377

Steen LA, Seebach JA Jr (1978) Counterexamples in topology, 2nd edn. Springer, Berlin/Heidelberg/New York

Stinchcombe MB (1992) Maximal strategy sets for continuous-time game theory. J Econ Theory 56:235–265

Sun Y (2006) The exact law of large numbers via Fubini extension and characterization of insurable risk. J Econ Theory 126:31–69

Thompson FB (1952) Equivalence of games in extensive form. RAND research memorandum, vol 759. Rand Corp., Santa Monica

Uhlig H (1996) A law of large numbers for large economies. Econ Theory 8:41–50

von Neumann J (1928) Zur Theorie der Gesellschaftsspiele. Mathematische Annalen 100:295–320

von Neumann J, Morgenstern O (1944) Theory of games and economic behavior. Princeton University Press, Princeton

von Stackelberg H (1934) Marktform und Gleichgewicht. Springer, Heidelberg

von Stengel B (2002) Computing equilibria for two-person games. In: Aumann RJ, Hart S (eds) Handbook of game theory, vol 3. Elsevier, Amsterdam, pp 1723–1759

Weiss W (1978) Countably compact spaces and Martin's axiom. Can J Math 30:243–249

Wichardt PC (2008) Existence of Nash equilibria in finite extensive form games with imperfect recall: a counterexample. Games Econ Behav 63(1):366–369

Zermelo E (1913) Über eine Anwendung der Mengenlehre auf die Theorie des Schachspiels. Proc Fifth Int Congr Math 2:501–504

Index

C. Alós-Ferrer, K. Ritzberger, *The Theory of Extensive Form Games*,
Springer Series in Game Theory, DOI 10.1007/978-3-662-49944-3

MIX
Papier aus verantwortungsvollen Quellen
Paper from responsible sources
FSC® C105338

If you have any concerns about our products,
you can contact us on
ProductSafety@springernature.com

In case Publisher is established outside the EU,
the EU authorized representative is:
Springer Nature Customer Service Center GmbH
Europaplatz 3, 69115 Heidelberg, Germany

Printed by Libri Plureos GmbH
in Hamburg, Germany